Additive Manufacturing Technologies From an Optimization Perspective

Kaushik Kumar
Birla Institute of Technology Mesra, India

Divya Zindani
National Institute of Technology Silchar, India

J. Paulo Davim
University of Aveiro, Portugal

A volume in the Advances in Logistics, Operations, and Management Science (ALOMS) Book Series

IGI Global
DISSEMINATOR OF KNOWLEDGE

Published in the United States of America by
IGI Global
Engineering Science Reference (an imprint of IGI Global)
701 E. Chocolate Avenue
Hershey PA, USA 17033
Tel: 717-533-8845
Fax: 717-533-8661
E-mail: cust@igi-global.com
Web site: http://www.igi-global.com

Library of Congress Cataloging-in-Publication Data

Names: Kumar, K. (Kaushik), 1968- editor. | Zindani, Divya, 1989- editor. | Davim, J. Paulo, editor.
Title: Additive manufacturing technologies from an optimization perspective / Kaushik Kumar, Divya Zindani, and J. Paulo Davim, editors.
Description: Hershey, PA : Engineering Science Reference, [2020] | Includes bibliographical references.
Identifiers: LCCN 2018058997| ISBN 9781522591672 (hardcover) | ISBN 9781522591689 (softcover) | ISBN 9781522591696 (ebook)
Subjects: LCSH: Three-dimensional printing. | Manufacturing processes--Technological innovations.
Classification: LCC TS171.95 .A45 2020 | DDC 621.9/88--dc23 LC record available at https://lccn.loc.gov/2018058997

This book is published in the IGI Global book series Advances in Logistics, Operations, and Management Science (ALOMS) (ISSN: 2327-350X; eISSN: 2327-3518)

British Cataloguing in Publication Data
A Cataloguing in Publication record for this book is available from the British Library.

All work contributed to this book is new, previously-unpublished material.
The views expressed in this book are those of the authors, but not necessarily of the publisher.

For electronic access to this publication, please contact: eresources@igi-global.com.

Advances in Logistics, Operations, and Management Science (ALOMS) Book Series

ISSN:2327-350X
EISSN:2327-3518

Editor-in-Chief: John Wang, Montclair State University, USA

MISSION

Operations research and management science continue to influence business processes, administration, and management information systems, particularly in covering the application methods for decision-making processes. New case studies and applications on management science, operations management, social sciences, and other behavioral sciences have been incorporated into business and organizations real-world objectives.

The **Advances in Logistics, Operations, and Management Science** (ALOMS) Book Series provides a collection of reference publications on the current trends, applications, theories, and practices in the management science field. Providing relevant and current research, this series and its individual publications would be useful for academics, researchers, scholars, and practitioners interested in improving decision making models and business functions.

COVERAGE

- Services management
- Decision analysis and decision support
- Finance
- Organizational Behavior
- Computing and information technologies
- Production management
- Networks
- Operations Management
- Risk Management
- Political Science

IGI Global is currently accepting manuscripts for publication within this series. To submit a proposal for a volume in this series, please contact our Acquisition Editors at Acquisitions@igi-global.com or visit: http://www.igi-global.com/publish/.

The Advances in Logistics, Operations, and Management Science (ALOMS) Book Series (ISSN 2327-350X) is published by IGI Global, 701 E. Chocolate Avenue, Hershey, PA 17033-1240, USA, www.igi-global.com. This series is composed of titles available for purchase individually; each title is edited to be contextually exclusive from any other title within the series. For pricing and ordering information please visit http://www.igi-global.com/book-series/advances-logistics-operations-management-science/37170. Postmaster: Send all address changes to above address.

Titles in this Series

For a list of additional titles in this series, please visit:
https://www.igi-global.com/book-series/advances-logistics-operations-management-science/37170

Strategic Thinking, Planning, and Management Practice in the Arab World
Fayez Albadri (Middle East University, Jordan) and Yacoub Adel Nasereddin (Middle East University, Jordan)
Business Science Reference • ©2019 • 312pp • H/C (ISBN: 9781522580485) • US $225.00

Global Supply Chains and Multimodal Logistics Emerging Research and Opportunities
Deepankar Sinha (Indian Institute of Foreign Trade, India)
Business Science Reference • ©2019 • 220pp • H/C (ISBN: 9781522582984) • US $165.00

Agile Approaches for Successfully Managing and Executing Projects in the Fourth ...
Hür Bersam Bolat (Istanbul Technical University, Turkey) and Gül Tekin Temur (Bahçeşehir University, Turkey)
Business Science Reference • ©2019 • 424pp • H/C (ISBN: 9781522578659) • US $235.00

Technology Optimization and Change Management for Successful Digital Supply Chains
Ehap Sabri (KPMG LLP, USA & University of Texas at Dallas, USA)
Business Science Reference • ©2019 • 323pp • H/C (ISBN: 9781522577003) • US $210.00

Strategy and Superior Performance of Micro and Small Businesses in Volatile Economies
João Conrado de Amorim Carvalho (Unidade de Ensino Superior Dom Bosco, Brazil) and Emmanuel M.C.B. Sabino (FORUM (Centro de Formação, Estudos e Pesquisas), Brazil)
Business Science Reference • ©2019 • 389pp • H/C (ISBN: 9781522578888) • US $225.00

Logistics and Transport Modeling in Urban Goods Movement
Jesus Gonzalez-Feliu (Ecole des Mines de Saint-Etienne, France)
Business Science Reference • ©2019 • 273pp • H/C (ISBN: 9781522582922) • US $185.00

Hierarchical Planning and Information Sharing Techniques in Supply Chain Management
Atour Taghipour (Normandy University, France)
Business Science Reference • ©2019 • 287pp • H/C (ISBN: 9781522572992) • US $185.00

For an entire list of titles in this series, please visit:
https://www.igi-global.com/book-series/advances-logistics-operations-management-science/37170

IGI Global
DISSEMINATOR OF KNOWLEDGE

701 East Chocolate Avenue, Hershey, PA 17033, USA
Tel: 717-533-8845 x100 • Fax: 717-533-8661
E-Mail: cust@igi-global.com • www.igi-global.com

Table of Contents

Section 4
Optimization

Detailed Table of Contents

Additive manufacturing (AM) is acquiring attention in the field of manufacturing. The technique facilitates building of parts through the addition of materials using a computerized three-dimensional solid model. However, the process does not require any coolants, cutting tools, or other resources that are used in conventional manufacturing. The numerous advantages over conventional manufacturing have created interest towards the applications of additive manufacturing in the field of engineering. The governing fundamental principles of additive manufacturing offer a wide spectrum of advantages which includes design, geometric flexibility, near-net-shape capabilities, and fabrication using various materials, reducing the cycle time for manufacturing and overall savings in both energy and costs. The chapter provides a step-by-step procedure for generation of a component through 3D printing and a brief discussion on advanced AM techniques. These can produce high-quality products at high speed and can be used as industrial manufacturing techniques.

Section 2
Importance and Application

Three-dimensional or 3D printing technology is a growing interest in medical fields like tissue engineering, dental, drug delivery, prosthetics, and implants. It is also known as the additive manufacturing (AM) process because the objects are done by extruding or depositing the material layer by layer, and the material may be like biomaterials, plastics, living cells, or powder ceramics. Specially in the medical field, this new technology has importance rewards in contrast with conventional technologies, such as the capability to fabricate patient-explicit difficult components, desire scaffolds for tissue engineering, and proper material consumption. In this chapter, different types of additive manufacturing (AM) techniques are described that are applied in the medical field, especially in community health and precision medicine.

Additive manufacturing (AM) is the most advanced recently trending manufacturing technique that employs 3D printers to create 3D objects by layer upon layer fabrication from the base to the top. The required trajectory of the fabricating tool to create the layer can be well programmed by CAD software available in the market. The 3D CAD model in the computer can be manipulated and customized for different design needs of the product. These manipulations in model and quick fabrication process make the system a flexible and an effective one. This chapter discusses the AM application in educational system by describing the individual AM processes, their limitations, advantages, feasibility in general conditions, and planning for future generations to get accustomed to this technology from the early education in schools to the specialized education in universities. The technology enables students to convert 2D objects into 3D on the CAD software and feel them physically by 3D printing. AM also enables teachers to demonstrate their ideas easily to students.

Section 3
Design and Analysis

Heart valve prostheses are well known and can be classified in two major types or categories: biological and mechanical. Biological valves (i.e., Homografts and Heterografts) make use of animal tissue as the valving mechanism whereas mechanical valves make use of balls, disks, and other mechanical valving mechanism. Mechanical valves carry considerable risk and require lifelong medication. The design of these valves is usually done on a "one size fits all" basis, with only the diameter changing depending on the model being produced. The author seeks to present an application of additive manufacturing in the design process for mechanical valves. This is expected to provide patients with customized prostheses to match their physiology and reduce the risk associated with the implantation.

This chapter builds new knowledge for design engineers adopting fused deposition modeling (FDM) technology as an end manufacturing process, rather than simply as a prototyping process. Based on research into 2.5D printing and its use in real-world additive manufacturing situations, a study featuring 111 test pieces across the range of 0.4-4.0mm in thickness were analyzed in increments of 0.1mm to understand how these attributes affect the quality and print time of the parts and isolate specific dimensions which are optimized for the FDM process. The results revealed optimized zones where the outer wall, inner wall/s, and/or infill are produced as continuous extrusions significantly faster to print than thicknesses falling outside of optimized zones. As a result, a quick reference graph and several equations are presented based on fundamental FDM principles, allowing design engineers to implement optimized wall dimensions in computer-aided design (CAD) rather than leaving print optimization to technicians and manufacturers in the final process parameters.

Fused deposition modelling is an extrusion-based automated fabrication process for making 3D physical objects from part digital information. The process offers distinct advantages, but the quality of part lacks in surface finish when compared with other liquid or powder based additive manufacturing processes. Considering the important factors affecting the part quality, the chapter attempted to optimize the raster angle, air gap, and raster width to minimize overall part roughness. Experiments are designed using face-centered central composite design and analysis of variance provides the effects of processing parameters on roughness of part. Suitability of developed model is tested using Anderson-darling normality test. Desirability method propose that roughness of different part faces are affected differently with chosen parameters, and thus, hybrid approach of WPCA based TOPSIS is used to break the correlation between part faces and reduce the overall part roughness. Optimizing shows that lower raster angle, lower air gap, and larger raster width minimizes overall part roughness.

Preface

We, editors, would like to present the book *Additive Manufacturing Technologies From an Optimization Perspective* under Book Series *Advances in Logistics, Operations, and Management Science (ALOMS)*. The present emphasis on Industry 4.0 or I 4.0 was the major thrust for choosing the book title. With globalization of market and advances in science and technology, the life span of products has shortened considerably. For early realization of products and short development period, engineers and researchers are constantly working together for more and more efficient and effective solutions. The most effective solution identified has been usage of computers in both designing and manufacturing. This gave birth to the nomenclatures CAD (Computer Aided Designing) and CAM (Computer aided Manufacturing). This was the initiation that ensured short product development and realization period. In this context ''Additive Manufacturing / Rapid Prototyping / 3D Printing / Layered Manufacturing" etc. etc. has become the buzz word for all major disciplines and many scholars are working in these areas. This book provides an insight for all researchers, academicians, post graduate or senior undergraduate students working in this important area.

Additive Manufacturing refers to a process by which digital 3D design data is used to build up a component in layers by depositing material. The term "3D printing" is increasingly used as a synonym for Additive Manufacturing. However, the latter is more accurate in that it describes a professional production technique which is clearly distinguished from conventional methods of material removal. Instead of milling a work-piece from solid block, for example, Additive Manufacturing builds up components layer by layer using materials which are available in fine powder form. A range of different metals, plastics and composite materials are currently being used. Additive Manufacturing is now being used increasingly in Series Production. It gives Original Equipment Manufacturers (OEMs) in the most varied sectors of industry, the opportunity to create a distinctive profile for them based on new customer benefits, cost-saving potential and the ability to meet sustainability goals.

The strengths of Additive Manufacturing lie in those areas where conventional manufacturing reaches its limitations. The technology is of interest where a new approach to design and manufacturing is required so as to come up with solutions. It enables a design-driven manufacturing process - where design determines production and not the other way around. What is more, Additive Manufacturing allows for highly complex structures which can still be extremely light and stable. It provides a high degree of design freedom, the optimization and integration of functional features, the manufacture of small batch sizes at reasonable unit costs and a high degree of product customization even in serial production.

The first methods for Additive Manufacturing became available in the late 1980s and were used to produce models and prototype parts. Today, they are used for a wide range of applications and are used to manufacture production-quality parts in relatively small numbers if desired without the typical unfavourable short-run economics. This economy has encouraged online service bureaus for early product realization or physical products for actual testing.

The chapters in the book have been provided by Researchers and Academicians working in the field and have gained considerable success in the field. The chapters of the book are segregated in four sections namely Section 1: Review of the State – of - Art; Section 2: Importance and Application; Section 3: Design and Analysis; and Section 4: Optimization .

Section 1 contains Chapter 1, whereas Section 2 contains Chapter 2 to Chapter 5, Section 3 consists of Chapter 6 to Chapter 8, and Section 4 with Chapters 9 to Chapter 12.

Section 1 of the book starts with Chapter 1. In it, a step by step procedure for generation of a component through 3D printing is presented and a brief discussion on Advanced Additive Manufacturing Techniques has been elaborated. As the technology facilitates in a building of parts through the addition of materials layer by layer using a computerized three-dimensional solid model and does not require any coolants, cutting tools, or other resources, as used in conventional manufacturing, hence the state of art requires some training and expertise.

Section 2 is dedicated to Importance and Application of Additive Manufacturing and contains four chapters. The first chapter of the section (i.e. chapter 2) talks about different types of Additive Manufacturing (AM) techniques are described which are applied in medical field especially in community health and precision medicine. Additive Manufacturing is currently developing an impact on medical fields like tissue engineering, dental, drug delivery, prosthetics and implant areas as the process is capable of creating objects using biomaterials, plastics, living cells or powder ceramics. This technology is also gaining importance reward over conventional technologies, such as the capability to fabricate patient-explicit difficult components, desire scaffolds for tissue engineering, proper material consumption.

Chapter 3 explores application in educational system by describing the individual AM processes, their limitations, advantages, feasibility in general conditions and planning for future generation to get accustomed to this technology from the early education in schools to the specialized education in universities. The technology enables students to convert 2D objects into 3D on the CAD software and feel them physically by 3D printing. AM also helps teacher to demonstrate their ideas to students with much ease.

The next chapter (i.e. Chapter 4) intrudes into the applications of Additive Manufacturing for crack repairs in Ti alloys. AM offers several advantages and has become a preferred freeform fabrication, processing, manufacturing, maintenance and repair technique for metals, thermoplastics, ceramics and composites. When using laser, it bears several names which include, Laser Additive Manufacturing, Laser Additive Technology, Laser Metal Deposition, Laser Engineered Net Shape, Direct Metal Deposition and Laser Solid Forming. These technologies use a laser beam to locally melt the powder or wire and the substrate that fuse upon solidification. Titanium (Ti) alloys offer very attractive properties much needed in Aerospace, Medical etc. industries. The chapter also discusses Metal cracking industrial challenges, crack detection and repair methods, challenges and milestones for Additive Manufacturing repair of cracks in Ti alloys.

The last chapter of the section (i.e. in Chapter 5) is a bio-design study of the thoracic aorta in adults. In recent decades, vascular surgery has seen the arrival of endovascular techniques for the treatment of vascular diseases such as aortic diseases (aneurysms, dissections, and atherosclerosis). The 3D printing process by addition of material gives an effect or of choice to the digital chain, opening the way to the manufacture of shapes and complex geometries, impossible to achieve before with conventional methods. In this chapter a bio-design protocol was established based on medical imaging, extraction of the shape and finally the 3D modelling of the aorta, secondly a bio-printing method based on 3D printing that could serve as regenerative medicine has been proposed. A simulation of the bio-printing process was carried out under the software Simufact additive whose purpose is to predict the distortion and residual stress of the printed model. The binder injection printing technique in a Powder Bed Printer (PBP) bed was used. The results obtained are very acceptable compared with the results of the error elements found.

From here the book starts with Section 3 which groups Design and Analysis part.

The first chapter of the section (i.e. Chapter 6) illustrates the design of a heart valve Heart valve. Heart valve prostheses are well known in the art, and can be classified in two major types or categories – biological and mechanical. Biological valves i.e. Homografts and Heterografts make use of animal tissue as the valving mechanism whereas mechanical valves make use of balls, disks and other mechanical valving mechanism. Mechanical valves carry considerable risk and require lifelong medication. The design of these valves is usually done on a "one size fits all" basis, with only the diameter changing depending on the model being produced. The chapter presents an application of additive manufacturing in the design process for mechanical valves. This is expected to provide patients with customized prostheses so as to match their physiology and reduce the risk associated with the implantation.

Chapter 7 provides new knowledge for design engineers adopting fused deposition modelling (FDM) technology, an Additive Manufacturing Technique, as an end manufacturing process, rather than simply as a prototyping process. Based on research into 2 ½ D printing and its use in real-world additive manufacturing situations, the chapter does a study featuring 111 test pieces across the range of 0.4-4.0 mm in thickness. They were analyzed in increments of 0.1mm to understand how these attributes affect the quality and print time of the parts, and isolate specific dimensions which are optimized for the FDM process. The results revealed optimized zones where the outer wall, inner wall/s and/or infill are produced as continuous extrusions, significantly faster to print than thicknesses falling outside of optimized zones. As a result, a quick reference graph and several equations were presented based on fundamental FDM principles, allowing design engineers to implement optimized wall dimensions in computer aided design (CAD), rather than leaving print optimization to technicians and manufacturers in the final process parameters.

The next chapter and last chapter of the section (i.e. Chapter 8) deals with advanced laser additive manufacturing (AM) process with capacity of directly producing parts through CAD model. Laser metal deposition, is an AM process used in the repair and fabrication of solid components but sometimes associated with thermal induced stresses which usually leads to cracks in deposited parts. Titanium aluminide alloys fall under the group of intermetallic compounds known for high temperature applications and display of superior physical and mechanical properties, which made them most sort after in the aeronautic, energy and automobile industries. This chapter looks at some AM processes with more emphasis laser metal deposition (LMD) technique and effect of LMD processing parameters and preheating of substrate on the physical, microstructural and mechanical properties of components produced through AM process.

From here the book starts with Section 4 which groups the Optimization part.

The first chapter of the section, Chapter 9, elaborates effect of printing temperature on layer sticking while dimensional accuracy is achieved. When the AM process is considered, there are also other parameters, such as nozzle size, flow rate of material, print-speed, print-bed temperature, cooling rate, pattern of printing etc. There are also dependencies which would be addressed in between these parameters; for example, if the printing temperature is increased, it is not clear if the viscosity of the material will increase or decrease. In this chapter theoretical modelling and experimental testing was performed to prove the relationship. This type of formulation can be later adapted into a slicer program, so that the program automatically selects some of the printing parameters to achieve desired dimensional accuracy and layer sticking.

Chapter 10 uses an integrated MCDM approach, a recent optimization tool, to study the various process parameters of fused deposition modelling (FDM). FDM fabricate the quality prototype directly from the CAD data and is depends on processing parameters, hence optimization is essential. The integrated MCDM approach consists of modified Fuzzy with ANP methods. Experimentation were performed considering three process parameters namely layer height, shell thickness and fill density and corresponding response parameters namely ultimate tensile strength, dimensional accuracy and manufacturing time were determined. Thereafter, optimization of FDM process parameters is done using proposed method. The result showed promising results providing higher ultimate TS, good DA, and lesser MT as well as improves the performance and efficiency of FDM.

Chapter 11 studies the effects of different manufacturing parameters on the materials properties of the finished parts. The chapter, hence, explains the effects of 6 different process parameters on the impact resistance. In particular, print temperature, print speed, infill ratio, infill pattern, layer height and print orientation parameters were studied and their effects on impact resistance were measured experimentally. Moreover, the optimum values of the process parameters for impact resistance were found. This chapter provides an important guideline for 3D manufacturing in terms of impact resistance of the printed parts. Furthermore, by using this methodology the effects of different 3D printing process parameters on the other material properties can be determined.

Chapters 12, the last chapter of the section and the book, attempted to optimize the raster angle, air gap and raster width to minimize overall part roughness. Fused deposition modelling, an extrusion based additive manufacturing process offers distinct advantages but the quality of part lacks in surface finish when compared with other liquid or powder based additive manufacturing processes. Considering the important factors affecting the part quality, experiments were designed using face centered central composite design and analysis of variance provides the effects of processing parameters on roughness of part. Suitability of developed model is tested using Anderson-darling normality test. Desirability method propose that roughness of different part faces are affected differently with chosen parameters and thus hybrid approach of WPCA based TOPSIS is used to break the correlation between part faces and reduce the overall part roughness. Optimizes shows that lower raster angle, lower air gap and larger raster width minimizes overall part roughness.

First and foremost, the editors would like to thank God for providing the power of thinking and means of expression. The Editors would also like to thank all the Chapter Contributors, the Reviewers, the Editorial Advisory Board Members, Book Development Editor and the team of Publisher IGI Global for their availability for work on this editorial project. Also, the editors, have sincere gratitude towards all who directly or indirectly helped them in this project.

Kaushik Kumar
Birla Institute of Technology Mesra, India

Divya Zindani
National Institute of Technology Silchar, India

J. Paulo Davim
University of Aveiro, Portugal

Section 1

Review of the State of the Art

Chapter 1
Recent Advancement in Additive Manufacturing

Satyanarayana Kosaraju
Gokaraju Rangaraju Institute of Engineering and Technology, India

Krishna Mohan B.
Gokaraju Rangaraju Institute of Engineering and Technology, India

Swadesh Kumar Singh
Gokaraju Rangaraju Institute of Engineering and Technology, India

ABSTRACT

Additive manufacturing (AM) is acquiring attention in the field of manufacturing. The technique facilitates building of parts through the addition of materials using a computerized three-dimensional solid model. However, the process does not require any coolants, cutting tools, or other resources that are used in conventional manufacturing. The numerous advantages over conventional manufacturing have created interest towards the applications of additive manufacturing in the field of engineering. The governing fundamental principles of additive manufacturing offer a wide spectrum of advantages which includes design, geometric flexibility, near-net-shape capabilities, and fabrication using various materials, reducing the cycle time for manufacturing and overall savings in both energy and costs. The chapter provides a step-by-step procedure for generation of a component through 3D printing and a brief discussion on advanced AM techniques. These can produce high-quality products at high speed and can be used as industrial manufacturing techniques.

DOI: 10.4018/978-1-5225-9167-2.ch001

INTRODUCTION

Manufacturing plays a vital role in the creation of wealth and progresses in the quality of life. The intricacies of manufacturing are applied to (1) system design and organization, (2) technological logistics, and (3) operational planning and control. Manufacturing technology can be broadly classified into two conventional or non-conventional processes. The conventional processes come in existence since prior to 1950, and the non-conventional processes have been adopted and implemented since the 1950s.

Additive manufacturing (AM), formerly known as rapid prototyping, is a process of joining materials to build parts from 3D model data (digital), usually layer by layer, as opposed to subtractive or conventional machining. AM, also known universally as 3D printing, evolved in 1980s with advances in Computer aided design (CAD). 3D printing allows a design a computerized three - dimensional model that can be easily transformed to a finished end product without the aid of any additional tools. This leads to the possibility of producing parts that have complex shape and would be difficult to obtain using conventional and nonconventional material removal processes. The benefits of additive manufacturing can lead to novel innovations in design, a necessary process for the manufacturing and assembly of any product. Application of AM are predominantly in the field of engineering and biomedical engineering. Medical devices named ZipDose®, HPAM™, printers which include bio-printer and inkjet printer, and prosthetics like hip, knee implants receiving interest day by day (Singh & Ramakrishna, 2017).

In the following sections a detailed discussion on classifications of AM process, steps involved in AM, and various applications of AM are described. In addition to that advanced AM methods like Multi Jet Fusion (MJF), Ink based Direct Write, and Continuous Liquid Interface Production (CLIP) are also discussed.

ADDITIVE MANUFACTURING PROCESS AND THEIR CATEGORIES

AM processes are categorized into seven families. Each category is characterized by the principles according to which material is added. They include Vat Photo-polymerization, Powder Bed Fusion (PBF), Material Extrusion, Directed Energy Deposition (DED), Binder Jetting, Material

Jetting, and Sheet Lamination. Table 1 illustrates AM processes and their methodologies.

VARIOUS STAGES OF THE AM PROCESS

Most common sequence of steps to produce the products for any AM process which are

1. Creation of CAD Model
2. Transformation of CAD Model to STL File
3. Transfer of STL file onto AM machine
4. Machine Settings
5. Printing or Building of Part
6. Post-processing of part

There are some variations in processes involved in particular steps which depend on the technology that is being used. Figure 1 shows the work flow of part printed using 3D printer built on FDM process.

Creation of CAD Model

For any AM process digital models are must which are generally designed by computer aided design (CAD) software. These CAD Models are the basic requirement for the AM Technology. Large range of CAD programs like Solidworks, CATIA, and PTC Creo are available in market to design 3D

Table 1. AM processes and their methodologies

AM Process	Process Methodology
Vat Photo-polymerization	Resins (photo polymers) are cured by light of specific wave length
Powder Bed Fusion (PBF)	Energy is used to fuse powder material which was spread as a thin layer, precisely, using some fusing agents
Material Extrusion (ME)	Molten material is precisely extruded through Nozzle
Directed Energy Deposition (DED)	Powder or filament material is fused by applying thermal energy from sources like laser or electron beam
Binder Jetting (BJ)	Strata of Powder material bonded using binding material
Material Jetting (MJ)	Material droplets are sprayed precisely.
Sheet Lamination	Parts are built by binding Sheets

Figure 1. Work flow of part printed using 3D printer built on FDM process

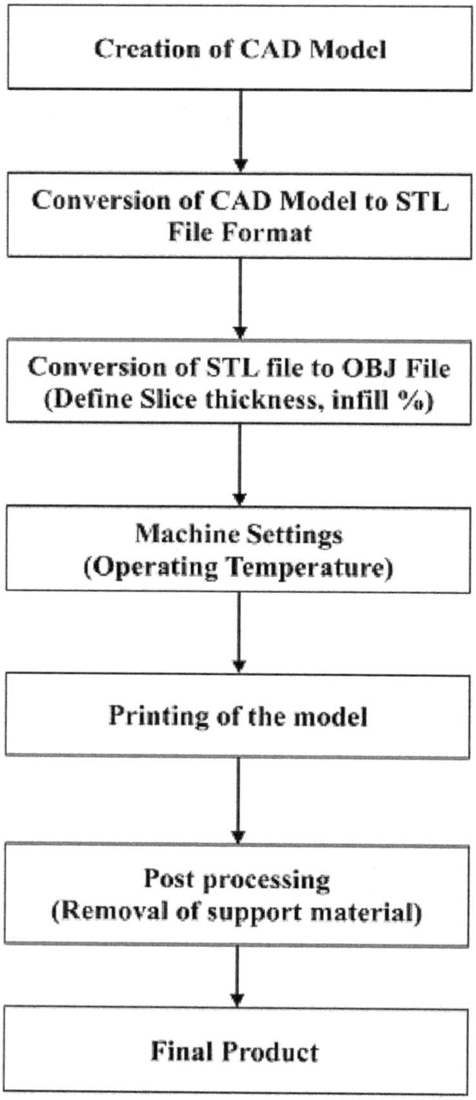

Models. In medical field, Medical imaging data (CT scan) is first converted into IGES files using software like Materialise- Mimics, 3-D Slicer, 3D Doctor, etc. These IGES files are then imported into CAD software to create 3D models of Tissue and organs. In addition to the above-mentioned methods a new technology has evolved which is known as reverse engineering. It uses 3D scanners to directly scan a part and create a 3D model of the part. Figure 2 shows a part being designed using Solidworks CAD package.

Transformation of CAD Model to STL File

Irrespective of creation of CAD model methods all 3D CAD models must be converted into STL file format. STL is universally considered to be standard format for AM systems. STL is derived from stereolithography, a commercial AM technology from 3D Systems. STL format represents a set of triangular facets on the surface of 3D CAD model. Most of CAD software are provided with STL file interface option. The CAD-STL interface executes surface tessellation, a method of approximating a surface using triangles, and converts data to either a Binary or ASCII STL file format. Software like Materialise- MAGICS convert CAD models into STL files and also repair STL files if there are any errors. Corrupted STL files produce inferior quality parts with poor resolution and geometric inaccuracies. The dimension of triangles' facets can be set as parameter in CAD. Smaller the dimensions of the triangle's smoother are the surfaces and time for building the model

Figure 2. Part designed using Solidworks

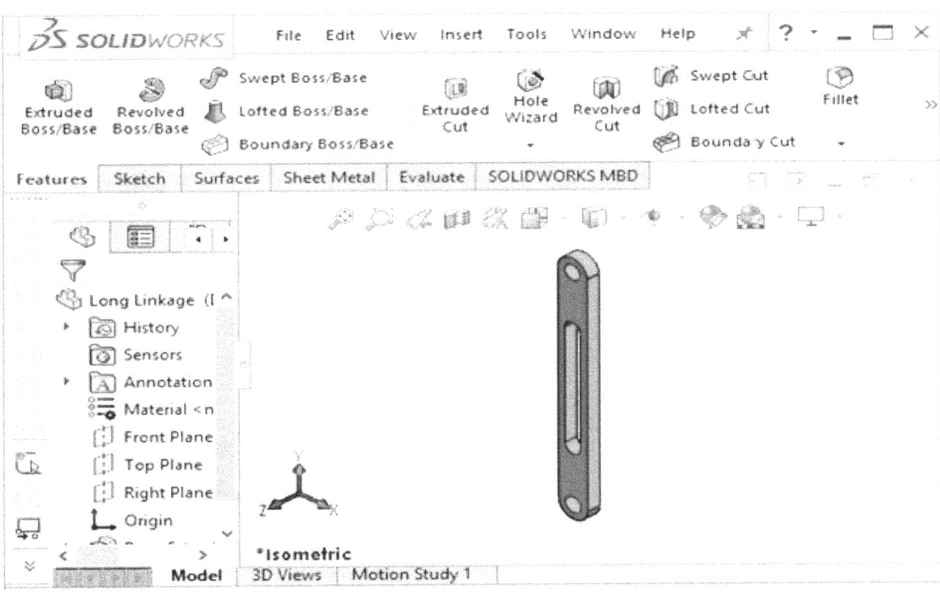

is more. So, choosing the dimensions of these triangles plays the important role in building the model.

Transfer of STL File Onto AM Machine

STL files are imported into AM system software or a dedicated slicer program as shown in Figure 3. The GUI of AM system software allows the user to view the part which is going to be printed. The operator can set the position of the part, change the orientation of build, and set build location on the platform. Multiples of same part or multiple parts can be built using an AM machine simultaneously. STL files can also be scaled, by maintaining the aspect ratio, to accommodate allowances like shrinkage allowance. Once the STL file is opened or imported into an AM system software like MakerBot Replicator, the operator has to define printer settings. Figure 4 shows the windows of printer settings. The operator has to set the resolution and decide whether supports are to be used. Under advanced options the operator can set the slicer parameters (Quality, Temperature and speed) like amount of infill, number of shells, layer height (slice thickness), extruder temperature, build plate temperature, speed while extruding and speed while travelling. When slicing is done, the STL file is transformed into G-code (NC part program). The AM system will read the G-code and build the part layer by layer.

Figure 3. STL file of a 3D part imported into system software of MakerBot

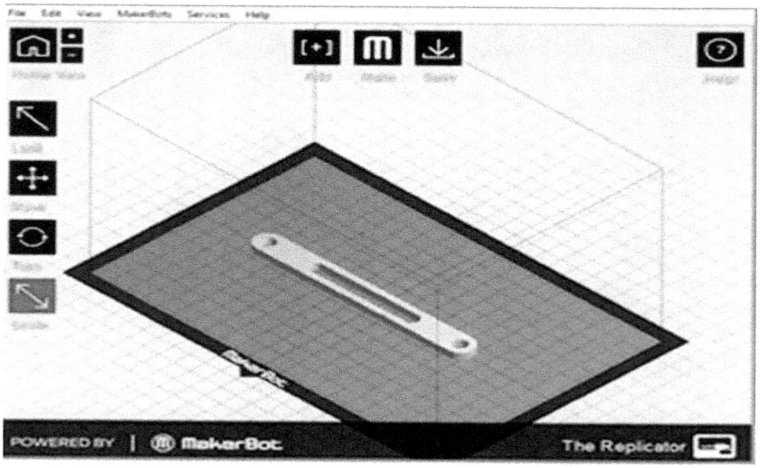

Figure 4. Windows of MakerBot Replicator software showing slicing options

Machine Settings

3D printing machines are assembly of complex parts. These machines should be maintained properly to achieve better results. Parts like nozzles should be cleaned regularly so that they are free of blockages. Leveling of build plate should be done frequently. Proper hygiene is to be maintained while handling AM machines which print parts related to medical field. The powder materials used in AM should be handled carefully as some of them have a limited shelf life. Excess build material should be inspected regularly and if the material properties change the build material should be replaced.

Printing or Building of Part

Printing or building of 3D parts is a layer by layers process and is common for most of the 3D printing techniques. But, method of material deposition varies from one process to the other.

In case of Vat Photopolymerization process photopolymer resin solidifies into a solid when it is exposed to UV light or light with certain wavelength

In case of Powder Bed Fusion (PBF) processes, powder particles (metallic or nonmetallic) are fused together using an energy source (lasers or electron beams). The powder is spread as thin layer as a part is being printed.

In case of Material Extrusion technologies molten material is extruded through nozzle which follows a predetermined path to print the part layer-by-layer.

In AM Machines employing DED, the raw material is melted using thermal energy (by focusing laser or electron beam) and the molten material is precisely deposited as strata to print a part. The built material mostly used in this process is metal in powder form. Metal parts are mostly printed using this process. So, it commonly referred as metal deposition process.

In AM Machines employing Binder jetting, strata of Powder material are bonded using binding material. The binding agent is precisely sprayed on thin layer of power

Material jetting process is like 2-D ink jetting. Generally, materials like photopolymers which solidifies when subjected by Ultra violet light or at optimum temperatures will be employed to print parts layer by layer. Multi-materials can be printing using this process.

Table 2 provides the summary of AM processes, technology they employ, common manufacturers and materials they use. Supporting material should be removed before submitting the printed part for post processing.

Post-Processing of Part

In post processing, printed parts are processed into final products and are ready for application purposes. Post processing procedures also vary according to printer technology and some printed parts does not require any post processing. Generally, SLA printed parts require curing with UV before using. Some metallic parts need sintering to relieve internal stresses. Support material are also removed at the post-processing stage. Post-processing techniques like polishing, tumbling, coloring, and high-pressure air cleaning are employed on printed parts.

The supports are provided for stability while building the part using AM Processes. In case of fused deposition modeling process the material used for supports is same as the build material or similar to build material. As a support material, High impact polystyrene (HIPS) is used along with Acrylonitrile Butadiene Styrene (ABS) and PolyVinyl Alcohol is used along with Polylactide (PLA). HydroFill is used as support material for both ABS and PLA. Supports are removed from the printed part by using physical means (using pliers) or by using solvents. HIPS is dissolvable in Limonene whereas PLA and HydroFill are dissolvable in water. The surface of the FDM parts is rough due to the presence of strips (representing different layers of material deposition). So, most of the FDM parts need surface finishing. First step

Table 2. AM processes, methodology they employ, list manufacturers and raw materials they use

AM Process	Methodology	List of Manufacturers	Raw Materials
Vat Photopolymerization	Stereolithography	DWS systems Formlabs 3D Systems	Clear, Greyscale, Color Resins
	Digital Light Processing	MoonRay B9 Creator	Biocompatible, Color Resins
	Continuous Liquid Interface Production	EnvisionTEC Carbon3D	Variety of Resins
PBF	Selective Laser Sintering	EOS Stratasys	Nylon, Poly Ether Ether Ketone, Thermoplastic polyurethane filaments
	Selective laser melting/ Direct metal laser sintering	EOS 3D Systems Sinterit	CoCr, Al, Ti, steels, Ni alloys,
	Electron-Beam Melting	Arcam	Ti, CoCr
	Multi Jet Fusion	HP	Nylon
ME	Fused Deposition Modeling	Stratasys Ultimaker MakerBot Markforged	fiber reinforced Nylon, Nylon, Acrylonitrile Butadiene Styrene, Polylactic Acid filaments
MJ	Material jetting	Stratasys (Polyjet), 3D Systems (MultiJet)	Rubber, Acrylonitrile Butadiene Styrene
			Inks
	NanoParticle Jetting	Xjet	steel, ceramics
	Drop On Demand	Solidscape	Wax
BJ	Binder jetting	3D Systems, Voxeljet	Gypsum, PMMA
		ExOne	Ceramics, steels, CoCr, WC
DED	Laser Engineered Net Shaping	Optomec	Ti, Al, Cu, steels
	Electron-beam additive manufacturing	Sciaky Inc	Ti, Al,CuNi, 4340 steels

in surface finishing is sanding. Sand papers of grit ranging from 150-2000 are used for sanding. Sanded parts are then cleaned and voids or gaps are filled. Epoxy resin is used to fill voids or gaps on PLA FDM printed parts whereas voids in ABS FDM parts are filled with mixture of ABS filament and acetone. After filling the voids FDM parts are polished with Blue Rogue before coloring. Before coloring the FDM parts are primed using aerosol primer. Acrylic paints are commonly used for coloring.

Barrel finishing or tumbling process can be used to improve the surface finish of FDM printed parts. Printed parts are placed in a drum along with media like ceramic, synthetic, etc. and rotated (Boschetto & Bottini, 2015; Cordes, 2019).

The surface finish of ABS FDM parts can be improved by Acetone vapor bath. When the parts are exposed to Acetone vapors the outer surface of the part will become like a smooth slurry of ABS and when the Acetone evaporates the slurry cool down and solidifies giving better surface finish (Lalehpour & Barari, 2016).

Laser based polishing can be used to improve the surface finish of parts printed using SLS AM Process. Titanium and Cobalt Chrome based alloy parts are processed, by re-melting a very thin outer surface layer, using Laser in inert gas atmosphere (Gora et al., 2016).

APPLICATIONS OF AM IN BIOMEDICAL FIELD

AM has tremendous scope in biomedical engineering field. The AM applications include printing models of organs & tissues, surgical instruments & tool, and designing splints, implants & prosthesis (Javaid & Haleem, 2018). AM is used to produce porous scaffolds to mimic bone performance. Ti–6Al–4V scaffolds with pore dimensions in range of 200- 500 μm are (Alabort, Barba, & Reed, 2019). 3-D zirconium dioxide ceramic are printed for orthopedic applications by DW printing method (Li, Li, & Li, 2015). AM processes like Electron beam melting (EBM), Laser powder bed fusion (LPBF) are used to print Metal implants used for Total joint arthroplasty (Narra, Mittwede, Wolf, & Urish, 2019). Hydroxyapatite scaffolds with multiscale porosity with symmetric architecture are fabricated by DW (Michna, Wu, & Lewis, 2005).Porous structures based on biodegradable resin materials are printer using stereolithography for tissue engineering application (Melchels, Feijen, & Grijpma, 2010).

AM can assist physicians and surgeons in diagnostics, planning, monitoring and analysis. Using AM technology cardiologists and cardiac surgeons can print and analyze 3D models of organs before performing surgical procedure (Haleem, Javaid, & Saxena, 2018).

Powder Bed Fusion: Multi Jet Fusion

Based on powder bed fusion AM process, HP developed an advanced 3D printing process called multi-jet fusion (MJF) ("Technical white paper, HP Multi Jet Fusion technology: A disruptive 3D printing technology for a new era of manufacturing ", 2017, May). MJF employs set of fusing agents, detailing agents and energy (infrared radiation) to precisely fuse thin layers of powder to create complicate parts. Figure 5 shows schematic of high speed MJF synchronous printing architecture. Dual carriages which are arranged mutually perpendicular to each other scan across the Working area. The first carriage coats the built area with fresh power material, and the second carriage spray fusing & detailing agents and fuses the area of interest. The main advantage of this process over other processes is that each process (recoating and printing) can be optimized independently. With MJF 3D Printer users can print a part, or multiple parts-built layer-by-layer. The process commences

Figure 5. HP Multi Jet Fusion synchronous printing architecture
("Technical white paper, HP Multi Jet Fusion technology: A disruptive 3D printing technology for a new era of manufacturing ", 2017, May)

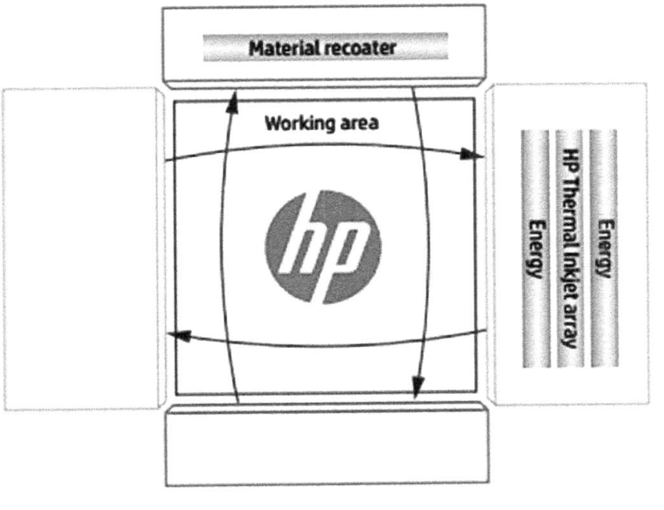

by coating the built area with thin layer of fresh material powder by the first carriage, as shown in Figure 6 (a). Figures 6 (b-e) represent scanning process done by the fusing carriage (second carriage). At the beginning of scanning process required amount of energy is applied to the fresh layer of powder material to maintain optimum material temperature (6 (b)). A fusing Agent "F" is selectively sprayed over power where particles are to be fused together and then a Detailing Agent "D" is selectively sprayed over area where the fusing action is be screened (Figures 6(c) and (d)). After selectively printing fusing agent and the detailing agent, the whole layer is exposed to heat energy so that selected areas are fused. The solidified material bonds to the layer below it which was fused in the previous cycle. Figure 6(f) shows the fused and unfused areas. The freshly fused layer is again recoated and the cycle continues till the part is complete. The strength of a part along Z-axis is more for parts printed with MJF when compared to other similar processes.

MJF approach rectifies the problem of leakage due to porous nature of parts printed with other AM technologies. It is capable of printing thirty million drops per second on build area and can achieve high dimensional precision (±0.2%) when compared with other technologies. High quality, dimensional precision, and rate of printing are the best attributes of MJF which encourage the manufactures to consider it for mass production (Morales-Planas, Minguella-Canela, Lluma-Fuentes, Travieso-Rodriguez, & Garcia-Granada, 2018). Parts fabricated based on MJF are isotropic in nature, but some mechanical and tribological properties depend on the print orientation (O'Connor, Dickson, & Dowling, 2018; Palma et al., 2019).

Figure 6. Working principle of Multi Jet Fusion printing process
("Technical white paper, HP Multi Jet Fusion technology: A disruptive 3D printing technology for a new era of manufacturing ", 2017, May)

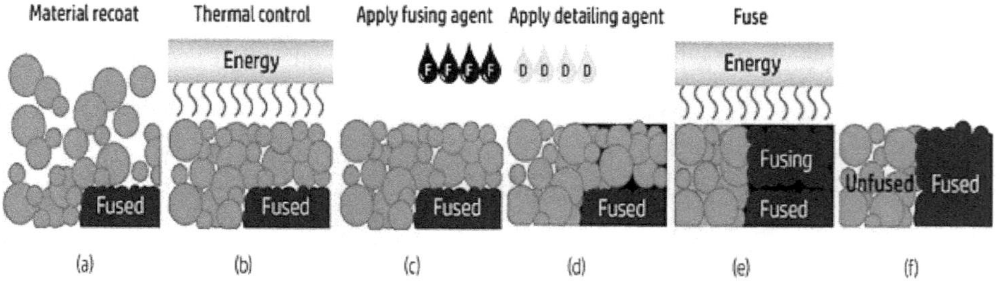

Figure 7. Classification of DW processes

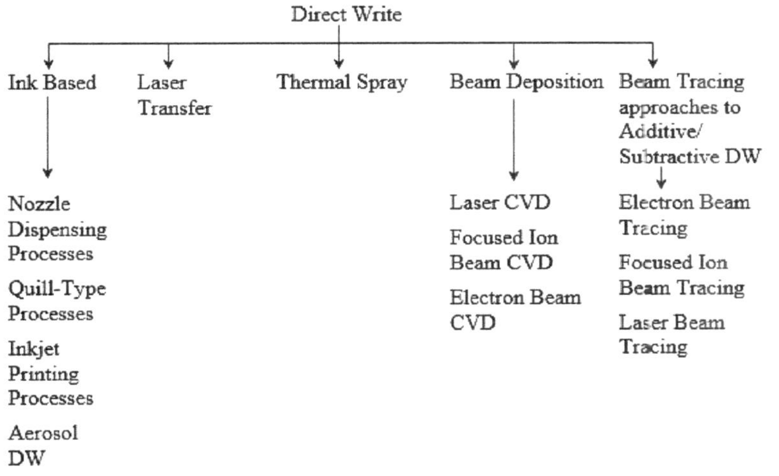

Direct Write Processes: Ink Based Direct Write

The term "Direct Write" (DW) is an additive manufacturing technique which can create 2D or 3D structures directly without using tooling or masks. DW techniques can create structures varying from macro to nano-scale using a freeform deposition. Figure 7 shows different classifications of DW processes based on the techniques through which the material is added (Gibson, Rosen, & Stucker, 2010).

Among all DW processes, Ink-based DW are more flexible and less expensive, processes. Figure 8 shows the equipment which was used to fabricate 3D Zirconium dioxide (ZrO_2) scaffolds for tissue engineering applications (Li et al., 2015).

The Ink based DW setup consists of three blocks. The first block represents CAD/ CAM interface which designs the part and transform it into digital information (NC part program). The second block consists of extrusion mechanism which eject inks at precise rates. The third block consists XYZ translation mechanism which drives the extrusion delivery system and the build platform. 3D ZrO_2 scaffolds are fabricated and are sintered at various temperatures (1100 to 1350 C range) to achieve the desired degree of densification (Li et al., 2015). Similar process was used to produce Hydroxyapatite scaffold (Michna et al., 2005). The 3D scaffolds printed using

Figure 8. Ink based DW setup
Source: (Li et al., 2015)

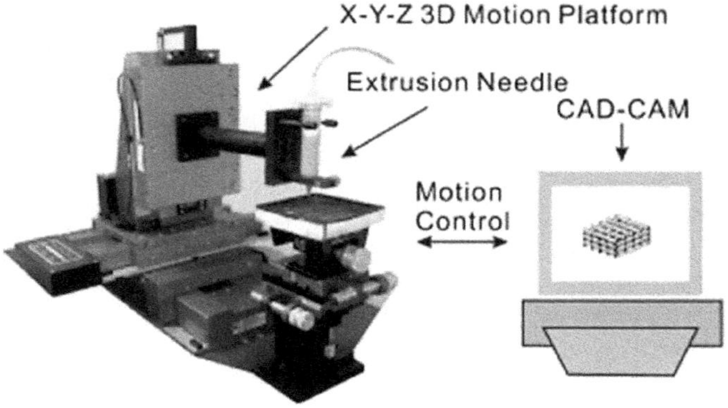

DW have better internal architecture and are suitable of tissue engineering applications. The ink used in DW should have excellent flow properties so that it can be flow through the extrusion mechanism, retain shape upon printing, and should have good cohesive and adhesive properties (Gibson et al., 2010).

Vat Photopolymerization Process: Continuous Liquid Interface Production (CLIP)

Current AM methods which include FDM, SLS, and SLA are extremely slow processes since they employ layer-by-layer printing (He, Yang, & Pan, 2019). A part of several hundred millimeters in height will take hours to fabricate using these methods and are not feasible for mass production. An additive manufacturing technique to be considered as realistic for industrial production, printing rate must increase without compromising part accuracy or resolution. CLIP process is much faster process (print rate 500mm/hr) than normal stereolithography process (Tumbleston et al., 2015).

CLIP is based on a vat photopolymerization AM technology where the curing of resin is continuous without interruption. The built part is drawn out of the resin pool at constant rate. Figure 9 illustrates the schematics of a 3D printer that uses CLIP process. CLIP process starts with projection of continuous sequence or video of UV images (from image processing unit) through an oxygen-permeable, UV transparent window below a photopolymer bath. An oxygen inhibition layer or a "dead zone (DZ)" is created between liquid resin and already cured part. When the solid part is moved out of the

Figure 9. Illustration of continuous printing based on CLIP process
Source: (He et al., 2019)

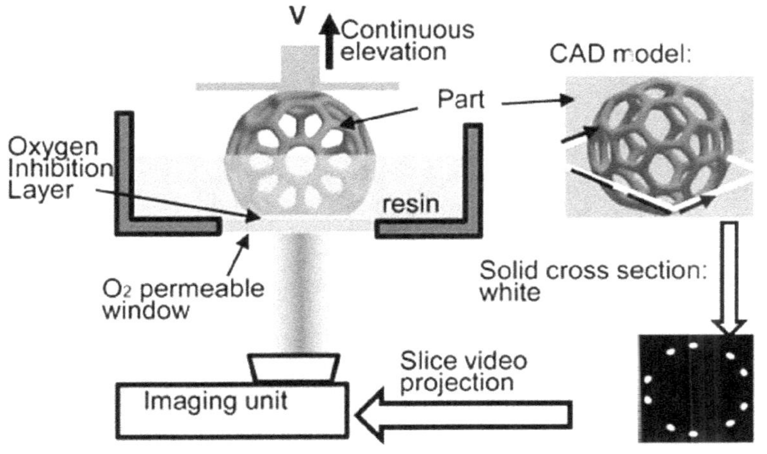

resin bath continuously from the top it grows at a rate equal to the speed at which it is drawn out. The photopolymerization is continuous which makes this process very fast when compared to other stereolithography approaches (Tumbleston et al., 2015).

Presence of oxygen-inhibited DZ is essential to the CLIP. Figure 10 shows that when the system is subjected to constant photon flux the DZ thickness is more when pure oxygen is present and vanishes in presence of nitrogen. In the presence of air the thickness is about half (of Pure oxygen).When the amount of photon flux increases the dead zone thickness decreases (Tumbleston et al., 2015). CLIP uses Teflon AF 2400 window material because of its excellent oxygen permeability, UV transparency, and chemical inertness. Continuous part production is not possible without a suitable DZ.

CONCLUDING REMARKS

The need for fast and exact manufacturing process combined with positive eco-friendly impact have been directed to additive manufacturing, which being referred as "third industrial revolution." For continues improvement especially for complex structural applications in the field of engineering and medical quantification studies are necessary. Such studies would allow manufacturers not only to optimize additive manufacturing techniques and

Figure 10. Relationship between Dead zone thickness and incident photon flux
Source: (Tumbleston et al., 2015)

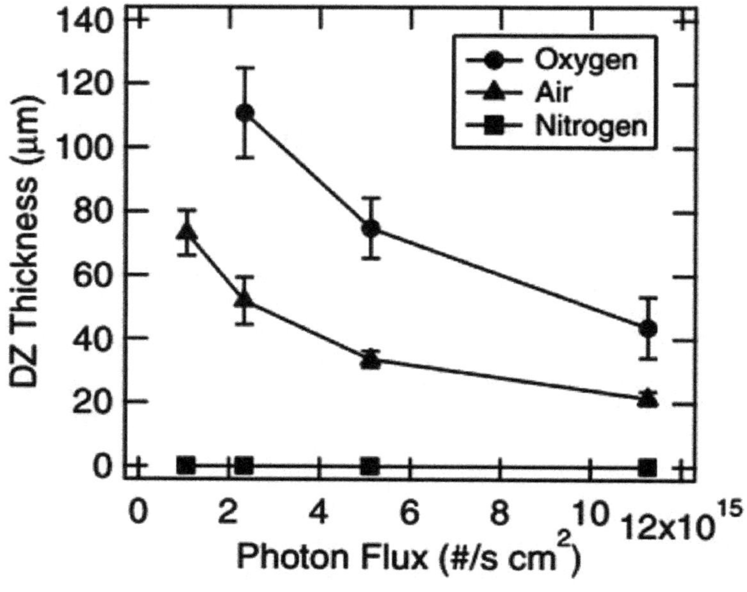

materials but also to grow effective methods for examining their products. The present utilization of added substance to produce complex shapes, for example, therapeutic gadgets, turbine edges, and even complex auxiliary parts. Even though AM have limitations like the rate of creation, surface completion, cost of generation, however with improving innovation AM can replace the ordinary machining procedures. As of now it made a havoc in the biomedical field. A most vital and creating pattern for added substance producing is its utilization for individual consumer purposes.

FUTURE SCOPE AND DEVELOPMENT

With the evolution of AM Process like SLS, Direct metal laser sintering, Selective metal sintering, Multi jet fusion, Continuous Liquid Interface Production, electron beam melting AM is no more a Rapid prototyping method but a method for future industrial production. AM is best suited for low or medium volume production. It has to evolve even more to consider it for high volume production leads to minimize the logistic and storage costs. Products can be produced near to the customers rather than at main production units there by reducing distribution costs. There is no need to produce goods in large quantities. Storage costs can be reduced by producing goods according to demand using AM. With AM total number of parts in car assembly can be reduced there by reducing the cost and weight of the vehicle. By optimizing the topology and using AM the weight of the parts can be reduced without compromising the strength. AM can reduce the wastage of material and also the amount of energy consumed in production of parts. In future 3D printers will be used as home appliances. Parts can be repaired or replaced by directly printing them at home. At the time of natural disasters AM can be used to build the infrastructure like building bridges in quick time. Potential of AM in the field of Medical, Defense, Automobile, Aviation is immense. Development of new AM materials can change the scope of AM.

Numerical procedures and simulations of the AM process for predictive process structure property relationships integrated with CAD/E/M tools is another thrusted area in AM research. It is important to have computational approaches for analysing materials and material combinations, designing materials and their combinations that can correlate material to processes to structure. Understanding the AM process, and microstructure improvement during AM process is critical to better and robust process control and getting custom fitted microstructures. As AM process is a layer-by-layer process it offers the chance of altering the process parameters as to make structured microstructures in the parts. At present work is in progress to know the effect of varying process parameters and how that can be applied in making parts with structured microstructures in different areas of application.

REFERENCES

Alabort, E., Barba, D., & Reed, R. C. (2019). Design of metallic bone by additive manufacturing. *Scripta Materialia, 164*, 110–114. doi:10.1016/j.scriptamat.2019.01.022

Boschetto, A., & Bottini, L. (2015). Surface improvement of fused deposition modeling parts by barrel finishing. *Rapid Prototyping Journal, 21*(6), 686–696. doi:10.1108/RPJ-10-2013-0105

Cordes, J. (2019). *Mass Finishing for FDM Parts*. Retrieved from http://articles.stratasys.com/finishing-processes/mass-finishing-for-fdm-parts

Gibson, I., Rosen, D. W., & Stucker, B. (2010). *Additive Manufacturing Technologies*. New York, NY: Springer. doi:10.1007/978-1-4419-1120-9

Gora, W. S., Tian, Y., Cabo, A. P., Ardron, M., Maier, R. R. J., Prangnell, P., ... Hand, D. P. (2016). Enhancing Surface Finish of Additively Manufactured Titanium and Cobalt Chrome Elements Using Laser Based Finishing. *Physics Procedia, 83*, 258–263. doi:10.1016/j.phpro.2016.08.021

Haleem, A., Javaid, M., & Saxena, A. (2018). Additive manufacturing applications in cardiology: A review. *The Egyptian Heart Journal, 70*(4), 433–441. doi:10.1016/j.ehj.2018.09.008 PMID:30591768

He, H., Yang, Y., & Pan, Y. (2019). Machine learning for continuous liquid interface production: Printing speed modelling. *Journal of Manufacturing Systems, 50*, 236–246. doi:10.1016/j.jmsy.2019.01.004

Javaid, M., & Haleem, A. (2018). Additive manufacturing applications in orthopaedics: A review. *Journal of Clinical Orthopaedics and Trauma, 9*(3), 202–206. doi:10.1016/j.jcot.2018.04.008 PMID:30202149

Lalehpour, A., & Barari, A. (2016). Post processing for Fused Deposition Modeling Parts with Acetone Vapour Bath. *IFAC, 49*(31), 42-48. doi:10.1016/j.ifacol.2016.12.159

Li, Y.-y., Li, L., & Li, B. (2015). Direct write printing of three-dimensional ZrO2 biological scaffolds. *Materials & Design, 72*, 16–20. doi:10.1016/j.matdes.2015.02.018

Melchels, F. P. W., Feijen, J., & Grijpma, D. W. (2010). A review on stereolithography and its applications in biomedical engineering. *Biomaterials, 31*(24), 6121–6130. doi:10.1016/j.biomaterials.2010.04.050 PMID:20478613

Michna, S., Wu, W., & Lewis, J. A. (2005). Concentrated hydroxyapatite inks for direct-write assembly of 3-D periodic scaffolds. *Biomaterials*, *26*(28), 5632–5639. doi:10.1016/j.biomaterials.2005.02.040 PMID:15878368

Morales-Planas, S., Minguella-Canela, J., Lluma-Fuentes, J., Travieso-Rodriguez, J., & Garcia-Granada, A.-A. (2018). Multi Jet Fusion PA12 Manufacturing Parameters for Watertightness, Strength and Tolerances. *Materials (Basel)*, *11*(8), 1472. doi:10.3390/ma11081472 PMID:30126216

Narra, S. P., Mittwede, P. N., Wolf, S. D., & Urish, K. L. (2019). Additive Manufacturing in Total Joint Arthroplasty. *The Orthopedic Clinics of North America*, *50*(1), 13–20. doi:10.1016/j.ocl.2018.08.009 PMID:30477702

O'Connor, H. J., Dickson, A. N., & Dowling, D. P. (2018). Evaluation of the mechanical performance of polymer parts fabricated using a production scale multi jet fusion printing process. *Additive Manufacturing*, *22*, 381–387. doi:10.1016/j.addma.2018.05.035

Palma, T., Munther, M., Damasus, P., Salari, S., Beheshti, A., & Davami, K. (2019). Multiscale mechanical and tribological characterizations of additively manufactured polyamide 12 parts with different print orientations. *Journal of Manufacturing Processes*, *40*, 76–83. doi:10.1016/j.jmapro.2019.03.004

Singh, S., & Ramakrishna, S. (2017). Biomedical applications of additive manufacturing: Present and future. *Current Opinion in Biomedical Engineering*, *2*, 105–115. doi:10.1016/j.cobme.2017.05.006

Technical white paper, HP Multi Jet Fusion technology: A disruptive 3D printing technology for a new era of manufacturing. (2017, May). Retrieved from https://www8.hp.com/us/en/printers/3d-printers/resource/3dtechpaper.html

Tumbleston, J. R., Shirvanyants, D., Ermoshkin, N., Janusziewicz, R., Johnson, A. R., Kelly, D., ... DeSimone, J. M. (2015). Continuous liquid interface production of 3D objects. *Science*, *347*(6228), 1349–1352. doi:10.1126cience.aaa2397 PMID:25780246

Section 2

Importance and Application

Chapter 2
Importance of 3D Printing Technology in Medical Fields

Ranjit Barua
O. M. Dayal Group of Institutions, India

Sudipto Datta
IIEST Shibpur, India

Amit Roychowdhury
IIEST Shibpur, India

Pallab Datta
IIEST Shibpur, India

ABSTRACT

Three-dimensional or 3D printing technology is a growing interest in medical fields like tissue engineering, dental, drug delivery, prosthetics, and implants. It is also known as the additive manufacturing (AM) process because the objects are done by extruding or depositing the material layer by layer, and the material may be like biomaterials, plastics, living cells, or powder ceramics. Specially in the medical field, this new technology has importance rewards in contrast with conventional technologies, such as the capability to fabricate patient-explicit difficult components, desire scaffolds for tissue engineering, and proper material consumption. In this chapter, different types of additive manufacturing (AM) techniques are described that are applied in the medical field, especially in community health and precision medicine.

DOI: 10.4018/978-1-5225-9167-2.ch002

INTRODUCTION

Three-dimensional printing is also known as an additive manufacturing technique where the things are made by depositing or fusing the materials layer by layer-for example ceramics, metal, plastic, powders or living cells [Schubert et al., 2014]. This process is also referred solid free-form technology (SFF) or rapid prototyping (RP) 6Various 3D printers are like as a conventional inkjet printers. Though, the finish product differs in that a 3D object is produced. 3D printing has taken many places in medical field among which bio-printing is use for cellular-scaffold printing and other biomedical application i.e. regenerative medicine. Numerous 3D printing techniques are available now a day with varying parameters i.e. printing speeds, printing methods, extrusion pressures, and printing materials [Banks, 2013]. With the help of Computer Aided Design software any imaginable objet can print by this technology. Basic principles of 3D printing processes are follows: i. Designing the CAD file through printer specific software, ii. Uploading the printer specific software to the 3D printer, iii. Running the program and iv. Print the final object. In this technology, the radiographic 2D images like MRI, CT, X-rays can be converted to 3D complex objects with customized medical structures [Sun et al., 2005].

TYPES OF 3D PRINTING TECHNOLOGY

Various methods of three-dimensional printing technology are available, stereo-lithography (SLA) technology is one of them [Hornbeck, 1997]. In this technique the SLA materials are generally light sensitive, solidification is initiated by laser undergoing photon [Mertz, 2013]. On the other hand other 3D printing technologies are now offered as well as selective laser sintering (SLS), fused deposition modelling (FDM), three-dimensional bio-printing (3DP), digital light processing (DLP), and laminated object manufacturing (LOM). Table 1 represents the processes and materials details which are using different types of 3D printing technologies. Whatever the printing technology is different but the fundamental hypothesis is that the item containing with the limited layers, also the more consisting the information of layers and also requires to elevate the declaration [Gross, 2014]. Different methods are available to attach the layers, several materials can be simply melted and

Table 1. Different types of 3D printing technology

Name	Process	Material	Reference
SLS	Selective laser sintering process, it is similar as SLA method, though materials are solidified by infrared laser.	Ceramic, plastics, metal materials and wax.	Tay et al., 2003
FDM	Fused deposition modeling process, this printing technology is the most fundamental 3D printing technology and most commonly used.	PLA, ABS types polymer and some foods	Anitha et al., 2001
LOM	Laminated object manufacturing process, in this technology the materials are fused by hot roller.	Metallic materials and ceramic	Melchels et al., 2010
SLA	Stereo-lithography process, in this process the light responsive materials are solidified in to a lean layer.	Thermoplastics	Mueller et al., 1999
DLP	Digital light processing technique, it is also similar to SLA process, though it is more rapidly because the total layer manufacture after scanning the laser .	Photopolymer	Singh, 2011; Utela et al.,2008
3D Bio-printing	In this technique biomaterials are use as a printing ink, basically this technology is use in tissue engineering process.	Alginate, Hydrogel.	Datta et al., 2108

gelatinous or several of materials can be molded by laser and also be easily gelatinous. An absolute set of 3D printing technique containing with the printer specific CAD software for designing any customized moulds and also printer machine, the major part of the 3D printing process. Printer may be hot extrusion base or inkjet base [Science and Society, 2013; Hoy, 2013].

Thermal Inkjet Printing (TIJ)

In thermal based inkjet printing the printing occurs without any contact between printer and the substrate top. Electromagnetic, thermal or piezoelectric techniques are utilized for deposition of the ink droplets on the substrate top according to the digital control [Nakamura et al., 2005]. Heating and mechanical pressure are applied for deposition of the ink droplet on the substrate. In these printers small air bubbles are created when heating is applied on the nozzle of volume 10 – 150 pico liters which results in pressure pulse creation for ejection on the ink from the nozzle [Ricci et al., 2008]. The size of the droplets can be changed by changing parameters like frequency of the pulse, viscosity and the temperature gradient applied. For tissue engineering and biomedical fields these printers are suitable [Vacanti, 2006]. Because of high digital accuracy and adaptability these printers are used in biomedical application like tissue engineering and regenerative medicine [Langer et al., 2013, Tschumperlin et al., 2013]. Just because of their versatility, digital exactness, fine manage and also the benevolent outcome on mammalian cells, this technique is currently applying in medical applications to print uncomplicated 2D as well as 3D organs and tissue which is also known as bio-printing technology [Fang et al., 2012]. Thermal inkjet printers may also provide evidence of model for other complicated uses, such as gene transfection during tissue construction and drug delivery [Delaere et al., 2009].

Selective Laser Sintering (SLS)

Laser based Selective Sintering is such a technique where the printer uses powder type materials as the substrate for creating to print the new substances [Atala et al., 2006]. In this process a laser draws the shape of the substance in the powder and also fusing it mutually. After that a fresh layer of powder is laid down and the process repeat again and one by one structuring each layer, to make the final object. Laser sintering also uses to make plastic, ceramic and metallic objects. However, in this process the degree of feature is restricted only by the exactness of the laser and the fine quality of the powder, so it is probable to construct particularly in depth and fragile structures with this category of printer.

Fused Deposition Modeling (FDM)

Fused deposition modeling process is such a technique where the printers are very much regular and economical than the Laser based Selective Sintering type printing process [Anitha et al., 2001]. In FDM process, printer is generally uses a print head comparable to an inkjet printer. On the other hand, as an alternative of ink, beads of heated polymer material like PLA, ABS are exited from the print head as it changes position, constructing the item in lean layers [Atala et al., 2006]. This procedure is repeated many times, letting accurate mechanism of the quantity along with the position of all deposit to make every layer [Anitha et al., 2001]. In FDM process the heating is done on the material by inside extruder heater and ultimate the melted material deposits below as per desire design layer by layer.4 Though each layer of plastic becomes cool and harden, also then step by step making the solid substance as the layers make. Despite the fact that relying on the difficulty along with the expenditure of FDM printer, machine may have improved features like as an example of multiple print heads. FDM printers can also use as a variety of polymers [Stanton et al., 2002]. Though, 3D Fused Deposition Modelling printed objects are frequently create from the similar thermoplastics which are practiced in machining, that's why the printed products have similar mechanical properties and durability, stability.

3D Bio-Printing

3D printing technology is one of the most cutting edge technique which is used in medical application specially in biomedical field, where it is known as bio-printing and it have a huge prospective for opportunity in upcoming regenerative treatment. A basic principle bio-printing technique is shown in Figure 1. First of all, exact data of tissues and organs is collected for making the exact replica. After designing the model, with the help of server, it convey the order into electrical signal for controlling the printer and then printing the objects, however the printer must be ability to sustain the cell viability during the time of printing procedure (Figure 1ii and iii). Typically, a tissue is combination of several types of cells and also the cells are assorted with various substances for better printing (Figure 1iii) [Sekine et al., 2013; Sirringhaus et al., 2000]. Now a day, a number of hard tissues are fabricated with bionic materials by bio-printer, also use in medical trials. An existent complex tissue is not able to build from 3D bio-printer at the moment. Upcoming days, 3D bio-printer also be able to print vital organs for repairing and reconstructing the injured body part, also to replicate some useful drug screening, tissues for therapy and biomedical research. In addition, 3D bio-printing can be used for modified treatment which will help to reduce the cost of treatment. Biodegradable and biocompatible materials are combined with 3D bio-printing to decrease the incompatibilities caused by materials [Datta et al., 2108]. For that reason, 3D bio-printing will lead to a cutting edge technology of medical application.

3D PRINTING IN MEDICAL APPLICATIONS

3D printing or additive manufacturing technology has been introduced in medical field since the beginning21st century, earlier the technology was mainly used for implants for dentals and custom prosthetics [Weng et al., 2012]. At that time, the medical applications for 3D printing process have evolved significantly. Many researchers have described the uses of 3D printing to fabricate organs, ears, bones, windpipes, exoskeletons, stem cells, eyeglasses, a jaw bone, blood vessels, cell cultures, vascular networks, and tissues, in accumulation to new dosage forms and delivery of drugs devices. Modern medical applications of 3D printing can be prepared into a number of extensive categories, like making prosthetics, dental implants, tissue and organ fabrication, anatomical models, drug pharmaceutical research with reference to dosage forms, drug discovery and delivery [Banks, 2013]. Medical applications of 3D printing are as follows.

Modified Implants and Prostheses

CT scan, MRI, X rays are used for creation of any geometry dental implants into 3D printing technique .stl files [Schubert et al., 2014, Banks, 2013]. Like this fashion, 3D printing technology is utilized effectively in medical health care area to build together usual along with intricate modified prosthetic

Figure 1. A basic bio-printing process

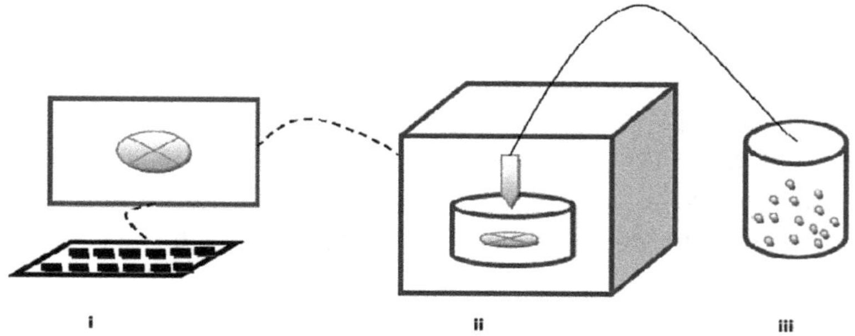

limbs and hip implants, surgical implants [Sun et al., 2005]. This method has been used to fabricate spinal and oral implant area like dental [Banks, 2013]. Up to that time, formerly for clinical use of implants the validation of the implants is necessary which is time consuming [Sun et al., 2005]. The capability to rapidly fabricate modified prostheses and the implants solves a comprehensible and determined difficulty in orthopedics, where typical implants are usually not satisfactory for few patients, mainly in multifaceted cases [Banks, 2013]. Formerly, surgeons had to execute bone graft surgeries or use scalpels and also drill to adjust implants by shaving pieces of metal and plastic to a preferred size, shape and fit [Schubert et al., 2014, Mertz et al., 2013]. This is similarly factual in neural surgery like skulls asymmetrical shapes, which is tough to normalize an implant of cranial. While in case of injury in head victims, where the head bone is to detached and give the brain a little place to swell, the accuracy of the plate of the cranial should be high which is later fixed [Mertz et al., 2013]. Even though few plates milling are done and are produced many by 3D printing which allows the modification of the design easily.

Cell Printing

In human body there have more than 200 type cells, and these cells create different types and complex tissues and also organs. For complicated structure, it is really difficult to makes replicate this functional organs and tissues *in vitro* system. Cell printing technique can work out this difficulty. In this process, the ink droplets are depositing on live cells to fabricate the tissues as preferred. In spite of upcoming potential for an organ transplantation, at the present time there have not any other biologically lively organ fabricated by bio-printing process. One more thing is most essential in this fabrication process, every organ and tissue of the human body is very complicated in structure, also the mixing of different type cells involve neither *in vitro* tests of functionality and viability or *in vivo* tests of communications with other organ and tissues. On the other hand, this technology has been extensively used in the manufacturing industry especially in biomedical area. It can be used in medical application like bio-therapy field with vessel-free and hard structures. It has printed small blood vessels and beating cardiac valves, representative that cell-printing is possible [Matsumoto et al., 2014]. Later they established a company named

Organovo, committed to bio-printer enlargement and marketing. In cell printing, fibroblasts cell have been effectively printed by improving bio-printers [Page et al., 2013]. Cells are collected from endothelial cells and mature stem cells, after printing process, a number of cell-cell interactions are observed. It covers the way for upcoming 3D organ-printing [Mironov et al., 2009]. While cells are encapsulate into the materials, also used as an ink, it is called as 3D bio-printing process. Here have one disadvantage of cell encapsulated in materials, which is reduce the cell viability. Primarily hydrogel is the primary aspirant for cell printing. For maintaining the cell viability and cell-cell interactions, sometime hydrogel materials can be mixed into the ink. Advantage of hydrogel is, it can also work as scaffoldings and subtract [Odde wt al., 1999]. It have particular physical and chemical properties, also it can be degraded *invivo* a little bit after grafting. 3D bio-printing process, the initial materials used as scaffoldings were electrospun fibers, it is necessary for substituting blood vessels [Choi et al., 2011]. At this time ployanhydride, collagen and fibronectin have been extensively used. It is still a cutting edge technology nowadays [Chien et al., 2013; Wilson et al., 2003]. There are some hydrogels which are used productively in 2D fabrication via 3D printing technology, also the functions of the fabricated products could also be detected: agarose, culture medium [Barron et al., 2004], collagen [Parzel et al., 2009], alginate [Ahmed, 2015], matrigel [Schuurman et al., 2013; Aguado et al., 2011], fibrin [Ratcliffe et al., 1984], polyvinyl alcohol (PVA) [Chung et al., 2013] and k-70 series [Van Den et al., 2000]. In this process, after extruded from the printer nozzle tips, hydrogels would be polymerized by cross-linker [Zhang et al., 2014]. It is a really big challenge for protecting cells, also it is necessary to maintain their solution during hydrogel's stabilization. Consequently, a lot of new altered-hydrogels and techniques are being discovered [Detsch et al.. 2013; Bartolo et al., 2009]. For containing maximum amount of water in the printed structures hydrogels are mainly used. The biological and mechanical properties of hydrogel require to be customized for printing and cell survival. Cell death occurs post printing during crosslinking of range 2- 45% because of environmental exposures [Mironov et al., 2007; Hassan et al., 2013]. There have some chemical and physical ways to crosslink hydrogels. For example, water soluble gelatin is a protein which has also good biocompatibility [Murphy et al., 2014]. With methacrylamide group adapted, the gelatin (gelMA) can be cross-linked by using ultraviolet rays [Seitz et al., 2005]. The adding together of hyaluronic acid (HA) progresses gelMA printable properties [Pati et al., 2014]. Cross-linking chemical calcium which helps to crosslink alginate and also the alginate-gelatin blends have been investigated as most likely materials for extrusion printing living cells [Duan et al., 2013]. A particular bio-printing has elevated resolution and can be used for material-cell blends precise control and cell patterning.

Organs/Tissues Bio-Printing

The damage of the tissues and the organs occurs due to many factors like accident, defect during birth, age and many diseases. The primary option for these is organ transplantation from other living dead donors. The availability of organs from the donors is not always possible because of non-availability of donors all the time. One more thing is that, in case of medical treatment, organ transplant surgery and follow-up is also very costly, estimate price more than $300 billion in the year 2012.10 An added difficulty is that organ transplantation involves the a lot of complex job of judgment a donor who should be tissue-equal with the recipient. This problem can only be minimized by using cells/tissues of the same patient taken to develop the required replacement organ [Schubert et al., 2014; Mertz et al., 2013]. The tissue rejection risk will also be minimized and the lifelong immune suppressants should no longer be needed. The extra benefit of the 3D printing technology is giving scaffold support which is not possible is regenerative medicine method, for an example fast digital speed control, and highly precise cell placement, concentration of cells, drop-volume, resolution and printed cells diameter. By 3D printing we can develop biomaterials, cell-laden biomaterials separately or in or direct tissue like structures [Datta et al., 2018]. Wide range of materials are now available for scaffold creation, porosity of the material, tissue strength and types and hydrogels for soft tissues development.

Anatomical Models Making for Surgical Preparation

The personage complexities and alternations of the human anatomical body construct the 3D-printed developed models use is perfect for surgery planning. It becomes easier for the physicians to have a real model of the patient and understanding of the anatomy of the patient becomes easier than 2D images like the CT scan, MRI and X-rays. The cost and accessibility is also reduced for cadavers in case of 3D printed developed models.3High level of pathology is also required for cadavers so the lesson of human anatomy can also be demonstrated more easily than surgical patients [Sun et al., 2005]. The surgical trainings which are out of reach for physicians are solved by the 3D printed models designed. Polypeptide chains models by 3D printing are commonly used currently because of their huge number of degree of freedom and insertion of bond rotational barriers it can crinkle inside secondary structures. Same as the models could be exploited to assist the considerate of other varieties of biochemical or biological structures. Students can better understand the molecular anatomical structures when 3D printed structures are involved as from the pre and the post conception studies. The biochemical and structural maintenance of the cells for growth and development of the cells are provided by the Extracellular matrix (ECM), it also contains various proteins and glucans. The modification of ECM would influence cell function and state. With more resolution, 3D printing help to use print scaffolds which resemblances *in vivo* constructions, and also the environment of the tissues, and the cellular viability from the scaffolds are used to understand the cell growth in the printed scaffolds, drug delivery and tissue engineering. Model development is the initial 3D printing application, with material modification and optimization of the technique, a number of hard bio-active tissues have been printed. In the year of 2012, the first medical experiment of the 3D bio-printing was done at in Netherlands. A bionic jaw printed by 3D printing technique was operated to a dental patient. Since then, more demos have been done in Japan, US and other countries worldwide. The joined materials can exclusively synchronize with vocal part without affecting the vocality and the hearing of the patients. To straight printing the living tissues and also printed scaffolds were used for cell culture and tissue engineering. Cells attained advanced viability [Datta et al., 2018] and function well cells were cultured on the scaffolds 3D printed with compared to the 2D environment. For various types of cell cultures on scaffolds, Ploy caprolactone (PCL) have been used. Though, this is just the mixture of scaffolds and cells combination at 2D level, and which is called indirect 3D bio-printing. In fact, the communication between materials and cells in the indirect 3D printing method is still 2D because only one face of the cell bound is scaffolds.

MODIFIED 3D-PRINTED DOSAGE FORMS AND DRUG DELIVERY INSTRUMENTS

The process of 3D printing is also used in drug fabrication and drug delivery pharmaceutical research because of transformative assurance. Pros of 3D printing technology process include accurate droplet dimension control and drug dosage, maximum reproducibility, and also the capability to fabricate dosage forms with intricate drug-releasing profiles.

Exclusive Dosage Forms

Mostly for pharmaceutical manufacturing process inkjet printers and inkjet powder based 3D printers are used. Whatever the material is used powder or additional material the substrate is used to categorize the 3D inkjet printing process. In case of inkjet type drug fabrication process, inkjet printers are used to spray formulations of medications and binders in small droplet sat exact motions, sizes and speeds onto a substrate. The substrate normally used are of cellulose, micro-porous, bio-ceramics, paper uncoated or coated, scaffolds of glass, potato starch films, potato starch films, potato starch films, and metal alloys. Researchers have improved the technology by forming nano-particles and micro-particles of liquid film by spraying identical "ink" droplets onto a liquid film. Matrices of such types can be used for delivery of small amount of hydrophobic molecules along with the growth factor. For powder-based 3D printing drug fabrication process, here the printer sprays the ink onto the base of the powder. When the ink comes in contact with the powder the powder becomes hard and solid dosed of layer by layer is formed. Active compositions like binders additions and immobile compositions can be present in the ink. When the solid dosage is formed it is removed from the loose powder substrate present nearby.

Personalized Drug Dosing

3D printing technology allows drug modification easily, as the development of the drugs requires enlarge efficacy and limits the risk of many unwanted reactions [Banks et al., 2013; Sun et al., 2003]. Because of many advantages of oral tablets like easily manufactured, dosing accuracy, no pain and better patient observations. The process of drugs manufacturing for modified drugs process is still not obtained for this reason 3D printing technology is commonly used for better built in diversity of active ingredients like theophylline, ofloxacin, steroidal anti-inflammatory drugs, paclitaxel, acetaminophen, dexamethasone, caffeine, folic acid, vancomycin, tetracycline, and others.3D drug printing have built-in a diversity of active ingredients, like as: theophylline, ofloxacin, steroidal anti-inflammatory drugs, paclitaxel, acetaminophen, dexamethasone, caffeine, folic acid, vancomycin, tetracycline and others. Another way, inactive components used in 3D drug printing have consisted: surfactants (such as Tween 20), poly (lacticco-glycolic acid), glycerin, ethanol-dimethyl sulfoxide, cellulose, methanol, propylene glycol, Kollidon SR, and acetone.

Difficult Drug-Release System

The formation of medications with versatile drug-release system is one of the most researchable uses of 3D printing technique. Conventional condensed dosage forms are often prepared from a uniform mixture of active, as well as inactive ingredients, and are therefore often restricted to a simple drug-release system. On the other hand, 3D printers help to print binder onto a matrix powder bed in layers normally 200 micrometers thick, creating a fence involving with the active ingredients to facilitate controlled drug release. 3D-printed dosage forms can also be fabricated in compound geometries that are basically porous and burdened with multiple drugs all through, enclosed by barrier layers that modulated is charge. Another area implantable drug delivery device with new drug-release systems can also be produced using 3D printing. Disparate conventional general treatments that can influence non afflicted tissue, these devices can be implanted to make available direct treatment to the area implicated. Bone infections are one case where direct treatment with a drug implant is more advantageous than general treatment. Luckily, powder-based 3D-printedbone scaffolding can be formed in high-resolution models with compound geometries that mimic the natural bone extracellular matrix. The printing of medications with customized drug release profiles into such bone implant scaffolds are considered. For an example of the printing of a multilayered bone implant with a disparate drug-release system alternating between isoniazid and rifampic in a pulse discharge mechanism.

FUTURE ASPECTS OF 3D BIO-PRINTING

It requires a huge time and desires a delivery of cells during an organ-printing process. In such an extended episode of time, to make sure the precision of printing and as well as the cell viability, which have become a critical questions. Preventing the exposure of radiations from laser it is necessary that the nozzle jet can be attached with aluminium alloy. Right now, complex scaffoldings of tissue can be printed by using hydrogels by laser-mediated printer and also 3D inkjet bio-printer. Until now the method of fusion, high pressure and temperature also requires, but some solvents which could lead to the loss of cells. Printed tissue becomes the combination of hydrogels with cells possesses; the mechanical properties are related to natural constructs but with limited function and cell extension abilities. This is one of the vital boundaries which bio-printing is facing at the time. While so many reports confirmed that cells may be fabricated, though there have not still any report regarding the function of cells printing. Right now, it is necessary to maintain cells living in materials [Datta et al., 2018]. Cell states are reliant on the different substrates, there have many materials are developed for mimicking the ECM, local tissue consequent the bio-ink, which is projected to simulated the *in vivo* environment. As a result, to improve the materials for modification of cell is one of upcoming challenge work. One more question of 3D bio-printing is to fabricate the organs containing blood vessels. Every organ have desires network of capillaries and vessels to supply cytokines, oxygen and nutrients, in addition to eliminate the wastes which are harmful to cells. It is reported that heart muscles containing blood vessel structure was achieved from endothelial cells during tissue culture. To fabricate the functional 3D tissues with blood vessels is really a huge challenge faced by 3D bio-printing [Datta et al., 2018]. The main problem is faced by 3D bio-printing is sources of cell, as cells are the fundamental units of an organ. In this technique, stem cells may turn into the main source for bio-printing. The advantage of stem cells is, it have high short cell cycle time and viability, it can distinguish into new cell types in defined environment. This may decrease the number of cells mandatory in bio-printing also to save printing time, and make environment for printed organs to active. At the present time, preliminary trials have been done in 3Dbio-printing with stem cells where embryonic stem cells were used in 3D printing to form embryoid bodies (EBs). Human umbilical vein endothelial cells (hUVECs) and human mesenchymal stem cells (hMSCs) were operated to print fabricated patches to restore cardiac muscles. On the other hand, we have to go further for choosing the accurate stem cells for bio-printing the desire organs. Preferably 3D bio-printing interacts *in situ* printing of cells at the place of injury. This requires rapid printing, rapid model structure and enough cell sources. Achievement of such demands may lead to a revolution in regenerative medicine. For an example, aim mediate repair of the skin wound with 3D bio-printing would raise the recovery rate and decrease stable scarring. For optimizing the 3D bio-printing process of tissues and organs, it is requires to development of characterization and estimate the roles of the printed organs. Significant studies will comprise cell tracing tests, practical marker recognition, cell viability and also animal model experiments. To complete 3D bio-printing of huge tissues structural scaffolds are to be used. Bio compatible material for example decomposable poly caprolactone (PCL) is used. Concerning of safety assessment systems for trying bio-printed object is one more problem to be measured. Suitable dictatorial backgrounds are now to be advanced. Whatever, bio-printing is a innovative and intricate technology, the method involves cell expansion, cell preparation, materials, safety, graft, observation after grafting, and efficiency valuation.

CONCLUSION

3D printing has become a constructive and potentially transformative tool in a number of dissimilar fields, including medical application. Because printer resolution, performance and accessible materials have rapidly growing up different applications. Investigators are trying hard to find out new methods and applications of 3D printing technologies in medical fields.3D printing in medical applications are more accurate, significant and exciting but still there need further improvement and innovative research for the 3D printing technology for high level medical applications like organ printing and tissue printing.

REFERENCES

Aguado, B. A., Mulyasasmita, W., Su, J., Lampe, K. J., & Heilshorn, S. C. (2011). Improving viability of stem cells during syringe needle flow through the design of hydrogel cell carriers. *Tissue Engineering. Part A*, *18*(7-8), 806–815. doi:10.1089/ten.tea.2011.0391 PMID:22011213

Ahmed, E. M. (2015). Hydrogel: Preparation, characterization, and applications. *Journal of Advanced Research*, *6*(2), 105–121. doi:10.1016/j.jare.2013.07.006 PMID:25750745

Anitha, R., Arunachalam, S., & Radhakrishnan, P. (2001). Critical parameters influencing the quality of prototypes in fused deposition modelling. *Journal of Materials Processing Technology*, *118*(1-3), 385–388. doi:10.1016/S0924-0136(01)00980-3

Atala, A., Bauer, S. B., Soker, S., Yoo, J. J., & Retik, A. B. (2006). Tissue-engineered autologous bladders for patients needing cystoplasty. *Lancet*, *367*(9518), 1241–1246. doi:10.1016/S0140-6736(06)68438-9 PMID:16631879

Banks, J. (2013). Adding value in additive manufacturing: Researchers in the United Kingdom and Europe look to 3D printing for customization. *IEEE Pulse*, *4*(6), 22–26. doi:10.1109/MPUL.2013.2279617 PMID:24233187

Barron, J., Spargo, B., & Ringeisen, B. (2004). Biological laser printing of threedimensional cellular structures. *Applied Physics. A, Materials Science & Processing*, *79*(4-6), 1027–1030. doi:10.100700339-004-2620-3

Bartolo, P. J., Almeida, H., & Laoui, T. (2009). Rapid prototyping and manufacturing for tissue engineering scaffolds. *International Journal of Computer Applications in Technology*, *36*(1), 1–9. doi:10.1504/IJCAT.2009.026664

Chien, K. B., Makridakis, E., & Shah, R. N. (2013). Three-dimensional printing ofsoy protein scaffolds for tissue regeneration. *Tissue Engineering. Part C, Methods*, *19*(6), 417–426. doi:10.1089/ten.tec.2012.0383 PMID:23102234

Choi, W. S., Ha, D., Park, S., & Kim, T. (2011). Synthetic multicellular cell-to-cell communication in inkjet printed bacterial cell systems. *Biomaterials*, *32*(10), 2500–2507. doi:10.1016/j.biomaterials.2010.12.014 PMID:21208654

Chung, J. H. Y., Naficy, S., Yue, Z. L., Kapsa, R., Quigley, A., Moulton, S. E., & Wallace, G. G. (2013). Bio-ink properties and printability for extrusion printing living cells. *Biomaterials Science*, *1*(7), 763–773. doi:10.1039/c3bm00012e

Datta, S., Barua, R., Sarkar, R., Barui, A., Roy Chowdhury, A., & Datta, P. (2018). Design and development of alginate: Poly-l-lysine scaffolds by 3D bio printing and studying their mechanical, structural and cell viability properties. *IOP Conf. Series: Materials Science and Engineering, 402*.

Datta, S., Sarkar, R., Vyas, V., Bhutoria, S., Barui, A., Roy Chowdhury, A., & Datta, P. (2018). Alginate-honey bioinks with improved cell responses for applications as bioprinted tissue engineered constructs. *Journal of Materials Research*, 1–11.

Delaere, P. R., & Hermans, R. (2009). Clinical transplantation of a tissue-engineered airway. *Lancet*, *373*(9665), 717–718, author reply 718–771. doi:10.1016/S0140-6736(09)60429-3 PMID:19249622

Detsch, R., Sarker, B., Grigore, A., & Boccaccini, A. R. (2013). Alginate and gelatine blending for bone cell printing and biofabrication. In *IASTED International Conference Biomedical Engineering*. Innsbruck, Austria: ACTA Press. 10.2316/P.2013.791-177

Duan, B., Hockaday, L. A., Kang, K. H., & Butcher, J. T. (2013). 3D bio-printing ofheterogeneous aortic valve conduits with alginate/gelatin hydrogels. *Journal of Biomedical Materials Research. Part A*, *101*(5), 1255–1264. doi:10.1002/jbm.a.34420 PMID:23015540

Fang, Y., Frampton, J. P., Raghavan, S., Sabahi-Kaviani, R., Luker, G., Deng, C. X., & Takayama, S. (2012). Rapid generation of multiplexed cell cocultures using acoustic droplet ejection followed by aqueous two-phase exclusion patterning. *Tissue Engineering. Part C, Methods*, *18*(9), 647–657. doi:10.1089/ten.tec.2011.0709 PMID:22356298

Gross, B. C., Erkal, J. L., Lockwood, S. Y., Chen, C., & Spence, D. M. (2014). Evaluation of 3D printing and its potential impact on biotechnology and the chemical sciences. *Analytical Chemistry*, *86*(7), 3240–3253. doi:10.1021/ac403397r PMID:24432804

Hassan, W., Dong, Y., & Wang, W. (2013). Encapsulation and 3D culture of humanadipose-derived stem cells in an in-situ crosslinked hybrid hydrogelcomposed of peg-based hyperbranched copolymer and hyaluronicacid. *Stem Cell Research & Therapy*, *4*(2), 32. doi:10.1186crt182 PMID:23517589

Hornbeck, L. J. (1997). Digital light processing for high-brightness high-resolution applications. In *Proceedings of Electronic Imaging '97*. International Society for Optics and Photonics. doi:10.1117/12.273880

Hoy, M. B. (2013). 3D printing: Making things at the library. *Medical Reference Services Quarterly*, *32*(1), 94–99. doi:10.1080/02763869.2013.749139 PMID:23394423

Langer, R., & Vacanti, J. P. (2013). Tissue engineering. Science, 1993, 260:920–926Trappmann B, Chen CS. How cells sense extracellular matrix stiffness: a material's perspective. *Current Opinion in Biotechnology*, *24*, 948–953. PMID:23611564

Matsumoto, K., Ishiduka, T., Yamada, H., Yonehara, Y., Arai, Y., & Honda, K. (2014). Clinical use of three-dimensional models of the temporomandibularjoint established by rapid prototyping based on cone-beam computed tomography imaging data. *Oral Radiology*, *30*(1), 98–104. doi:10.100711282-013-0127-3

Melchels, F. P., Feijen, J., & Grijpma, D. W. (2010). A review on stereolithography and its applications in biomedical engineering. *Biomaterials*, *31*(24), 6121–6130. doi:10.1016/j.biomaterials.2010.04.050 PMID:20478613

Mertz, L. (2013). Dream it, design it, print it in 3-D: What can 3-D printing do for you? *IEEE Pulse*, *4*(6), 15–21. doi:10.1109/MPUL.2013.2279616 PMID:24233186

Mironov, V., Prestwich, G., & Forgacs, G. (2007). Bio-printing living structures. *Journal of Materials Chemistry*, *17*(20), 2054–2060. doi:10.1039/b617903g

Mironov, V., Trusk, T., Kasyanov, V., Little, S., Swaja, R., & Markwald, R. (2009). Biofabrication: A 21st century manufacturing paradigm. *Biofabrication*, *1*(2), 022001. doi:10.1088/1758-5082/1/2/022001 PMID:20811099

Mueller, B., & Kochan, D. (1999). Laminated object manufacturing for rapidtooling and patternmaking in foundry industry. *Computers in Industry*, *39*(1), 47–53. doi:10.1016/S0166-3615(98)00127-4

Murphy, S. V., & Atala, A. (2014). 3D bio-printing of tissues and organs. *Nature Biotechnology*, *32*(8), 773–785. doi:10.1038/nbt.2958 PMID:25093879

Nakamura, M., Kobayashi, A., Takagi, F., Watanabe, A., Hiruma, Y., Ohuchi, K., ... Takatani, S. (2005). Biocompatible inkjet printing technique for designed seeding of individual living cells. *Tissue Engineering*, *11*(11-12), 1658–1666. doi:10.1089/ten.2005.11.1658 PMID:16411811

Odde, D. J., & Renn, M. J. (1999). Laser-guided direct writing for applications inbiotechnology. *Trends in Biotechnology*, *17*(10), 385–389. doi:10.1016/S0167-7799(99)01355-4 PMID:10481169

Page, H., Flood, P., & Reynaud, E. G. (2013). Three-dimensional tissue cultures: Current trends and beyond. *Cell and Tissue Research*, *352*(1), 123–131. doi:10.100700441-012-1441-5 PMID:22729488

Pati, F., Jang, J., Ha, D. H., Won Kim, S., Rhie, J. W., Shim, J. H., ... Cho, D. W. (2014). Printing three-dimensional tissue analogues with decellularizedextracellular matrix bioink. *Nature Communications*, *5*(1), 3935. doi:10.1038/ncomms4935 PMID:24887553

Pepper, M. E., Parzel, C. A., Burg, T., Boland, T., Burg, K. J. L., & Groff, R. E. (2009). Designand implementation of a two-dimensional inkjet bio-printer. *Proceedings of Annual International Conference of the IEEE Engineering in Medicine and Biology Society*, 6001–6005. 10.1109/IEMBS.2009.5332513

Ratcliffe, J. H., Hunneyball, I. M., Smith, A., Wilson, C. G., & Davis, S. S. (1984). Preparation and evaluation of biodegradable polymeric systems for the intra-articular delivery of drugs. *The Journal of Pharmacy and Pharmacology*, *36*(7), 431–436. doi:10.1111/j.2042-7158.1984.tb04419.x PMID:6146685

Ricci, J. L., Clark, E. A., Murriky, A., & Smay, J. E. (2012). Three-dimensional printing of bone repair and replacement materials: Impact on craniofacialsurgery. *The Journal of Craniofacial Surgery*, *23*(1), 304–308. doi:10.1097/SCS.0b013e318241dc6e PMID:22337431

Schubert, C., van Langeveld, M. C., & Donoso, L. A. (2014). Innovations in 3D printing: A 3D overview from optics to organs. *The British Journal of Ophthalmology*, *98*(2), 159–161. doi:10.1136/bjophthalmol-2013-304446 PMID:24288392

Schuurman, W., Levett, P. A., Pot, M. W., van Weeren, P. R., Dhert, W. J., Hutmacher, D. W., ... Malda, J. (2013). Gelatinmethacrylamidehydrogels as potential biomaterials for fabrication of tissue-engineered cartilage constructs. *Macromolecular Bioscience*, *13*(5), 551–561. doi:10.1002/mabi.201200471 PMID:23420700

(2013). Science and society: Experts warn against bans on 3D printing. *Science*, *342*(6157), 439. PMID:24163835

Seitz, H., Rieder, W., Irsen, S., Leukers, B., & Tille, C. (2005). Three-dimensional printing of porous ceramic scaffolds for bone tissue engineering. *Journal of Biomedical Materials Research. Part B, Applied Biomaterials*, *74*(2), 782–788. doi:10.1002/jbm.b.30291 PMID:15981173

Sekine, H., Shimizu, T., Sakaguchi, K., Dobashi, I., Wada, M., Yamato, M., ... Okano, T. (2013). In vitro fabrication of functional three-dimensional tissues with perfusable blood vessels. *Nature Communications*, *4*(1), 1399. doi:10.1038/ncomms2406 PMID:23360990

Singh, R. (2011). Process capability study of polyjet printing for plastic components. *J MechSciTechnol*, *25*, 1011–1015.

Sirringhaus, H., Kawase, T., Friend, R. H., Shimoda, T., Inbasekaran, M., Wu, W., & Woo, E. P. (2000). High-resolution inkjet printing of all-polymer transistor circuits. *Science*, *290*(5499), 2123–2126. doi:10.1126cience.290.5499.2123 PMID:11118142

Stanton, R. A., & Billmire, D. A. (2002). Skin resurfacing for the burned patient. *Clinics in Plastic Surgery*, *29*(1), 29–51. doi:10.1016/S0094-1298(03)00085-3 PMID:11827368

Sun, C., Fang, N., Wu, D.M., & Zhang, X. (2005). Projection micro-stereolithography using digital micro-mirror dynamic mask. *Sensor Actuat a-Phys, 121*, 113–120.

Tay, B. Y., Evans, J. R. G., & Edirisinghe, M. J. (2003). Solid freeform fabrication of ceramics. *International Materials Reviews, 48*(6), 341–370. doi:10.1179/095066003225010263

Tschumperlin, D. J., Liu, F., & Tager, A. M. (2013). Biomechanical regulation ofmesenchymal cell function. *Current Opinion in Rheumatology, 25*(1), 92–100. doi:10.1097/BOR.0b013e32835b13cd PMID:23114589

Utela, B., Storti, D., Anderson, R., & Ganter, M. (2008). A review of process developmentsteps for new material systems in three dimensional printing (3DP). *Journal of Manufacturing Processes, 10*(2), 96–104. doi:10.1016/j.jmapro.2009.03.002

Vacanti, C. A. (2006). The history of tissue engineering. *Journal of Cellular and Molecular Medicine, 10*(3), 569–576. doi:10.1111/j.1582-4934.2006.tb00421.x PMID:16989721

Van Den Bulcke, A. I., Bogdanov, B., De Rooze, N., Schacht, E. H., Cornelissen, M., & Berghmans, H. (2000). Structural and rheological properties of methacrylamide modified gelatin hydrogels. *Biomacromolecules, 1*(1), 31–38. doi:10.1021/bm990017d PMID:11709840

Weng, B., Liu, X., Shepherd, R., & Wallace, G. G. (2012). Inkjet printed polypyrrole/collagen scaffold: A combination of spatial control and electrical stimulation of PC12 cells. *Synthetic Metals, 162*(15-16), 1375–1380. doi:10.1016/j.synthmet.2012.05.022

Wilson, W. C., & Boland, T. (2003). Cell and organ printing 1: Protein and cellprinters. *The Anatomical Record. Part A, Discoveries in Molecular, Cellular, and Evolutionary Biology, 272*(2), 491–496. doi:10.1002/ar.a.10057 PMID:12740942

Zhang, K., Chou, C. K., Xia, X., Hung, M. C., & Qin, L. (2014). Block-cell-printing for live single-cell printing. *Proceedings of the National Academy of Sciences of the United States of America, 111*(8), 2948–2953. doi:10.1073/pnas.1313661111 PMID:24516129

Chapter 3
Additive Manufacturing:
A Tool for Better Education

Hridayjit Kalita
Birla Institute of Technology Mesra, India

Divya Zindani
National Institute of Technology Silchar, India

Kaushik Kumar
Birla Institute of Technology Mesra, India

ABSTRACT

Additive manufacturing (AM) is the most advanced recently trending manufacturing technique that employs 3D printers to create 3D objects by layer upon layer fabrication from the base to the top. The required trajectory of the fabricating tool to create the layer can be well programmed by CAD software available in the market. The 3D CAD model in the computer can be manipulated and customized for different design needs of the product. These manipulations in model and quick fabrication process make the system a flexible and an effective one. This chapter discusses the AM application in educational system by describing the individual AM processes, their limitations, advantages, feasibility in general conditions, and planning for future generations to get accustomed to this technology from the early education in schools to the specialized education in universities. The technology enables students to convert 2D objects into 3D on the CAD software and feel them physically by 3D printing. AM also enables teachers to demonstrate their ideas easily to students.

DOI: 10.4018/978-1-5225-9167-2.ch003

INTRODUCTION

The traditional methods of manufacturing required separate processes of casting, machining, joining, which consumed a lot of energy, labour power and time. Moreover, a lot of material is wasted in shaping the workpiece into its desired shape [ASTM, 2010; Kruth et al., 1998; Levy et al., 2003]. There is no flexibility in design change and dimensional errors are common. The arrangements of the equipment for the above-mentioned processes cannot be relocated (due to bulk equipment) to a new facility with ease and lots of hazards can be associated with the relocation such as noise hazards [Jaymes, 2012; Niosh, 2011]. Precautionary guidelines must be tight and safety clothing must be worn by every working personnel.

Additive manufacturing (AM) is the process of replicating a given CAD 3D model into a physical model by sequentially generating layers one above the other [ASTM, 2010] based on the cross-sectional area of the slice that has been cut across the 3D CAD model. It can be considered the "3rd industrial revolution" and recently gained tremendous popularity [Lolur & Dawes, 2014]. AM can eliminate all the above mentioned drawbacks of traditional manufacturing by shifting the manufacturing technique to an additive one [Huang et al., 2012] rather than the subtractive one as used in the traditional approach. This facilitates AM manufacturing process to fabricate solid objects having complex and sharp features [Levy et al., 2003; Kruth et al., 1998; ASTM, 2010] at minimum utilization of raw material [Huang et al., 2012]. It also implies that the material quantity to be used for a given 3D CAD solid model to be physically built, can be decided right in their designing phase and the same amount of material can be pre-ordered for production. The flexibility in design change during production adds up to feature for customized production [Huang et al., 2012]. In spite of all these advantages, AM processes lags behind in the strength of their components, fatigue resistance and material limitations. ABS and PLA are the major materials used for 3DP and since ABS extrusion emits an unpleasant odour resulting in requirement of proper ventilation and isolated space, 3DP integrated libraries (as will be explained in the below sections) adopt PLA which emit a pleasant sweet fragrance [Bharti et al., 2015].

Also known as the rapid prototyping [Kruth et al., 1998] and rapid manufacturing [Levy et al., 2003], the AM technology, integration with the educational system has been studied across various domains like architecture

[Celani, 2012; Paio et al.; 2012, Oxman, 2010], computing [Eisenberg, 2013; Ishengoma & Mtaho, 2014], engineering [Stier & Brown, 2000; Chong et al., 2018], library studies [Niaki & Nonino, 2017; Wei et al., 2017; Huang et al., 2013; Seuring & Muller, 2008], science [Horowitz & Schultz, 2014; Cook et al., 2015; Loy, 2014; McGahern et al., 2015; McMenamin et al., 2014; Horejsi, 2014], medicine [Oxman, 2010; Ishengoma & Mtaho, 2014] and technology [Schelly et al., 2015; Buehler et al, 2016; Buehler et al., 2014]. The need for this integration can be realised with the growing difference in adoption rate of AM technology and human skills to the industrial advancement and there is a chance that humans may lag behind on this technology and skills associated with it [Simpson et al., 2017; Despeisse et al., 2017; U.A.M.S. Group, 2016; Snyder et al., 2014; A.M., 2017].

With further improvement in the technology, the AM devices has become more compact, cost effective [Hoy & Brigham, 2013], safety oriented and ease in synchronization and transfer of .stl files from the Computer aided design (CAD) software to the printer [Bharti et al., 2015]. The designs can easily be manipulated and evaluated by users using various CAD software which presents a user-friendly environment. In this paper, a few significant AM techniques have been discussed and the concept of integrating the technology into educational system has been explored. The chapter is divided into 5 sections of which the first section discusses some of the available techniques in additive manufacturing and their feasibility to get incorporated in the educational institutions with proper safe handling measures. From section 2, the term 3DP is frequently used instead of AM. The 2nd section gives a detailed literature review on the present application of 3D printers in educational institutions and libraries. The 3rd section discusses some of the advantageous features and challenges of using 3D printing in educational institutions and how are they going to be tackled and used in different aspects of learning and teaching rather than only using it in research applications. The 4th section gives a brief overview of the probable sources of impacts on learning and collaborative culture in educational institutions. The effectiveness of the integration of additive manufacturing and educational system and scope for future enhancement forms the fifth section.

ADDITIVE MANUFACTURING TECHNIQUES, THEIR FEASIBILITY, AND SAFETY MEASURES

The methods involved in additive manufacturing differs in their implementation of bonding techniques where few of these depends on the thermal fusion of the layers of material one above the other while in other cases, cohesive and adhesive bonding is obtained by spraying binders onto powder formed ceramic or polymer particles. In some other techniques layers of sheets are placed one above the other with adhesives being applied onto each combination of surfaces for strong solid object. The initial design procedures for each additive manufacturing technique though remain the same, involving 3D CAD solid model construction and converting it into AM file (.stl) format, slicing of the 3D model based on the need of accuracy and type of AM machines to be involved and then loading the file into the AM machine. Few of the additive manufacturing techniques involved in 3D object fabrication are described below:

Fused Deposition Modelling (FDM) was patented in the year 1992 [Huang et al, 2013], which uses liquid thermoplastic material raised in temperature of up to 1 degree above the melting point so that it readily solidifies once it is extruded out of the nozzle and cold weld to the underlying layer. Ultra thin layer of liquid polymer is spread across the previously generated layer in the form of slice. In education system, this AM technique can be used to build small objects where the weight of the object is of least concern. In case of building large objects two nozzles are generally employed, one of which is used to construct the supporting element structure made of cheaper material which can easily be removed out once the object is perfectly capable of holding its own weight [Pham & Gault, 1998]. The process uses material such as wax, ceramic and metal [Kruth et al., 1998].

Inkjet Printing (IJP) is based on non impact dot matrix technology which involves a liquid phase material or ink consisting of solutes dissolved or dispersed in a solvent. It was first used for creating 2D image but later was developed for 3D object building. It was first patented in the year 1951 [Le, 1998] and is still being used and developed with time, till the present age. The liquid material is forced out of a nozzle made of piezoelectric elements which triggers a sudden quasi-volumetric contraction. The ejected material droplets impinges on the substrate and the solvent gets evaporated which dries off and accumulates the solute layer after layer to form a 3D object. The technique exhibits a quick process and the manufacturing of transistors, sensors and solar cells are commonly seen exploiting its use [Singh et al., 2010]. The disadvantage lies in the equipment cost and delicacy where the ink cartridges are too expensive and print heads too fragile.

Laminated object manufacturing (LOM) process is an additive technique which is relatively less toxic and expensive. Patented in 1988 [Huang et al., 2013], this process utilizes sheets of materials coated with adhesives on its surface. The layers are spread one above the other and a 2D profile on each layer of sheet is cut using laser. The focus and height of the laser is adjusted such that the penetration of the laser is only about the thickness of each sheet and prevents any damage to the underlying layer. The procedure begins with the base and gradually increasing in its height to the top forming a 3D object, the laser head being gradually moving up in the Z direction and adjusting the focus and penetration. Though LOM process has an accuracy defect in its Z direction resulting in dimensional stability issues and post processing time required for eliminating the waste [Kamrani & Nasr, 2010], it can be useful for educational system to demonstrate things which do not require much accuracy and are small in size.

Stereolithography (SLA) is an additive manufacturing technique which employs photosensitive monomer resin treated with light or UV laser for solidification. It was patented in the year 1986 and was first publicised by Hull [Hull, 1988]. The base monomer layer of the desired 3D object is first fused using laser, making a 2D solidified sectional layer which is lowered into the liquid monomer to get coated over the prepared surface and lifted by an amount, such that a blade wipes out the extra top liquid resin, maintaining a thickness of a single slice for treating the next layer. Similar sequential steps are performed for each layer which build up the part to the top and the "support material" is later manually removed from the object. Though SLA technique can prove to be a less time consuming process with good surface finish attainable, it is a costly affair as the photopolymer itself cost about $300 to $400. Also it is applicable to production of only small sized products and not exceeding a cube size of 2-foot.

Selective laser sintering (SLS) is an additive manufacturing process which utilizes the heat generated by laser to sinter the powdery particles to reach a temperature lower than its melting point so that distortions can be avoided and fuses only to particles on the previous layer. SLS was first patented in

1989 [Huang et al., 2013] and employs powdery form of ceramics, polymers, metals, glass or any other material as raw materials. Layers are being sintered one above the other by spreading new layers of powder on top of the already sintered structure. The un-sintered particles remain in place to support the solid structure till the end of the fabrication, then it is removed and reused for next fabrication. SLS technique is fast and provides better durability and functionality. It lags behind in their surface finish, as compared to the SLA and material changeover [Kamrani and Nasr, 2010].

Three dimensional printing (3DP) was first patented in the year 1993 [Huang et al., 2013] which is employed for fabrication of metal, ceramic and metal/ceramic parts and is fast and cost effective [Marks, 2011]. Powdered material is deposited on the substrate in which binders are sprayed at selected position and area for joining. The material is first misted by water droplets for stabilization to prevent disturbances caused by hitting of the binder. The unbounded particles are removed, and further processing is carried out by heating the part to high temperature for effectiveness in binding strength.

Feasibility in General Applications

Some of the difficulties faced while considering feasibility of AM in general applications are surface roughness, dimensional instability and cost of the device itself. High surface roughness and dimensional instability can be attributed to large size of the powdered particles and their arrangement on top of one another at the trajectory of the printing head. The 3D printers comes at a high cost, starting at $5000 to about $50,000 excluding the binders, powdered material and other extra accessories [Huang et al., 2013].

Some of the environmental feasibility factors in additive manufacturing include less pollution of aquatic, terrestrial and aerial ecosystems; less wastage of material and energy consumption [Luo et al., 1999]. The use of cutting fluids in conventional shaping of material such as machining by a tool contributes the highest, to the liquid pollutants. Moreover the scrap material removed as chips causes an additional wastage and less efficient and with it energy consumption also elevates. The measurements for energy consumption is taken in kilowatt hour (khW) per kilogram of the 3D object produced. Experiments have been performed considering different AM techniques for studying the energy consumption characteristic and were found to depend on the way of conducting the experiments. The energy fluctuations were caused due to higher energy rate consumption during warm up and cool down stages [Baumers et al., 2011].

Occupational hazards in AM are not as severe as in the conventional manufacturing processes such as machining but it possesses a different set of hazards such as disposal of waste materials in various AM processes, their usage and handling. A low viscosity liquid resin called Tuxedo TMG3-HCM is found to cause genetic mutation and alteration of cells to completely change its structure [American dye source, 2002]. Some other health effects are harsh reactions on skin, eye irritation and allergies caused by either inhaling the harmful chemicals or spilling of it onto the skin. Long exposure to these harmful chemicals can cause serious health issues, the fatality of which is not known completely yet. The materials used in AM processes are generally long hydro carbon chains which are non bio degradable and survive for an extended period of time after the operation has been shut down. These long chain carbons degrade with time to release harmful gases like carbon dioxide, carbon monoxide and nitrogen oxides. Some other harmful chemicals and gases evolved during the operation include CFC's, HCFC's, CCl_4, CH_3CCl_3, nickel and lead compounds [Drizo & Pegna, 2006]. The solvent used for dissolving the support materials in AM processes like SLA are found to cause some health issues like skin burns and respiratory uneasiness.

Proper handling and safety measures during operation of AM machines have to be considered by all personnel working on them to avoid any dermal contact or chemical spills. Protective masks, goggles and gloves must be provided to every operator during AM operations and they must be trained before handling and operating on high intensity laser heads.

RECENT APPLICATIONS OF 3D PRINTING IN EDUCATIONAL INSTITUTIONS

Though the conventional manufacturing processes cannot be replaced completely due to limitations as given in [Stein, 2012], additive manufacturing processes or 3DP can play a major role in design sector. Complex prototype building and modifications in between operations for any change in flexibility of the material or stiffness of the prototype [Huang et al., 2012], can be achieved in a very short duration of time. The technology is efficient and environmental friendly as compared to traditional manufacturing [Huang et al., 2012].

Due to these advantages, falling costs of the AM hardware [Hoy & Brigham, 2013] and realising the need to shorten the gap between the rate of adoption of the human skills and computational technology in industrial revolution (as already discussed in the introduction part), 3DP's efficient and productive application in educational institutions have been discussed. The application of the AM technology in educational institutions as tools for the students to help them in learning various regular subjects can be distinguished based on the education level of the students and requirement of the type of demonstration by teachers. From the below section, the additive manufacturing will be denoted as 3D printing in a general way.

3DP as Educational Tool in Primary, Middle, and Secondary Schools

In schools, 3DP technology have been employed for building physical models for better understanding of STEM subjects like mathematics and science [Bull et al., 2014]. The motive was to bring in the engineering and technology curriculum into school learning of the regular subjects. Table 1 shows a few applications of 3DP technology integrated into school curriculum.

As can be seen from the Table 1, most of the integration of 3DP into schools were based on introductory overview of the design and fabrication aspect of it and stressing more on its application in improving the concepts of STEM subjects.

Additive manufacturing proves to be an effective tool for teachers and students to explore new ideas in a physical realm, learning in the process instead of imagining from the textbook [Kostakis et al., 2015; Loy, 2014]. To carry out this integration teacher must be fit and dynamic enough to develop their professional skills and keep themselves updated [Kostakis et al., 2015; Bull et al., 2015]. The major issue is with the students who needs to build their attitude towards adaptation to new technology [Kostakis et al., 2015; Nemorin & Selwyn, 2016], their previous technological awareness and being comfortable with the inclusion of 3D printing into curriculum.

Table 1. Table for papers presenting 3DP integration in primary, middle and secondary schools

Papers	School and Country Adopting 3DP	Educational Level	Subject Area	Paper Summary
Cook et al., 2015		4th grade student (elementary)	Science, technology, engineering, Arts and mathematics	With self design and innovation, a teacher with his/her students learnt to be creative and built a 3D printed prosthetic hand using 3D printing in STEAM lab or makerspace.
Corum and Garofalo, 2015		5th grade student	mathematics	Students were able to understand the concept of 3-D object dimensions and learn to built physical models such as rectangular prism and cubes, using modelling software and die cutters.
Huleihil, 2017		6th grade students (middle school)	mathematics	Two student groups (reference and the intervention groups) were studied in their improvement in mathematics by letting them think, design and produce 3 dimensional geometries such as cubes, prism and cylinder. Comparing both the groups by conducting tests proved the result to be positive
Stansell and Tyler-Wood, 2016		Middle school	Science and mathematics	A transmedia study was carried out with few students accessing the 3D printers in evaluating their STEM projects end solution and the other group not having access to 3D printers. A significant development in mathematics concept was built in the students of the group having access to it.
Kostakis et al., 2015	Loannina, greece	High school	Artefact building	A 3 moth project run in two high schools to examine the role of open source 3D printers and a design platform in improving the design and skills in students by producing creative artefacts.
Makino et al., 2017	Japan	High school	Science, mathematics, technology and engineering	Students were able to create police whistle and study the sound frequencies by altering the length of the mouthpiece and radius.
Roscoe et al., 2014		Secondary school	Computational thinking	Participants were given access to open world game minecraft and 3D printers, to learn computer aided design and additive manufacturing by design thinking and building a miniature world from scratch.
Chery et al., 2015	Girls high school, Philadelphia, PA	10th grade (high school)	Chemistry	A study was carried out where half of the students were given the task to design the structure of atom using iPads with the help of an app called AutoDesk123 design and then 3D printed at Drexel University. Tests were performed before and after the programme for all the students and the result showed an improvement in the concept of students who have experimented their models against the students who have not.
Grant et al., 2016	California and florida	Middle and high school (K-12 schools in U.S.)	Fossils and science of paleontology	A curricular prototype of a giant extinct shark *Carcharocles Megalodon* has been presented to students in two high school and middle school in their study of fossils and science of paleontology, made by 3D printing.
Jacobs et al., 2016	Vertus charter school in Rochester, New York	9th grade students (high school)	Mastery based online coursework and hands on learning activity	A summer program around e-Nable involving a group of 12-16 students led by an educator had been implemented to assess students performance after each level of their learning experience in designing, student-teacher interaction activity and hands on activities such as 3D printing, game designing, machine designing, art, chess and slam poetry.

3DP as Educational Tool in Tertiary Education (Universities, etc.)

3DP adoption in universities and specialized studies can be identified based on the purpose of the 3DP product being produced, as in making artefacts to assist in STEM education [Bagley & Galpin, 2015; Hall et al., 2017] and producing test specimens for experiments [Golub et al., 2016]. Building prototypes and components by the students for their projects in a project-based learning environment [Martinez et al., 2016], hold the other option for 3DP applications in universities including robotic applications (building components for educational robots) [Ziaeefard et al., 2015] and 3D printer constructions [Mercuri & Meredith, 2014; Kayfi et al., 2015]. Table 2 describes some of the initiatives taken by various universities in incorporating 3D printing into their university curricula.

As we can see from the Table 2 most of the application of 3D printing is limited to engineering field with few in medical applications such as in building molecular structures for pharmacy applications, simulation models for application based on real patient situation and 3D printing of skeletal and muscle cells [McGahern et al., 2015; Hall et al., 2017; Kroger et al., 2016; Bagley & Galpin, 2015].

Most of the 3DP applications (from the table) are through project based learning and integration with the course curriculum. This enables students to utilize 3DP as a tool for experimentation, product development and concept verification leading to new ideas and innovation.

With 3D integration, student learning has become more active, interesting and passionate. Firstly, there is enough exposure for students to visualize their 3D problem [Chen et al., 2014] and understanding mechanisms and concepts. Secondly, student teacher relationship [Nemorin & Selwyn, 2016] has improved with the introduction of 3DP in the education system as it ensured maintained level of interest in students by adding a visual representation along with the teacher's lecture. With full fledge adoption of 3DP in the education system, the teacher would be called more often as mentor because the mentor would be learning the subject in a physical realm alongside with the students [Loy, 2014]. Thirdly, it gave the students the vision to quickly analyse their objectives or future goals as they get thorough the structure of the problem. Students get accustomed to their responsibilities [Fernandes & Simoes, 2016] and thinking of a design for a product which is environmentally sustainable [Luo et al., 1999].

Table 2. Table for papers presenting 3DP integration in tertiary education

Papers	University and Country Adopting 3DP	Educational Level	Subject Area	Paper Summary
Jaksic, 2014	Colorado state university, Pueblo	Undergraduate engineering (Mechatronics and industrial engineering)	3D printing	The paper explores utilization of inexpensive 3D printers in supporting engineering and non engineering programmes and challenges for a successful 3D printing.
Reggia et al., 2015	The University of Maryland, college park	First year engineering major	Design course (project based)	Students in a multidisciplinary group of 8-10 had to design and build an autonomous vehicle using 3DP and then test. Also CAD courses have been merged with 3D printers to give an experimental motive to the design software learning.
Minetola et al., 2015	Politecnico di Torino, Italy	Masters in Mechanical engineering	Design and 3DP technology	A survey to assess the outcome of incorporating 3DP technology into mechanical engineering curriculum. The participants got to design and create multiple polymeric components using Fused Deposition Method (FDM) which had to be assembled by the students itself for the final prototype.
Bilen et al., 2015		Undergraduate engineering	Design course	The students used 3D Printing to design and built rocket to accomplish a launch trajectory set by the faculty. Application of 3DP enabled students to make quick changes to their model for any design alteration.
Jankovic et al., 2016	University of Belgrade, Serbia	Masters studies in science	Fans and turbo chargers	Students were able to design axial fan impeller blade and fabricate using 3D printing.
Kroll and Artzi, 2011	Israel institute of technology, Israel	4th year aerospace engineering	Aerodynamics, experimental project	Two polymer based Rapid prototyping aircraft models were built to be tested in subsonic wind tunnels by the students. The 3DP models bought down the cost of wind tunnel experimentation and within the budget of academics.
Pieterse and Nel, 2016	University of Johannesburg (UJ), South Africa	Undergraduate mechanical engineering	Design and research projects	Students performed their design projects using Dimension elite 3D printers and research works through capstone research projects.
Go and Hart, 2016	Massachusetts institute of technology, MA	Graduate and advanced undergraduate	Additive manufacturing	In a 14 week course curriculum, students were introduced to the technical aspect of the AM, design procedures, machine controls along with lab tasks of fabrication using 3D printers, a bridge of maximum strength to weight ratio with end conditions.
Valero-Gomez et al., 2012	Open source	Engineering	Design and A.M.	A training course based on the open source community oriented project based learning had been implemented where the participants with the help of open source 3D printers and connected to other designers, built a PrintBot (printable mobile robot).

continued on following page

Table 2. Continued

Papers	University and Country Adopting 3DP	Educational Level	Subject Area	Paper Summary
Gatto et al., 2015	University of modena and reggio emilia, Italy	2nd year mechanical engineering undergraduate	Cost analysis, prototype testing, design and development	Students built an eye traker head mount by reverse engineering a digital head model using 3D printing in an inter-disciplinary project based learning.
Dahle and Rasel, 2016	State university of new york, USA	Engineering	Modelling and simulation	Students learned the simulation and development of MEMS device using 3D printer in a semester long design module.
Payne, 2015	University of north georgia, GA	Computer science and information system	Computer graphics course	With the integration of 3D printers in the computer graphics course students learned to scan, model and 3D printing.
Lin et al., 2012	Tsinghua University, china	Undergraduate engineering	Additive manufacturing	Students were able to operate on 3D softwares, design and fabricate their own desired prototypes.

3DP as Educational Tool in Libraries

3DP has revolutionized the consumer market by being capable of providing an easy and quick access to various customized products [Petrick & Simpson, 2013] desired by the consumers. In a way, this technology boom can be considered to bring a democratized environment for design [Lipson, 2012] and fabrication just like books have democratized information through ages and internet democratized communication world [Griffey, 2012]. Library holds a neutral zone in the entire university campus to provide a multidisciplinary and cross disciplinary interactive environment [Van Epps et al., 2015; Gonzalez & Benett, 2014]. This interactive environment integrated with 3DP can suffice the growing research need among scholars and university students. This can also solve the core motive of the future libraries of democratizing information and technology [Hoy & Brigham, 2013].

Future public libraries are sure to inevitably adopt 3DP as tool [Hoy & Brigham, 2013] for building innovative and productive collaborative environment where people will be able to gain easy design file access or bring their own design to convert it into 3D product. On one hand where there is great feasibility of adoption of 3DP in technical university and public libraries, the feasibility of medical libraries in adopting 3DP can be a matter of concern. Few possible reasons why 3DP should be integrated in medical library can be described [Hoy & Brigham, 2013] as below:

Table 3. Table for papers presenting 3DP integration in libraries

Paper	University and Country	Who Will Benefit?	Library Type	Paper Summary
Scalfani et al., 2013	University of Alabama	Students, faculties, staffs	University library	An open access 3D printing studio has been maintained in the library to be accessed by users after a two step training procedure. The user can avail the 3D printers and design workstations installed, for their project and research works.
Nowlan, 2015	University of Regina, Canada	same	University library	A 3DP library service has been implemented to better understand the outcomes of such adoption and modifications necessary for the sustainability of it. Training for the users is found to be necessary.
Gonzales and Bennett, 2014	University of Florida, United states	Undergraduates	Science and health science library	A 3D printing and scanning service consisting of 3D printers and scanners being installed in respective libraries, has been implemented to assess improvement in research and teachings from such integration.
Bharti et al., 2015	University of Florida, United states	Teachers and faculty	Science library	3D printing technology and a visualization wall has been integrated into Marston science library to provide assistance in teaching and research. Assessment of the integration proved to be successful.
Williams and folkman, 2017	State library of north carolina,	Information studies students and library professionals across the state	makerspace	Librarians were assisted to quickly adapt to the use of makers technology including trainings, workshops and conferences.
Groenendyk and Gallant, 2013	Dalhousie University, canada	Faculty and students	University library	A 3D printing and scanning technology has been implemented which served students and faculties in carrying out their research projects. Scanning devices make it possible for carrying out archival studies.

1. Medical libraries are not technology focused. There are few cases of using 3DP for organ, tissue, medical equipment development [Waldo, 2012; Thilmany, 2012], but it is very unlikely that these fabrication will be taking place in medical libraries as these procedures require expertise and medical professionals. Librarians though can learn the procedure

which remains same be it in a consumer grade or educational low budget 3D printer, to help medical users with their learning.

2. 3DP requirement in medical libraries can be realised in creating anatomical artefacts and models for clinical practice and education [Shapeways 3d Printing News and Innovation, 2011].
3. Medical libraries can provide a shared access to an expensive resource like 3D printers in universities to be availed by multidisciplinary students and faculties.

Table 3 gives us an account of few initiatives taken to integrate 3DP in libraries.

From the Table 3 it is found that digital library play a major role in providing a collaborative environment for the university community to communicate and learn from external sources. Students, faculties and librarians can avail this opportunity to carry out their own research works and teaching plans [Groenendyk & Gallant, 2013; Bharti et al., 2015]. Few of the examples give us an idea of the plan for successfully integrating 3DP into libraries [Gonzalez & Benett, 2014; Nowlan, 2015].

Libraries are spaces for gaining and sharing knowledge, which enables students to look for solutions to their problem by communicating with other users (faculty, librarians, etc) across different universities, innovate and build 3D parts and components for their projects. As Groenendyk and Gallant explain, the motive of the library is "*to take the knowledge-sharing, innovation-driven ideals of hackerspaces and bring these into an academic library setting*" [Groenendyk & Gallant, 2013].

Libraries are grounds of learning for both non-technical and technical students in an institution [Van Epps et al., 2015]. Non technical student who are new to this technology, can discuss and learn how to use 3DP devices from educators and technical users for designing artefacts and decorative items.

Librarians can go through training on the basics of 3DP so that communications with the library users [Groenendyk & Gallant, 2013; Williams & Folkman, 2017] becomes more relatable and understandable, collaboration with teachers is easy and maintenance of the 3DP equipment and troubleshooting it in case there is any malfunction is possible [Bharti et al., 2015, Groenendyk & Gallant, 2013].

3DP in library must consider the safety measures and health issues, the cause of which depends on materials to be used for printing and method of prototyping. This is the reason behind using Poly lactic acid (PLA) as replacements of ABS [Bharti et al., 2015; Bharti & Singh, 2017] as it decreases the production of ultra fine particles by about 10 times during fabrication process. Another issue with using 3DP in libraries is the intellectual property (IP) where the library users and librarians must be instructed for preventing any infringement of copyright materials and products [Finley, 2016; Jones, 2015; Chan & Enimil, 2015].

3DP in Special Educational Settings

3DP application in education for students with visual, motor and cognitive impairment or the combination of all three can be improved by specific settings in its use in the educational system [Horowitz & Schultz, 2014, Buehler et al., 2014]. These special settings include custom adaptive devices and educational aids developed for facilitating enhancement of the interests in students with the above mentioned issues, towards the STEM subjects [Buehler et al., 2016, Buehler et al., 2014]. Students with cognitive impairment were instructed to use tindercad software before going for construction of their own 3D CAD models [Buehler et al., 2016].

The students, due to lack of time and difficulties in performing tasks to create their own designs had to adopt and modify designs already available in the open source sites. The students had to go through a time consuming process of learning the software first before facing further challenges in developing their own 3D design and manipulations, due to which the interest level of students decreases and crave for innovation is lost. With less time available, student's enthusiasm for exploring potentials in 3DP has been affected by their concern for efforts required in learning the software as said "*they currently see the task of 3D design and printing to be someone else's work, and see themselves as consumers of that work*" [Buehler et al., 2016].

TRANSFER OF KNOWLEDGE FROM THE 3DP MANUFACTURER TO THE USERS

The different ways by which 3DP can be implemented in educational system is by giving trainings or introduce courses to students, teachers, librarians and educators. This is an active integration of 3DP [Junk & Matt, 2015]. After gaining knowledge, 3DP can be utilized in a number of different ways which can be in the form of assistance, artefacts or concept solver. This can be considered as passive integration of 3DP as the technology is used only to support facts and evidence for other projects and courses.

Teaching Students in Educational Institutions

3DP integration in universities include introductory courses, project-oriented courses, workshops and 3DP application in other projects as passive integration. The courses can be briefed and summarized as given in [Minetola et al.; 2015, Junk & Matt, 2015; Payne, 2015; de Sampaio et al., 2013; Liou et al., 2012] and it can be "detailed descriptive" and specific as given in [Radharamanan, 2017; Go & Hart, 2016]. 3DP can be implemented for creative experimentation [Loy, 2014; de Sampaio et al., 2013; Chiu et al., 2015], product development and entrepreneurship [de Sampaio et al., 2013; Liou et al., 2012], assistance of 3DP for understanding concepts in other technical courses [Liou et al., 2012], and assist multi and inter-disciplinary skills [de Sampaio et al., 2013]. The main motive behind all integration is to improve technical and non-technical skills by having the best and efficient use of 3DP technology.

A 14 day MIT-based course on additive manufacturing with 30 students participating had been organized. The first five days of the course concentrated in giving an introductory overview of different additive manufacturing techniques available such as SLA, SLS/SLM and FDM along with the different techniques involved in scanning and lab sessions held for exercising their skills. Two group exercises were introduced alongside the introductory classes in other special topics like bio-printing, computational designs, design assembly, design and economics of additive manufacturing, micro and nano-additive manufacturing and .electronic printing. The two group exercises enabled students to put in their own design ideas and creating their own innovation. In overall, the course was highly satisfactory with positive feedbacks from students in both the lecture and workshop sessions [Go & Hart, 2016].

3DP teaching has also been incorporated in the City university, Hong Kong into an engineering-based course [Chiu et al., 2015]. Gagne's model [Gagne, 1985] for structural framework of teaching was used to carry out these courses for 2013/14 and 2014/15 with student participation of 89 and 28 in number. The aim was to cover knowledge regarding the different stages in 3DP designing and fabrication which are pre-processing, fabricating and post-processing. Though 50% of the participating students were from engineering background, some were even from business studies background, liberal arts and social science background. The overall feedback from the student was positive and they experienced ease of designing with CAD software in presence of tutor. They improved in their attitude and innovation towards designing with 3DP. Though a range of differences in difficulty level in learning the concept and implementation was also observed which can be related to weaker technical background in non-technical and engineering students [Chiu et al., 2015].

A semester long course on AM technology have also been carried out in the departments of mechanical, industrial design and manufacturing at the Metropolitan state university in Denver. In a 16 days course students learned about the 3D scanning, solid modelling, design for AM techniques, mesh manipulating and designing, fabricating, sustainability issues and post processing. The challenges in carrying out the course lies in non availability of appropriate textbooks in the 3DP domain, continuous requirement of supervision and requirement of working in lab outside class hours [Paudel & Kalla, 2016]. Similar courses on AM was also introduced in Mercer university where it was incorporated as elective for students in engineering field for a 16 days programme.

In all the above- mentioned programmes, the basic 3D solid fabricating from the initial designing phase can be taught to the students in mainly three approaches [Junk & Matt, 2015]. In the first approach, students are told to download previously designed CAD 3D models from the database and transferring the file to the AM machine for fabrication. In this way, the students are able to solve for complications and design sustainability in transferring CAD data file into the machine and the limits of the AM machine itself. The students are able to explore the capabilities of AM in different designing domains. In the second approach, the students were allowed to build their own designs applying their innovation and modelling skills instead of using database and extract designs already available [Junk & Matt, 2015]. This enables them to tackle obstacles and complications related to following the design procedure and builds confidence and motivation in using the technology. In the third approach, the techniques involved using the 3D scanner for building the 3D model in CAD software, manipulating the scanned profile and printing the final model were used.

For teaching students about 3DP in schools, the examples can be taken of the two Greek high schools where the students were given knowledge about 3DP designing and printing by letting them participate in a collaborative project [Kostakis et al., 2015]. Two groups were chosen from the two high schools of age ranging from 15 to 18, who were made to work on a project which includes 700mins of introductory class about 3D CAD modelling and various designs to be considered while transferring digital models to AM devices. In the experimental phase of the project session students were instructed to build their own imaginations and design accordingly so that an errorless 3D model is created. They build artefacts and perfected their model by making modifications and necessary changes in design repeatedly to get an errorless 3D model. Between operations, students had to go through various obstacles and complications which needs to be solved with utmost delicacy, care and thoroughly. This art of handling complications in between design operations require instructor's attention, working with whom, the students can gain hands in experience and confidence in 3D designing and printing.

Programs have been tried in an elementary school in Baltimore for the 3rd and 4th grade students incorporating FDM 3D printers for prototyping. The program is basically to improve interest in arithmetic and all other STEM subjects by introducing 3DP in their project which helped students visualize their problems and concepts. The students improved in their problem solving ability and developing iterative designs, first in a low fidelity format to the final prototype building.

A lot of time is needed to be invested on the students while teaching modelling and printing techniques and instructor's effort in doing so will make a significant effect on the understanding of the subject by the students. Another drawback of 3DP is the time taken for the large 3D prints [Easley et al., 2017; Plemmons, 2014] to be accomplished which is very high and it will be very inefficient for the students to remain stuck in one project while missing all other class lectures.

Teaching Educators About 3DP in Educational Institutions

Apart from all the challenges in integrating 3DP in school and university course curriculum and workshops, a major problem lies in teaching of the educators who can be considered the real connectors for the ideas perceived by the students about the fundamentals of the 3DP technology. In the present scenario schools, the educators does not fully qualify to pass on their knowledge in the 3DP field to the students as they lack the expertise in design thinking and engineering designs [Bull et al., 2014]. Current teacher's training program does not include this expertise, so there is a huge role to be played in teaching the educators about 3DP integration in designing and school courses.

Programs have been initiated as given in the literature [Maloy et al., 2017; Moorefield-Lang, 2014] to teach the educators about 3DP technology, to teach the active educators (who are already related to this field) as professional development training and to teach the librarians.

Early childhood educators [Sullivan & McCartney, 2017] were given the preference along with science educators [Vemer & Merksamer, 2015; Irwin et al., 2014] who had received first, the teachings concerning 3DP technology in their teacher's training program. The program was aimed at students and educators integration towards achieving a common design goal using 3DP technology, learning together in the process and critically evaluating new technology, analyse the technology hurdles in incorporating it into early childhood classrooms [Sullivan & McCartney, 2017].

3DP workshops can be attended by childhood educators, science teachers-biology, chemistry, physics teachers where 3DP technology, its construction and principles can be passed on so that educators can build their own 3D printers for fabricating instruments and equipments in their respective lab. Similar involvement of biology educators [Irwin et al., 2014] were observed during 2 day workshop on Reprap 3D printers at the Michigan Technological University. Middle school teachers can be given access to 3DP workshop training to build artefacts that would enable them to explain their concerned subjects to students more appropriately and with ease. Teachers have first found difficulties in using Tinkercad design software but found it easy to understand in the presence of technical support [Maloy et al., 2017].

Active educators who already have the basic knowledge about the recent computing technologies in the middle and high school levels have been trained in a program of 3 workshop days where recent technologies based on 4 major advancement in application of computational technology such as internet of things, robotics, mobile application and 3DP were explained and analysed for any limitations and challenges in integrating the technology in the teaching. The program was held at King Saud University and the feedback was that 3DP score for the willingness of the educators to integrate 3DP in school courses was the minimum among all the other emerging technologies and the percentage of acceptance was about 57% [Al-Mouh et al., 2016].

There is not much evidence and literature on workshops being carried out for librarians and they do not get any assistance regarding acquiring skills on 3DP. A little self curious attitude towards the subject or self learning, self experimentation, exploring the field by travelling to other university libraries and educators, travelling museums and peer learning will help librarians to acquire enough material to decide about integration of the technology in their own libraries. An initiative to build skills on 3DP technology was attempted at the university of north Carolina, Greensboro for all librarians across the states [Williams and Folkman, 2017] and was successful as it stressed upon the environment of "okay to fail" attitude which encouraged the participants to actively get involved with the hands in experience on 3DP designing and fabricating.

Probable Sources of Impact of 3DP on Educational Domain

Considering the discussion above, the impact of 3DP on educational culture in the near future can be portrayed in terms of learning habits, teaching skills, student-teacher relationship, contribution to disabled learners, rate of transfer of knowledge and its utility in medical applications. Two probable sources of these impacts have been discussed in the form of 3DP artefacts and support systems as learning assistance.

3DP Artefacts as Aid in Learning

Producing artefacts using 3DP technology has changed the learning experience of the students in their STEM subjects. It also enabled educators to quickly and easily explain the complex concepts behind any 3D representation. As already discussed, biology teachers can be trained to build their own 3DP artefacts to explain very small size organism or organs that is difficult to visualize just by reading the textbook or referring to 2D diagrams. Mathematics teachers can utilize 3DP technology to construct complex 3D geometries that cannot be visualized during normal textbook readings and referring diagrams in 2D. Similar utilization of artefacts can be observed in the teaching of chemistry [Blauch & Carroll, 2014; Stone-Sundberg et al., 2015] and physics. In anatomy, for the study of the human bones, 3DP artefacts can be employed resembling exactly like real or plastinated bones that are presently very expensive to buy. Moreover, 3DP artefacts will ensure safety of the original specimen [AbouHashem et al., 2015].

Investigations have been carried out in anatomical field where groups of medical students were detailed about the upper limb using 3DP artefacts. The 3DP parts were accurate and in compliance with the actual body organ dimensions. Yet, the students were in agreement that although 3DP parts were perfect in understanding the mechanism, it can be best utilized with plastinated specimens to aid learning [Mogali et al., 2017]. A one-third scale model of lower limb posterior compartment had been fabricated using 3DP and presented in a limb anatomy class. The class was divided into two groups of students, one of which is using the dissection specimen and the other using the 3DP copy of the specimen. No significant difference between the knowledge gained by both the groups about the subject on anatomy had been observed [O'Reilly et al., 2016] which implies 3DP models can be an effective substitute when dissection specimen is not available or not sufficient [McMenamin et al., 2014].

In chemistry, 3D printed molecular structures has played a major role in the understanding of the subject in a more detailed way rather than using 2D diagrams and texts to imagine the object. Most importantly, it was observed that cost of the 3D printed chemistry models is 1/50th of the commonly available artefacts in market and using freeware CAD design software [Griffith et al., 2016].

Integration of 3DP technology in the teaching of dentistry have also been studied and applied. A 3DP model for prosthodontic practices was introduced to a group of 22 dental student in their fourth year [Kroger et al., 2016]. The experience was healthy with high positive feedbacks from the students as they claimed the worthiness of 3DP in practicing removal of crowns and installing new crowns. Clinical courses can be well explained and worked on with the introduction of 3DP technology in a more practical environment.

3DP as Support Systems

3DP can be used to develop technology as support systems [Horowitz & Schultz, 2014; Buehler et al., 2014; Buehler et al., 2016] for visually, cognitively and motor impaired condition. It can also be employed for educating blind or hearing-impaired people in gaining a better knowledge of the subjects in science, mathematics, history and literacy by incorporating 3DP specimens [Kolitsky, 2014; Cavanaugh & Eastham, 2017] and graphics to assist them touch, feel and understand concepts in a physical realm.

In the subject of history, the introduction of tactile 3DP textbook has enabled students to enhance their capability to memorize things that are too obscure to understand by traditional means of education [Jo et al., 2016]. It also helped educators to reduce their time of teaching and in better transferring of knowledge. Tactile graphics is another provision which enabled students to touch and understand ancient artifacts or weapons and develop eagerness and excitement in learning the subject [Kane & Bigham, 2014]. The main issues with these tactile graphic was its brittleness and more time to 3D print.

EFFECTIVENESS IN THE INTEGRATION OF AM TECHNOLOGY WITH THE EDUCATION SYSTEM AND FUTURE SCOPE

As seen from the above section, it can be concluded that 3DP integration with engineering and technological universities has reached the stage of maturity as it perfectly suits with the design research and manufacturing advancements (as observed in Table 2 and Table 3). Apart from the design and engineering, other STEM subjects have also adopted this technology to strengthen the grasp of the subjects by the students (Table 1). Artefacts and models are being manufactured using 3DP which mimic the actual object, and helps students in understanding some of the complex aspects of the STEM subjects. Non STEM subjects, however has not much advanced in adopting 3DP for in-class teaching.

In universities, engineering and design departments have opened their dedicated services of integrating 3DP technology into other disciplines. Centralized libraries have been provided by universities which incorporate 3DP technology to be accessed by institute community including the non-technical users for advance exposure [Table 3].

In the present scenario, the expertise in AM technology is less and educators themselves are not fully aware of the technology and lack basic design thinking and computational skill [Bull et al., 2014]. 3DP educational components needs to be incorporated into the training curriculum of pre-service educators and workshops and courses need to be arranged for in-service educators to enhance skills in 3DP applications and to provide better guidance to the students. This issue of "training the teacher" have often been overlooked by the current papers. At present, there seems to be a lack of books on "teaching" regarding integration of 3DP into educational curriculum [Paudell & Kalla, 2016] whether it be courses on 3DP skill development or teaching through artefacts the STEM subjects. Additional attention has to be given on this topic.

SUMMARY

From the above discussions, we have come to know about the different initiatives that has been taken to integrate AM technology into educational curriculum. We have discussed about the various challenges and hurdles towards achieving this integration and how are they tackled in different universities and schools. The chapter begins with its first section dedicating to a few commonly used AM technology, their feasibility and safety issues. The main issue with integrating AM in education has been found which does not limit to the capability of the technology but also extends to the IQ (design thinking) level of the educators and how are they able to cope up with the modern technology. A number of initiatives to integrate this technology had already been successfully accomplished in different schools and universities with the aim to give students exposure to AM technologies by self designing models and artefacts that help elevate the understanding of the STEM subjects. Moreover, different design procedures, computational skills and design innovations are also being taught to the students and educators for their own design needs. After attending workshops on AM, science, mathematics and history educators can build their own miniature artefacts of their interest to explain their subject to the students in a more systematic and interactive way. It also enables the students to understand complex geometries, micro organisms, cellular structures and ancient historical artefacts that are too hard to imagine and visualize under normal circumstances. In concerning departments of science and technology, students were given a full-fledged course structure concentrating on the different CAD design procedures, file formatting, file transfer, fabricating using AM techniques and post processing which elevates knowledge and hands in experience of the students. Some other advantages of AM technology applications in universities and schools are fabrication of artefacts and models for visually, cognitively and motor impaired people which enables them to understand subject in mathematics, science and history by physically touching and feeling the specimen.

REFERENCES

AbouHashem, Y., Dayal, M., Savanah, S., & Strkali, G. (2015). The application of 3D printing in anatomy education. *Medical Education Online*, *20*(1), 29847. doi:10.3402/meo.v20.29847 PMID:26478143

Al-Mouh, N., Al-Khalifa, H. S., Al-Ghamdi, S. A., Al-Onaizy, N., Al-Rajhi, N., Al-Ateeq, W., & Al-Habeeb, B. (2016). A professional development workshop on advanced computing technologies for high and middle school teachers. *2016 15th Int. Conf. Inf. Technol. Based High. Educ. Train*, 4–7. .2016.776069610.1109/ITHET

A.M. UK. (2017). *Additive Manufacturing UK: National Strategy, 2018–2025*. Author.

American Dye Source. (2002). *Water soluble thiophene polymer*. American Dye Source Inc. Retrieved from http://www.adsdyes.com/products/pdf/polythiophene/ADS2000P_DATA.pdf

ASTM. (2010). *F2792-10e1 Standard terminology for additive manufacturing technologies*. ASTM International. Retrieved from http://enterprise.astm.org/filtrexx40.cgi?+REDLINE_PAGES/F2792.htm

Bagley, J. R., & Galpin, A. J. (2015). Three-dimensional printing of human skeletal muscle cells: An interdisciplinary approach for studying biological systems. *Biochemistry and Molecular Biology Education*, *43*(6), 403–407. doi:10.1002/bmb.20891 PMID:26345697

Baumers, M., Tuck, C., Bourell, D. L., Sreenivasan, R., & Hague, R. (2011). Sustainability of additive manufacturing: Measuring the energy consumption of the laser sintering process. *IMechE Part B: J Eng Manuf*, *225*(12), 2228–2239. doi:10.1177/0954405411406044

Bengtson, J., & Bunnett, B. (2012). Across the table: Competing perspectives for managing technology in a library setting. *Journal of Library Administration*, *52*(8), 699–715. doi:10.1080/01930826.2012.746877

Bharti, N., Gonzalez, S., & Buhler, A. (2015). 3D Technology in Libraries: Applications for Teaching and Research. *4th Int. Symp. Emerg. Trends Technol. Libr. Inf. Serv.*, 161–166. 10.1109/ETTLIS.2015.7048191

Bharti, N., & Singh, S. (2017). Three-dimensional (3D) printers in libraries: Perspective and preliminary safety analysis. *Journal of Chemical Education, 94*(7), 879–885. doi:10.1021/acs.jchemed.6b00745

Bilen, S. G., Wheeler, T. F., & Bock, R. G. (2015). MAKER: applying 3D printing to model rocketry to enhance learning in undergraduate engineering design projects. *ASEE Annu. Conf. Expo.* 10.18260/p.24448

Blauch, D. N., & Carroll, F. A. (2014). 3D printers can provide an added dimension for teaching structure-energy relationships. *Journal of Chemical Education, 91*(8), 1254–1256. doi:10.1021/ed4007259

Buehler, E., Comrie, N., Hofmann, M., McDonald, S., & Hurst, A. (2019). Investigating the implications of 3D printing in special education. *ACM Trans. Access. Comput., 8.* . doi:10.1145/2870640

Buehler, E., Hurst, A., & Hofmann, M. (2014). Coming to Grips: 3D Printing for Accessibility. *ASSETS' 14 Proc. 16th Int. ACM SIGACCESS Conf. Comput. Access.*, 291–292. 10.1145/2661334.2661345

Buehler, E., Kane, S. K., & Hurst, A. (2014). ABC and 3D: opportunities and obstacles to 3D printing in Special education environments. *ASSETS' 14 Proc. 16th Int. ACM SIGACCESS Conf. Comput. Access.*, 107–114, 10.1145/2661334.2661365

Bull, G., Chiu, J., Berry, R., Lipson, H., & Xie, C. (2014). Advancing children's engineering through desktop manufacturing. In Handb. Res. Educ. Commun. Technol. (4th ed.). Springer Science +Business Media. doi:10.1007/978-1-4614-3185-5_54

Bull, G., Haj-Hariri, H., Atkins, R., & Moran, P. (2015). An educational framework for digital manufacturing in schools, 3D print. *Addit. Manuf., 2*, 42–49. doi:10.1089/3dp.2015.0009

Cavanaugh, T., & Eastham, N. (2017). The 3D printer as assistive technology. *Soc. Inf. Technol. Teach. Educ. Int. Conf.*, 95–102. Retrieved from https://www.learntechlib.org/p/ 177280/

Celani, G. (2012). Digital fabrication laboratories: Pedagogy and impacts on architectural education. *Nexus Network Journal, 14*(3), 469–482. doi:10.100700004-012-0120-x

Chan, J. R., & Enimil, S. A. (2015). Copyright Considerations for Providing 3D Printing Services in the Library. *Bulletin of the American Society for Information Science and Technology*, *42*, 26–31. doi:10.1002/bul2.2015.1720420109

Chen, M., Zhang, Y., & Zhang, Y. (2014). Effects of a 3D printing course on mental rotation ability among 10-year-old primary students. *International Journal of Psychophysiology*, *94*(2), 240. doi:10.1016/j.ijpsycho.2014.08.925

Chery, D., Mburu, S., Ward, J., & Fontecchio, A. (2015). Integration of the arts and technology in GK-12 science courses. *2015 IEEE Front. Educ. Conf.*, *1–4*. doi:10.1109/FIE.2015.7344165

Chiu, P. H. P., Lai, K. W. C., Fan, T. K. F., & Cheng, S. H. (2015). A pedagogical model for introducing 3D printing technology in a freshman level course based on a classic instructional design theory. *2015 IEEE Front. Educ. Conf.*, 1–6. 10.1109/FIE.2015.7344287

Chong, L., Ramakrishna, S., & Singh, S. (2018). A review of digital manufacturing-based hybrid additive manufacturing processes. *International Journal of Advanced Manufacturing Technology*, *95*(5-8), 2281–2300. doi:10.100700170-017-1345-3

Cook, K. L., Bush, S. B., & Cox, R. (2015). Creating a prosthetic hand: 3D printers innovate and inspire and maker movement. *Sci. Child, 53*, 80–86. Retrieved from http://stats.lib.pdx.edu/proxy.php?url=http://search.ebscohost.com/login.aspx?direct=true&db=ehh&AN=111061979&site=ehost-live

Cook, K. L., Bush, S. B., & Cox, R. (2015). Engineering encounters: Creating a prosthetic hand. *Science and Children*, *53*(4), 80–86. doi:10.2505/4c15_053_04_80

Corum, K., & Garofalo, J. (2015). Using digital fabrication to support student learning, 3D Print. *Addit. Manuf.*, *2*, 50–55. doi:10.1089/3dp.2015.0008

Dahle, R., & Rasel, R. (2016). 3-d printing as an effective educational tool for MEMS design and fabrication. *IEEE Transactions on Education*, *59*(3), 210–215. doi:10.1109/TE.2016.2515071

de Sampaio, C. P. de O., Spinosa, R.M., Tsukahara, D.Y., da Silva, J.C., Borghi, S.L.S., Rostirolla, F., & Vicentin, J. (2013). 3D printing in graphic design education: educational experiences using fused deposition modeling (FDM) in a Brazilian university. *Proc. 6th Int. Conf. Adv. Res. Virtual Rapid Prototyp.*, 25–30.

Despeisse, M., Baumers, M., Brown, P., Charnley, F., Ford, S.J., Garmulewicz, A., … Rowley, J. (2017). Unlocking value for a circular economy through 3D printing: a research agenda. *Technol. Forecast. Soc. Change, 115*, 75–84, .2016.09.021 doi:10.1016/j.techfore

Drizo, A., & Pegna, J. (2006). Environmental impacts of rapid prototyping: An overview of research to date. *Rapid Prototyping Journal, 12*(2), 64–71. doi:10.1108/13552540610652393

Easley, W., Buehler, E., Salib, G., & Hurst, A. (2017) Fabricating engagement: using 3D printing to engage underrepresented students in STEM learning. *ASEE Annu. Conf. Expo.* 10.18260/1-2--28347

Eisenberg, M. (2013). 3D printing for children: What to build next? Int. J. Child-Comput. *Interact., 1*, 7–13. doi:10.1016/j.ijcci.2012.08.004

Elrod, R. E. (2016). Classroom innovation through 3D printing. *Library Hi Tech News, 33*(3), 5–7. doi:10.1108/LHTN-12-2015-0085

European Commission. (2014). *Additive Manufacturing in FP7 and Horizon 2020: Report from the EC Workshop on Additive Manufacturing*. Author.

Fernandes, S. C. F., & Simoes, R. (2016). Collaborative use of different learning styles through 3D printing. *2016 2nd Int. Conf. Port. Soc. Eng. Educ.*

Finley, T. K. (2016). The impact of 3D printing services on library stakeholders: A case study. *Public Services Quarterly, 12*(2), 152–163. doi:10.1080/15228959.2016.1160808

Gagné, R. M. (1985). *The Conditions of Learning and Theory of Instruction* (4th ed.). New York, NY: Holt, Rinehart & Winston.

Gatto, A., Bassoli, E., Denti, L., Iuliano, L., & Minetola, P. (2015). Multi-disciplinary approach in engineering education: Learning with additive manufacturing and reverse engineering. *Rapid Prototyping Journal, 21*(5), 598–603. doi:10.1108/RPJ-09-2014-0134

Go, J., & Hart, A. J. (2016). A framework for teaching the fundamentals of additive manufacturing and enabling rapid innovation. *Addit. Manuf., 10*, 76–87. doi:10.1016/j.addma.2016.03.001

Golub, M., Guo, X., Jung, M., & Zhang, J. (2016). *3D Printed ABS and Carbon Fiber Reinforced Polymer Specimens for Engineering Education. In REWAS 2016 Towar. Mater. Resour. Sustain* (pp. 281–285). Cham: Springer. doi:10.1007/978-3-319-48768-7_43

Gonzalez, S. R., & Bennett, D. B. (2014). Planning and implementing a 3d printing service in an academic library. *Issues Sci. Technol. Librariansh.*, *78*, 1–11. doi:10.5062/F4M043CC

Grant, C. A., MacFadden, B. J., Antonenko, P., & Perez, V. J. (2016). 3-d fossils for K–12 education: A case example using the giant extinct shark carcharocles megalodon. *Paleontol. Soc. Pap.*, *22*, 197–209. doi:10.1017cs.2017.15

Griffey, J. (2012). Absolutely Fab-ulous. *Library Technology Reports*, *48*(3), 21–24.

Griffith, K. M., de Cataldo, R., & Fogarty, K. H. (2016). Do-It-Yourself: 3D models of hydrogenic orbitals through 3D printing. *Journal of Chemical Education*, *93*(9), 1586–1590. doi:10.1021/acs.jchemed.6b00293

Groenendyk, M., & Gallant, R. (2013). 3D printing and scanning at the Dalhousie University Libraries: A pilot project. *Library Hi Tech*, *31*(1), 34–41. doi:10.1108/07378831311303912

Group, U. A. M. S. (2016). *Additive Manufacturing*. Leading Additive Manufacturing in the UK.

Hall, S., Grant, G., Arora, D., Karaksha, A., McFarland, A., Lohning, A., & Dukie, S. (2017). A pilot study assessing the value of 3D printed molecular modelling tools for pharmacy student education. *Curr. Pharm. Teach. Learn.*, *9*(4), 723–728. doi:10.1016/j.cptl.2017.03.029 PMID:29233449

Horejsi, M. (2014). Teaching STEM with a 3D printer. *Science Teacher (Normal, Ill.)*, *10*. doi:10.1126cience.1153539

Horowitz, S. S., & Schultz, P. H. (2014). Printing space: Using 3D printing of digital terrain models in geosciences education and research. *Journal of Geoscience Education*, *62*(1), 138–145. doi:10.5408/13-031.1

Hoy, M. B., & Brigham, T. (2013). Emerging technologies. *Medical Reference Services Quarterly*, *32*(1), 93–99. doi:10.1080/02763869.2013.749139 PMID:23394423

Huang, S. H., Liu, P., Mokasdar, A., & Hou, L. (2013). Additive manufacturing and its societal impact: A literature review. *International Journal of Advanced Manufacturing Technology*, *67*(5-8), 1191–1203. doi:10.100700170-012-4558-5

Huleihil, M. (2017). 3D printing technology as innovative tool for math and geometry teaching applications. *IOP Conf. Ser.: Mater. Sci. Eng., 164.*

Hull, C. (1988). Stereolithography: Plastic prototype from CAD data without tooling. *Modern Casting*, *78*, 38.

Irwin, J. L., Pearce, J. M., Anzalone, G., & Oppliger, D. E. (2014). The RepRap 3-D printer revolution in STEM education. *ASEE Annu. Conf. Expo.*, 24.1242.1- 24.1242.13.

Ishengoma, F. R., & Mtaho, A. B. (2014). 3D printing: Developing countries perspectives. *International Journal of Computers and Applications*, *104*, 30–34. doi:10.5120/18249-9329

Jacobs, S., Schull, J., White, P., Lehrer, R., Vishwakarma, A., & Bertucci, A. (2016). e-NABLING education: curricula and models for teaching students to print Hands. *2016 IEEE Front. Educ. Conf.* 10.1109/FIE.2016.7757460

Jaksic, N. I. (2014). New inexpensive 3D printers open doors to novel experiential learning practices in engineering education. *ASEE Annu. Conf. Expo.,* 24.932.1-24.932.23. Retrieved from https://peer.asee.org/22865

Janković, N. Z., Slijepčević, M. Z., Čantrak, D. S., & Gađanski, I. I. (2016). Application of 3D printing in M.Sc. Studies – axial turbocompressors. *Int. Conf. Multidiscip. Eng. Des. Optim*, 96–99. 10.1109/MEDO.2016.7746545

Jaymes, C. (2012). Adverse health effects of oil mist in machine tool industries. *SD Editorials*. Retrieved from http://www.streetdirectory.com/travel_guide/159029/health/adverse_health_effects_of_oil_mist_in_machine_tool_industries.html

Jo, W. (2016). Introduction of 3d printing technology in the classroom for visually impaired students. *Journal of Visual Impairment & Blindness*, *110*(2), 115–121. doi:10.1177/0145482X1611000205

Jones, B. M. (2015). 3D printing in libraries: a view from within the american library association: privacy, intellectual freedom and ethical policy framework. *Bulletin of the American Society for Information Science and Technology*, *42*, 36–41. Retrieved from https://search.proquest.com/docview/1812277903?accountid=10297%0Ahttp://resolver.ebscohost.com/openurl?ctx_ver=Z39.88-2004 &ctx_enc=info:ofi/enc:UTF-8&rfr_id=info:sid/ProQ%3Aabiglobal&rft_val_fmt=info:ofi/fmt:kev:mtx:journal

Junk, S., & Matt, R. (2015). New approach to introduction of 3D digital technologies in design education. *Procedia Cirp, 36*, 35–40. .2015.01.045 doi:10.1016/j.procir

Junk, S., & Matt, R. (2015). Workshop rapid prototyping - a new approach to introduce digital manufacturing in engineering education. *2015 Int. Conf. Inf. Technol. Based High. Educ. Train.*, 1–6. 10.1109/ITHET.2015.7217965

Kamrani, A. K., & Nasr, E. A. (2010). *Engineering design and rapid prototyping*. New York: Springer. doi:10.1007/978-0-387-95863-7

Kane, S. K., & Bigham, J. P. (2014). Tracking @stemxcomet: Teaching Programming to Blind Students via 3D Printing, Crisis Management, and Twitter. *SIGCSE' 14 Proc. 45th ACM Tech. Symp. Comput. Sci. Educ.*, 247–252, 10.1145/2538862.2538975

Kayfi, R., Ragab, D., & Tutunji, T. A. (2015). Mechatronic system design project: a 3D printer case study. *2015 IEEE Jordan Conf. Appl. Electr. Eng. Comput. Technol*, 1–6. 10.1109/AEECT.2015.7360570

Kolitsky, M. (2014). 3D printed tactile learning objects: Proof of concept. *J. Blind. Innov. Res.*, *4*. doi:10.5241/4-51

Kostakis, V., Niaros, V., & Giotitsas, C. (2015). Open source 3D printing as a means of learning: An educational experiment in two high schools in Greece. *Telematics and Informatics*, *32*(1), 118–128. doi:10.1016/j.tele.2014.05.001

Kröger, E., Dekiff, M., & Dirksen, D. (2016). (n.d.). 3D printed simulation models based on real patient situations for hands-on practice. *European Journal of Dental Education*, 1–7. doi:10.1111/eje.12229 PMID:27470072

Kroll, E., & Artzi, D. (2011). Enhancing aerospace engineering students' learning with 3D printing wind-tunnel models. *Rapid Prototyping Journal*, *17*(5), 393–402. doi:10.1108/13552541111156522

Kruth, J. P., Leu, M. C., & Nakagawa, T. (1998). Progress in additive manufacturing and rapid prototyping. *CIRP Ann-ManufTechn*, *47*(2), 525–540. doi:10.1016/S0007-8506(07)63240-5

Le, H. P. (1998). Progress and trends in ink-jet print technology. *The Journal of Imaging Science and Technology*, *42*, 49–62.

Letnikova, G., & Xu, N. (2017). Academic library innovation through 3D printing services. *Library Management*, *38*(4/5), 208–218. doi:10.1108/LM-12-2016-0094

Levy, G. N., Schindel, R., & Kruth, J. P. (2003). Rapid manufacturing and rapid tooling with layer manufacturing (LM) technologies: State of the art and future perspectives. *CIRPAnn-Manuf Techn*, *52*(2), 589–609. doi:10.1016/S0007-8506(07)60206-6

Lin, F., Zhang, L., Zhang, T., Wang, J., & Zhang, R. (2012). Innovative education in additive manufacturing in China. *23rd Annu. Int. Solid Free. Fabr. Symp.*, 14–44. Retrieved from http://www.scopus.com/inward/record.url?eid=2-s2.0-84898487345&partnerID=40&md5=669226e0bb4642ee5e7d2d578c02fb63

Liou, F. W., Leu, M. C., & Landers, R. G. (2012). Interactions of an Additive Manufacturing Program with Society. *23rd Annu. Int. Solid Free. Fabr. Symp.*, 45–61.

Lipson, H. (2012). Design in the Age of 3D Printing. *Mechanical Engineering (New York, N.Y.)*, *134*(10), 30–35. doi:10.1115/1.2012-JAN-1

Lolur, P., & Dawes, R. (2014). 3D Printing of Molecular Potential Energy Surface Models. *Journal of Chemical Education*.

Loy, J. (2014). eLearning and eMaking: 3D Printing blurring the Digital and the Physical. *Education in Science*, *4*(1), 108–121. doi:10.3390/educsci4010108

Luo, Y. C., Ji, Z. M., Leu, M. C., & Caudill, R. (1999). *Environmental performance analysis of solid freeform fabrication processes. In The 1999 IEEE Int Symp on Electron and the Environ* (pp. 1–6). IEEE.

Makino, M., Suzuki, K., Takamatsu, K., Shiratori, A., Saito, A., Sakai, K., & Furukawa, H. (2017). 3D printing of police whistles for STEM education. *Microsystem Technologies*, *1–4*. doi:10.100700542-017-3393-x

Maloy, R., Trust, T., Kommers, S., Malinowski, A., & LaRoche, I. (2017). 3D modeling and printing in History/Social studies classrooms: Initial lessons and insights. *Contemporary Issues in Technology & Teacher Education, 17*, 229–249. Retrieved from https://citejournal.s3. amazonaws.com/wp-content/uploads/v17i2socialstudies1.pdf

Marks, D. (2011). 3D printing advantages for prototyping applications. *Articles Base*. Retrieved from http://www.articlesbase.com/technologyarticles/3d-printing-advantages-for-prototyping-applications- 1843958.html

Martinez, M. O., Morimoto, T. K., Taylor, A. T., Barron, A. C., Pultorak, J. D. A., Wang, J., . . . Okamura, A. M. (2016). 3-D Printed Haptic Devices for Educational Applications. *2016 IEEE Haptics Symp.*, 126–133. 10.1109/HAPTICS.2016.7463166

McGahern, P., Bosch, F., & Poli, D. (2015). Enhancing learning using 3D printing: An alternative to traditional student project methods. *The American Biology Teacher, 77*(5), 376–377. doi:10.1525/abt.2015.77.5.9

McMenamin, P. G., Quayle, M. R., McHenry, C. R., & Adams, J. W. (2014). The production of anatomical teaching resources using three-dimensional (3D) printing technology. *Anatomical Sciences Education, 7*(6), 479–486. doi:10.1002/ase.1475 PMID:24976019

Mercuri, R., & Meredith, K. (2014). An educational venture into 3D printing. *2014 IEEE Integr. STEM Educ. Conf*, 1–6. 10.1109/ISECon.2014.6891037

Minetola, P., Iuliano, L., Bassoli, E., & Gatto, A. (2015). Impact of additive manufacturing on engineering education – evidence from Italy. *Rapid Prototyping Journal, 21*(5), 535–555. doi:10.1108/RPJ-09-2014-0123

Mogali, S. R., Yeong, W. J., Kuan, H., Tan, J., Jit, G., Tan, S., . . . Ferenczi, M. A. (2017). Evaluation by medical students of the educational value of multi-material and multi-colored three-dimensional printed models of the upper limb for anatomical education. *Anat. Sci. Educ.*

Moorefield-Lang, H. M. (2014). Makers in the library: Case studies of 3D printers and maker spaces in library settings. *Library Hi Tech, 32*(4), 583–593. doi:10.1108/LHT-06-2014-0056

Nemorin, S. (2016). The frustrations of digital fabrication: An auto/ethnographic exploration of "3D Making" in school. *International Journal of Technology and Design Education*. doi:10.100710798-016-9366-z

Nemorin, S., & Selwyn, N. (2016). Making the best of it? Exploring the realities of 3D printing in school. *Research Papers in Education*, *32*(5), 578–595. doi:10.1080/02671522.2016.1225802

Nevarez, H. E. L., Pitcher, M. T., Perez, O. A., Gomez, H., Espinoza, P. A., Hemmitt, H., & Anaya, R. H. (2016). Work in progress: designing a university 3D printer open lab 3D model. *ASEE Annu. Conf. Expo*. 10.18260/p.27219

Niaki, M.K., & Nonino, F. (2017). Additive manufacturing management: a review and future research agenda. *Int. J. Prod. Res., 7543*.

NIOSH. (2011). *Noise and hearing loss prevention*. Centers for Disease Control and Prevention. Retrieved from http://www.cdc.gov/niosh/topics/ noise/

Nowlan, G. A. (2015). Developing and implementing 3D printing services in an academic library. *Library Hi Tech*, *33*(4), 472–479. doi:10.1108/LHT-05-2015-0049

O'Reilly, M. K., Reese, S., Herlihy, T., Geoghegan, T., Cantwell, C. P., Feeney, R. N. M., & Jones, J. F. X. (2016). Fabrication and assessment of 3D printed anatomical models of the lower limb for anatomical teaching and femoral vessel access training in medicine. *Anatomical Sciences Education*, *9*(1), 71–79. doi:10.1002/ase.1538 PMID:26109268

Oxman, R., & Oxman, R. (2010). The new structuralism design, engineering and architectural technologies. *Archit. Des., 80*, 15–25.

Paio, E., Eloy, S., Rato, V. M., Resende, R., & de Oliveira, M. J. (2012). Prototyping Vitruvius, New Challenges: Digital Education, Research and Practice. *Nexus Network Journal*, *14*(3), 409–429. doi:10.100700004-012-0124-6

Paudel, A. M., & Kalla, D. K. (2016). *Direct digital manufacturing course into mechanical engineering technology curriculum. ASEE Annu. Conf. Expo*. doi:10.18260/p.26848

Payne, B. R. (2015). Using 3D printers in a computer graphics survey course. *Journal of Computing Sciences in Colleges*, *31*, 44–251. doi:10.1017/CBO9781107415324.004

Petrick, I. J., & Simpson, T. W. (2013). 3D printing disrupts manufacturing. *Research Technology Management*, *56*(6), 12–16. doi:10.5437/08956308X5606193

Pham, D. T., & Gault, R. S. (1998). A comparison of rapid prototyping technologies. *International Journal of Machine Tools & Manufacture*, *38*(10-11), 1257–1287. doi:10.1016/S0890-6955(97)00137-5

Pieterse, F. F., & Nel, A. L. (2016). The advantages of 3D printing in undergraduate mechanical engineering research. *2016 IEEE Glob. Eng. Educ. Conf.*, 25–31. 10.1109/EDUCON.2016.7474526

Plemmons, A. (2014). Building a culture of creation. *Teach. Libr., 41*, 12–16. Retrieved from http:// search.proquest.com/docview/1548229289?accountid=8194%5Cnhttp://primo. unilinc.edu.au/openurl/ACU/ACU_SERVICES_PAGE?url_ver=Z39.88-2004&rft_val_fmt=info:ofi/fmt:kev:mtx:journal&genre=article&sid=ProQ:ProQ%3Aeducation&atitle=Building+a+Culture+of+Cre.

Radharamanan, R. (2017). Additive manufacturing in manufacturing education: a new course development and implementation. *ASEE Annu. Conf. Expo.*

Reggia, E., Calabro, K. M., & Albrecht, J. (2015). A scalable instructional method to introduce first-year engineering students to design and manufacturing processes by coupling 3D printing with CAD assignments. *ASEE Annu. Conf. Expo*. 10.18260/p.23447

Roscoe, J. F., Fearn, S., & Posey, E. (2014) Teaching computational thinking by playing games and building robots. *2014 Int. Conf. Interact. Technol. Games*, 9–12. 10.1109/iTAG.2014.15

Scalfani, V. F., & Sahib, J. (2013). A model for managing 3D printing services in academic libraries. *Issues Sci. Technol. Librariansh., 72*, 1–9.

Schelly, C., Anzalone, G., Wijnen, B., & Pearce, J. M. (2015). Open-source 3-D printing technologies for education: Bringing additive manufacturing to the classroom. *Journal of Visual Languages and Computing*, *28*, 226–237. doi:10.1016/j.jvlc.2015.01.004

Seuring, S., & Muller, M. (2008). From a literature review to a conceptual framework for sustainable supply chain management. *Journal of Cleaner Production*, *16*(15), 1699–1710. doi:10.1016/j.jclepro.2008.04.020

Shapeways 3d Printing News and Innovation. (2011). *3D Printing Bone on a Budget!* Retrieved from http://www.shapeways.com/blog/archives/995-3DPrinting- Bone-on-a-budget!html

Simpson, T. W., Williams, C. B., & Hripko, M. (2017). Preparing industry for additive manufacturing and its applications: summary & recommendations from a National Science Foundation workshop. *Addit. Manuf., 13*, 166–178. 10.1016/j.addma.2016.08.002

Singh, M., Haverinen, H. M., Dhagat, P., & Jabbour, G. E. (2010). Inkjet printing: Process and its applications. *Advanced Materials, 22*(6), 673–685. doi:10.1002/adma.200901141 PMID:20217769

Snyder, T. J., Andrews, M., Weislogel, M., Moeck, P., Sundberg, J. S., Birkes, D., ... Graft, J. (2014). 3D systems' technology overview and new applications in manufacturing, engineering, science, and education, 3D print. *Addit. Manuf., 1*, 169–177. doi:10.1089/3dp.2014.1502 PMID:28473997

Stansell, A., & Tyler-Wood, T. (2016). Digital fabrication for STEM projects: a Middle school example. *IEEE 16th Int. Conf. Adv. Learn. Technol.*, 483–485. 10.1109/ICALT.2016.44

Stein, A. (2012). Disadvantages of 3D printers. *eHow TECH*. Retrieved from http://www.ehow.com/facts_7652991_disadvantages-3d-printers.html

Stier, K., & Brown, R. (2000). Integrating rapid prototyping technology into the curriculum. *J. Ind. Technol., 17*, 1–6. Retrieved from http://www.scopus.com/inward/record.url?eid=2-s2.0-3242746763&partnerID=40&md5=6ee1529a2624ddb9053d19f9f18949d2

Stone-Sundberg, J., Kaminsky, W., Snyder, T., & Moeck, P. (2015). 3D printed models of small and large molecules, structures and morphologies of crystals, as well as their anisotropic physical properties. *Crystal Research and Technology, 50*(6), 432–441. doi:10.1002/crat.201400469

Sullivan, P., & McCartney, H. (2017). Integrating 3D printing into an early childhood teacher preparation course: Reflections on practice. *Journal of Early Childhood Teacher Education, 38*(1), 39–51. doi:10.1080/10901027.2016.1274694

Thilmany, J. (2012). Printed Life. *Mechanical Engineering (New York, N.Y.), 134*(1), 44–47. doi:10.1115/1.2012-JAN-5

Valero-Gomez, A., Gónzalez-Gómez, J., González-Pacheco, V., & Salichs, M. A. (2012). Printable creativity in plastic valley UC3M. *Glob. Eng. Educ. Conf. (EDUCON), 2012 IEEE*, 1–9. 10.1109/EDUCON.2012.6201151

Van Epps, A., Huston, D., Sherrill, J., Alvar, A., & Bowen, A. (2015). How 3D printers support teaching in engineering, technology and beyond. *Bulletin of the American Society for Information Science and Technology*, *42*, 16–20. doi:10.1002/bul2.2015.1720420107

Verner, I., & Merksamer, A. (2015). Digital design and 3D printing in technology teacher education. *Procedia Cirp, 36*, 182–186. .2015.08.041 doi:10.1016/j.procir

Waldo, C. (2012). 10 Ways 3D Printing Is Changing the Medical World. *3D Printer*. Retrieved from http://www.3dprinter.net/10-ways-3d-printing-is-changing-themedical-world

Wei, Y., Tay, D., Panda, B., Paul, S. C., Mohamed, N. A. N., Tan, M. J., & Leong, K. F. (2017). 3D printing trends in building and construction industry: A review. *Virtual and Physical Prototyping*, *12*(3), 261–276. doi:10.1080/17452759.2017.1326724

Williams, B. F., & Folkman, M. (2017). Librarians as makers. *Journal of Library Administration*, *57*(1), 17–35. doi:10.1080/01930826.2016.1215676

Ziaeefard, S., Ribeiro, G. A., & Mahmoudian, N. (2015). GUPPIE, underwater 3D printed robot a game changer in control design education. *2015 Am. Control Conf.*, 2789–2794. 10.1109/ACC.2015.7171157

Zuberbier, D. P., Agarwala, R., Sanders, M. M., & Chin, R. A. (2016). An academic library's role in improving accessibility to 3-D printing. *ASEE Annu. Conf. Expo*. 10.18260/p.26551

Chapter 4
Additive Manufacturing for Crack Repair Applications in Metals:
A Case of Titanium (Ti) Alloys

Tawanda Marazani
https://orcid.org/0000-0003-2268-5867
University of Johannesburg, South Africa

Daniel Makundwaneyi Madyira
University of Johannesburg, South Africa

Esther Titilayo Akinlabi
University of Johannesburg, South Africa

ABSTRACT

Additive manufacturing (AM) builds intricate parts from 3D CAD model data in successive layers. AM offers several advantages and has become a preferred freeform fabrication, processing, manufacturing, maintenance, and repair technique for metals, thermoplastics, ceramics, and composites. When using laser, it bears several names, which include laser additive manufacturing, laser additive technology, laser metal deposition, laser engineered net shape, direct metal deposition, and laser solid forming. These technologies use a laser beam to locally melt the powder or wire and the substrate that fuse upon solidification. AM is mainly applied in the aerospace and biomedical industries. Titanium (Ti) alloys offer very attractive properties much needed in these industries. This chapter explores AM applications for crack repairs in Ti alloys. Metal cracking industrial challenges, crack detection and repair methods, challenges, and milestones for AM repair of cracks in Ti alloys are also discussed.

DOI: 10.4018/978-1-5225-9167-2.ch004

INTRODUCTION

The 1737 Tsar Bell cracking incident is a paradigm that metal cracking is a perennial metal industry challenge (Mann, 2011). It was only after two decades during the World War II (1939-1945) that the cracking of metals became a subject of scientific research after hydrogen cracking related failure caused unbearable costs in the metal industry (Hart, 1999). Since then, there has been growing research in the field of metal cracking which include crack detection, classification, propagation, and repair techniques (Dexter, et al., 2013; Kumar, 2009; Locknstitch-Inc; Marazani, Madyira, & Akinlabi, 2017; Hashemite-University). Mechanical, chemical, thermal and metallurgical effects or combined effects were realised to be the main sources of crack initiation and propagation. Unattended crack propagation can result in complete fracture which leads to loss of function or failure, violates safety practice requirements, can lead to injuries and death, increased downtime and operational costs (Callister, 2007; Young, Budynas, & Sadegh, 2012). Research studies on metal cracking recommended the immediate arrest of cracks upon detection before they grow and propagate to fracture. Of the many developed crack repair technologies, laser additive technology (LAT) has increasingly become the most preferred advanced, quick, time saving, economic, and customizable component repair technique in industries where net shape results are required (Marazani, Madyira, & Akinlabi, 2017). LAT is an automated technique, which directly converts 3D CAD data into functional components, which are built layer upon layer via a material deposition process (Mahamood & Akinlabi, 2017; Stucker, 2015). The technology focuses its laser beam on the target surface locally melting the deposited material in either wire or powder form and the base material, which then fuse together upon cooling and solidification producing net shape results with no need for tooling, milling or post machining. Further, LAT can build complex geometries, is a fast process capable of producing a smaller heat affected zone (HAZ) and dilution zone (DZ), all unattainable by conventional means. These features or benefits have made LAT find wide use in the aerospace industry for the built-up and repair of high performance, high value critical parts and in the biomedical industries for free-form fabrication of medical implants. These industries extensively use titanium and its alloys, particularlyTi-6Al-4V due to its very attractive attributes. The application of LAT for repair of cracks in metals is a recent success with victory only registered in wider and Vee shaped cracks. Very few and recent failed research attempts to extend the application of LAT onto the repair of narrow rectangular cracks in metals is in the public domain. This has been faced with challenges associated with groove inaccessibility, powder impedance, lack of sidewall laser irradiation, lack of intralayer and interlayer fusion, entrapped unmelted powders and porosity that all affected the integrity of the repaired substrates. Only very recently, were successful attempts on this subject made on Ti alloy components. The current book chapter seeks to explore AM crack repair applications in Ti alloys. The following sections present a brief background on additive manufacturing, laser additive manufacturing, its advantages, applications, challenges and the breakthroughs made in its applications on the repair of narrow rectangular cracks in Ti alloy components.

ADDITIVE MANUFACTURING

Additive Manufacturing (AM) is a process wherein digital 3-dimensional model data is used to build up components in successive layers through material deposition contrary to subtractive techniques like machining (Mahamood, Akinlabi, Shukla, & Pityana, 2014; Mahamood, Akinlabi, Shukla, & Pityana, 2013; Ahuja, Schaubb, Karga, Lechnerb, & Merkleinb, 2014; Klahn, Leutenecker, & Meboldt, 2014; ASTM-International, n.d.; Sinha, n.d.). Areas where AM finds applications have remarkably increased in the past decade. AM now allows for both complete design and industrial revolution, in the aerospace, chemical, energy, automotive and transportation, medical (biomedical implants), tooling (plastic processing), and consumer goods industries (European-Powder-Metallurgy-Association, 2015).

According to (European-Powder-Metallurgy-Association, 2015), AM offers numerous advantages, which include:

- Increased design freedom as compared to conventional casting and machining.
- Capability of producing near net shape and net shape results.
- Ability to produce intricate parts of complex geometry.
- No tools needed, unlike in subtractive technologies.
- Fast method capable of producing complex shapes in a shorter time compared to traditional machining, moulding or forging.

There exist numerous AM technologies and to date, a wide range of terms and acronyms are in use. The terms, 3D printing, rapid prototyping or rapid manufacturing, or freeform fabrication or solid freeform fabrication, laser beam melting, direct energy deposition or direct metal deposition are used in exchange with Additive Manufacturing which is the most popular term (Marazani, Madyira, & Akinlabi, 2017; European-Powder-Metallurgy-Association, 2015). AM technologies that use laser as their energy source are called laser additive manufacturing (LAM) processes (Mahamood & Akinlabi, 2017). Laser is an acronym for light amplification by stimulated emission of radiation (Ahsan, 2011; Majumdar & Manna, 2003; Cottam, 2012). Laser is a lucid, confluent monochromatic electromagnetic radiation beam, which propagates linearly with minimal divergence and occurs at various wavelengths, energy and beam-configurations with numerous fields of applications (Majumdar, Laser Gas Alloying of Ti-6Al-4V, 2011).

Laser Additive Manufacturing (LAM)

In LAM, parts are built in successive layers by sharply focusing the collimated laser beam together with the shielding gas assisted powder onto the substrate (Kobryn & Semiatin, 2002). Both the powder and the substrate are melted at once, and upon cooling and solidification form the component predefined by the CAD data information (Mahamood, Akinlabi, Shukla, & Pityana, 2014; Mahamood, Akinlabi, Shukla, & Pityana, 2013; Pityana, Mahamood, Akinlabi, & Shukla, 2013). LAM is also known as laser metal deposition (LMD), laser engineered net shape (LENS), direct metal deposition (DMD) or laser solid forming (LSF) (Yu, Rombouts, Maes, & Motmans, 2012).

Advantages of Laser Additive Manufacturing (LAM)

LAM presents numerous advantages in materials and components processing, manufacturing, maintenance and repair. Its ability to produce less heat enables it to yield a smaller heat affected zone (HAZ) and a smaller dilution zone (DZ) which offers minimum distortion and less thermal damage in the substrate (Graf, Gumenyuka, & Rethmeiera, 2012). This enables safeguarding of both the metallurgical and mechanical integrities of the built or repaired parts. LAM enables building of new components with no need for tooling (Mahamood,

Akinlabi, Shukla, & Pityana, 2013; Farayibi, Abioye, Murray, Kinnell, & Clare, 2015). Parts of complex geometry mostly unattainable through conventional means are easily built using LAM. The technology is a good contender for the processing of functionally graded materials (Mahamood, Akinlabi, Shukla, & Pityana, 2013; Pityana, Mahamood, Akinlabi, & Shukla, 2013). Compared to traditional manufacturing processes which generate scrap and are energy intensive, LAM is very economic (no material loss), capable of producing net shape results, and is a very fast process and hence an energy serving technology (Mahamood, Akinlabi, Shukla, & Pityana). Further, LAM can effectively process metal matrix composites (MMCs), remarkably reduces the raw material to actual aerospace part ratio and consequently their production time, two factors which are key to cost reduction (Rottwinkel, Nölke, Kaierle, & Wesling, 2014). The ability of LAM to build components of complex geometry reduces parts assemblies and this helps in reducing aircraft weight (Mahamood, Akinlabi, Shukla, & Pityana, 2013; Pityana, Mahamood, Akinlabi, & Shukla, 2013).

Applications of Laser Additive Manufacturing

LAM has gained prominence in the biomedical and aerospace industries owing to its flexibility and cost saving nature (Ahsan, 2011). LAM finds applications in the building of functional prototypes, FGMs, fabrication of short-run components, repair of worn out parts, subtractive and additive manufacturing, joining, cutting, powder metallurgy, and surface coating for improved wear and corrosion resistance (Kobryn & Semiatin, 2002; Pityana, Mahamood, Akinlabi, & Shukla, 2013; Akinlabi, 2015; Baloyi, Popoola, & Pityana, 2014). LAM also finds use in the manufacture and repair of airplane wings, landing gears, engine turbines blades and body parts (Baloyi, Popoola, & Pityana, 2014; Titanium Company, 2014). LAM has been used for manufacturing and repair of titanium (Ti) alloy components (Kobryn & Semiatin, 2002). Ti alloys are highly biocompatible and many medical implants are LAM manufactured. Ti medical implants find wide applications in the biomedical industry (Mahamood, Akinlabi, Shukla, & Pityana, 2014; Baloyi, Popoola, & Pityana, 2014). In the medical field, LAM can be used to build skeletal implants and dental fixtures. LAM has also been widely used for commercial alloy powders, carbides and intermetallics deposition (Akinlabi,

2015), particulate reinforced metal matrix composite (PR-MMC) coatings, and in high quality alloying (Ahsan, 2011). LAM was applied for Titanium Matrix Composites (TMCs) (Mahamood, Akinlabi, Shukla, & Pityana).

LASER ADDITIVE MANUFACTURING FOR METAL CRACK REPAIRS

In the recent years, LAM has gained extended use in the repair of cracks in high performance and critical metal components, mainly in the aerospace industry where net shape results are a strict requirement. There have been an increasing number in the conducted research studies on this subject though it still remains rarely studied. The few available studies primarily focused on stainless steel and Ti alloys. The challenges and milestones of this subject area, including own work with the obtained results is presented and discussed in this section.

Laser Additive Manufacturing Crack Repair Applications in Ti Alloy Components

Graf et al (Graf, Gumenyuka, & Rethmeiera, 2012) investigated the use of LAM for metal cracks repair. The authors used the V, U and the top open U-slots as shown in Figure 1. The choice of the groove depth was made under the assumption that intolerable defects extend not deeper than 10 mm.

The repair process was conducted using a 5-axis TRUMPF TruDisk 2.0 kW Yb:Yag laser machine with a 3-jet powder nozzle. Helium 5.0 and argon 5.0 were used as carrier and shielding gases respectively, at below 50 ppm oxygen. Powder grain sizes ranging between 45-125μm were used. Table 1

Figure 1. From left to right: V; U and top open U-slots
Source: (Graf, Gumenyuka, and Rethmeiera, 2012)

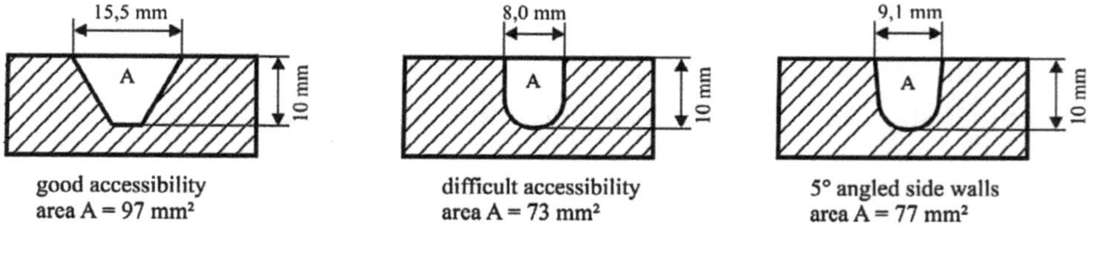

Table 1. LAM process parameters for the V, U and top open U-slots

Deposition Parameters	(i)	(ii)
Weld speed	0.5 m/min	1 m/min
Laser power	2 kW	1 kW
Laser spot size	2.2 mm	1 mm
Powder flow rate	9.4g/min	3.8 g/min

(Graf, Gumenyuka, & Rethmeiera, 2012)

presents the parameters used. The parameters were applied to each of the three groove geometries. For both materials, the V and U-slots with open top angle sidewalls allowed better powder delivery and accessibility. The V-slotted repaired plates were examined using X-ray diffraction and showed better sidewall fusion although lack of sidewall fusion and unmelted powders were evident at the bottom sharp edges shown in Figure 2.

The cross-section of the U-slot shown in Figure 3 (a) showed lack of sidewall fusion defects, which resulted from groove inaccessibility challenges since the laser beam could not be adjusted perpendicular to the sidewalls. It also shows cavities created by the irregular deposits. This further resulted in irregular sidewalls material deposition. Instead of being deposited within the groove, most of the powder was impeded on the upper lip of the slot creating irregular powder streams on groove sidewalls. Suggestions were made to widen the U-grooves or create inclines for uniform powder deposition and for successful implementation of the technique as shown in Figure 3 (b). The

Figure 2. LAM repaired stainless steel and Ti-6Al-4V substrates
Source: (Graf, Gumenyuka, & Rethmeiera, 2012)

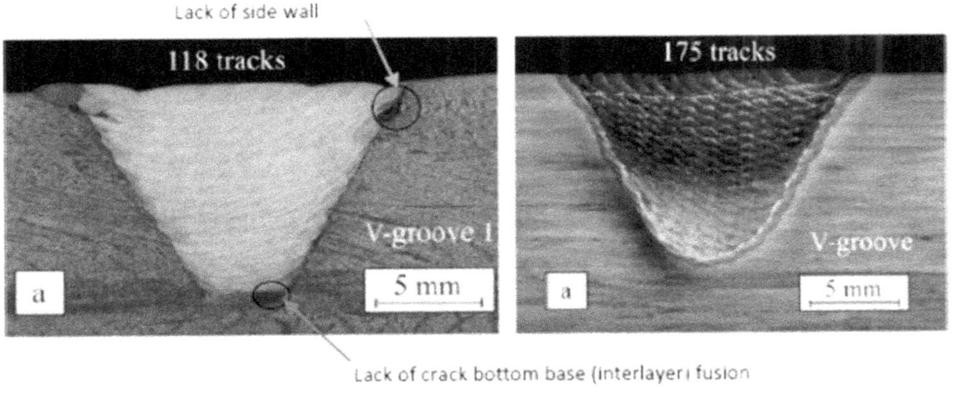

Figure 3. Left (a) U-groove; Right (b) 5° U-groove top open sidewalls
Source: (Graf, Gumenyuka, & Rethmeiera, 2012)

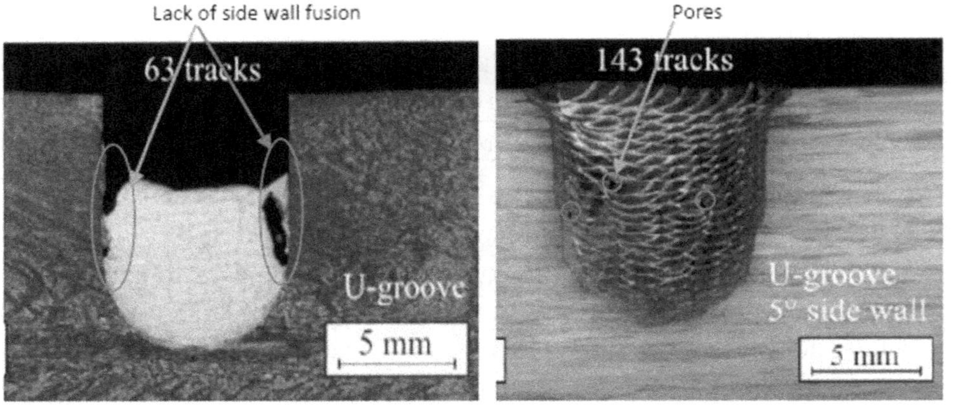

authors concluded the attempts to use LAM for repair of narrow U-slots as unsuccessful.

Rottwinkel et al. (Rottwinkel, Nölke, Kaierle, & Wesling, 2014) further studied V-grooves LAM repair of single-crystal (SX) cracks. The obtained results showed cracked sections of the repaired substrates as shown in Figure 4. Cracks were more visible in deposits made without preheating and attempts to eliminate them by preheating were fruitless. The repaired samples were analysed using scanning electron microscopy (SEM) which exposed the cracks as shown in Figure 5. The authors concluded that their AM research

Figure 4. Cracks on LMD repaired grooves without preheating
Source: (Rottwinkel, Nölke, Kaierle, & Wesling, 2014)

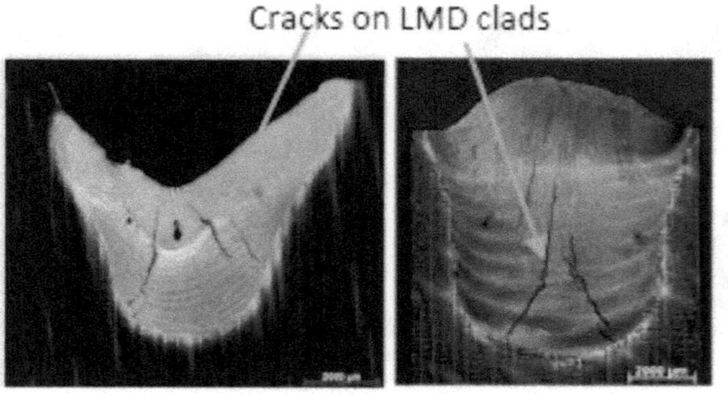

Figure 5. Cracks detected by SEM analysis
Source: (Rottwinkel, Nölke, Kaierle, & Wesling, 2014)

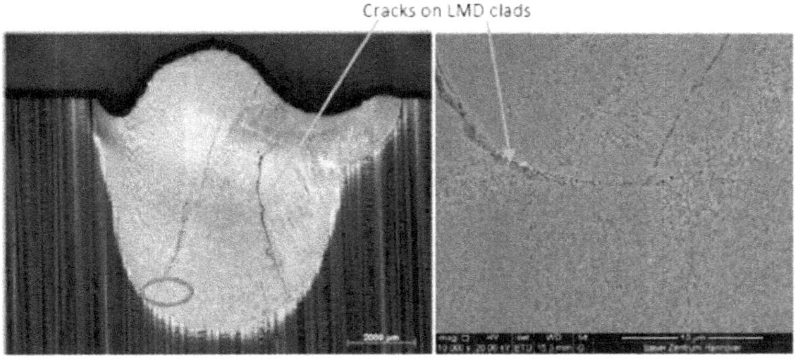

work was unsuccessful since the repaired substrates had cracks. They further recommended optimization of the key LAM parameters for improved results.

Studies by Pinkerton et al, (Pinkerton, Wang, & Li, 2008), used diode laser direct metal deposition for repair of cracks in the form of rectangular and V-slots on H13 steel components. Results obtained revealed porosity defects on both slots. Lack of vertical sidewall fusion was noticed on the vertical sidewalls of the rectangular slots. The lowest points of the V-slots had the highest porosity, which increased with increased laser power and powder delivery. Figure 6 shows the micrographs of the repaired substrates. In other studies of LAT applications to the repair of Inconel 718, (Chen, Zhang, Huang, Hosseini, & Li, 2016), liquation cracks propagating from the top to the bottom weld were found to form during the deposition process closer to the weak fusion zone-heat affected zone interface. Successful repair of wider cast iron cracks of V-slots (10 mm top gap and 3 mm deep) truncated at the bottom to avoid sharp edges which normally serve as porosity pockets in full V-slots was reported by Yu et al., (Yu, Choi, Shim, & Park, 2018). The repaired substrates were defect free and showed enhanced mechanical properties. The wider slot allowed for easier accessibility as opposed to narrow slots.

Cracking challenges were noted during laser additive repair of 316L stainless steel of truncated V-slots (10 mm wide on the top base and 5 mm deep) by Sun et al., (Sun, et al., 2019). It was only through careful optimization of process parameters that defect-free welds were achieved. The 10 mm grooves were wider, which allowed easier groove accessibility. Cracking and porosity have been the major challenges to the many attempts made to use LAM for repair of cracks in metals, (Zghair & Lachmayer, 2017; Barr, et al., 2018).

Figure 6. Cracks detected by SEM analysis
Source: (Pinkerton, Wang, & Li 2008)

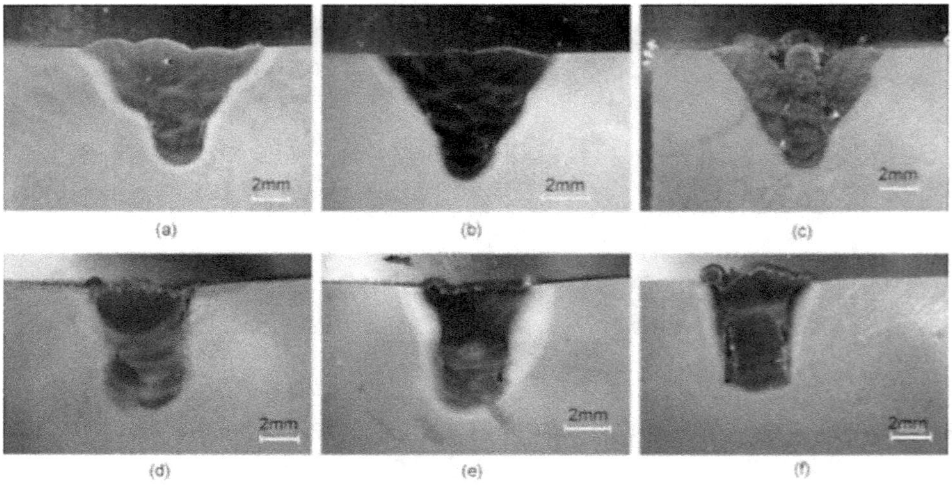

In all the consulted studies, no breakthrough was made in the repair of narrow rectangular metal cracks and V-slots of breadth less than 10 mm. In the present work, focus was put on establishing how to successfully repair 3.5 mm and 2.5 mm cracks of rectangular cross-section. The section that follows presents the method used and its noted challenges.

Difficulties for LAM Repair of Narrow-Slot Cracks in Titanium Alloy Substrates

The researchers investigated LAM repair of rectangular grooves (cracks) of sizes 2.5 and 3.5 mm, both 5 mm deep and 60 mm length. Wire electrical discharge machining (WEDM) was used to machine the slots. The parameters used during the preliminary laser additive repair process are shown in Table 2.

The LAM repair process was carried out at the Council for Scientific and Industrial Research National Laser Centre (CSIR NLC) in Pretoria, South Africa. The 3 kW Nd-YAG Roffin laser with a Kuka robot attachment, used for the experimental work is shown in Figure 7.

Ti-6Al-4V powder of diameters 45 to 90μm and 99.6% purity was used in the study. Argon was used as the shielding gas and before the deposition process, acetone was used to clean the substrates for better laser absorptivity.

Table 2. Preliminary LAM crack repair process parameters

Sample Designation	Parameters						
	Deposition Laser Power (kW)	Remelt Laser Power (kW)	Laser Spot Size (mm)	Powder Feed Rate rpm	Gas Flow Rate (l/min)	Laser Remelt Runs	Scanning Speed (m/min)
2.5 mm Crack: Focal Distance: 179 mm							
S1	2.4	None	2	2	10	None	2.5
S2	1.8	None	2	2	10	None	2.5
S3	2.8	None	2	2	10	None	2.5
3.5 mm Crack: Focal Distance: 197 mm							
S4	2.8	2.8	3	2	10	1	2
S5	1.8	1.8	3	2	10	1	2

Work was clamped on the working table and the repair process with four replicates followed the process parameters summarized in Table 2, resulting in an overall of twenty welds being made.

Physical Observed Deposition Difficulties

The main challenges included groove inaccessibility of the narrow slot, deposited bead sizing, top groove powder impedance and irregular sidewall powder delivery. It was noted that the deposited beads were broader than the grooves irrespective of them having been produced by spot sizes smaller than the slots. The 2 mm spot size generated an averaged 2.4 mm bead width while the 3 mm spot size produced a 3.4 mm bead width. These could not be completely deposited into the cracks (grooves) and instead were irregularly impeded on the top lips of the grooves creating bridges, which left full sample length tunnels or cavities. Processes done without remelt run were observed to bear macro voids with lack of bonding between the deposited interlayers. The preliminary matrix samples S4 and S5 samples were both given a single remelt run on the top most layers to investigate the effects that remelt could have to the repairs made. It was observed that the top band was melted and produced a smooth fuse between the deposited material and the top surface of the substrates. The reheat effect from the remelt process showed insignificant effects to the subsequent deposited bands, which remained visibly unfused and defective.

Figure 7. 3 kW ND-YAG Roffin laser attached with a Kuka robot

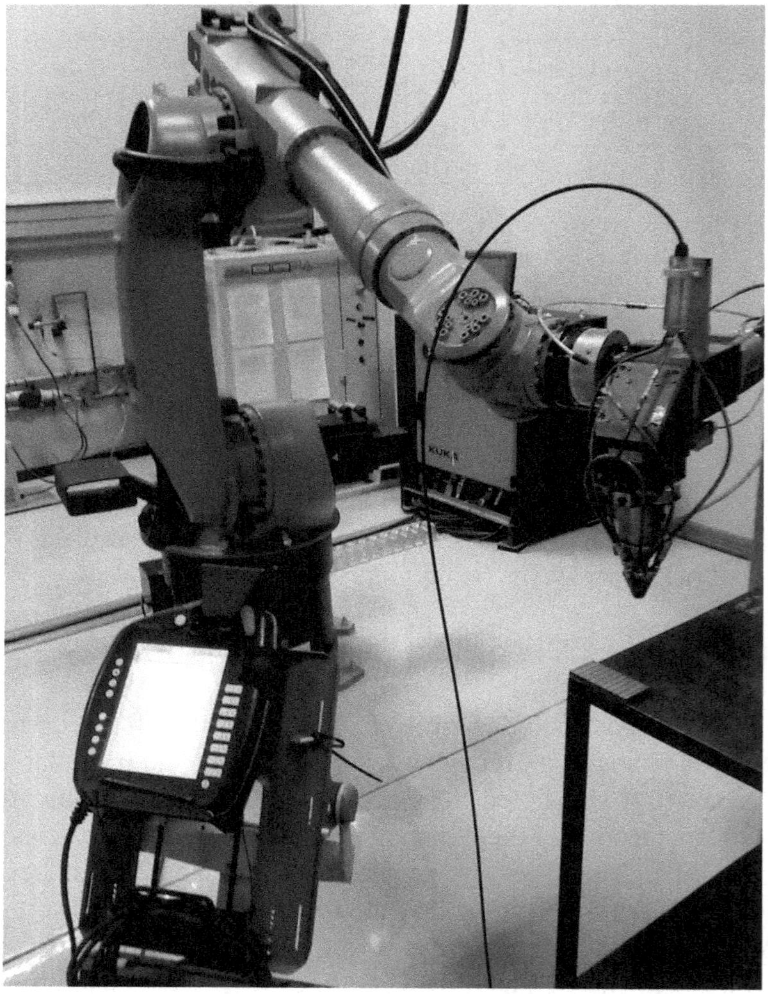

Metallographic Sample Preparation and Characterization

This was performed at the University of Johannesburg Metallurgy laboratory. Sample preparation was done as outlined by Struers (Struers, 2015) and Taylor and Weidmann (Taylor & Weidmann, 2015). The repaired samples were examined for defects using optical microscopy (OM). Optical images of the S-samples are shown in Figure 8. All the samples showed lack of sidewall

Figure 8. Optical images of the preliminary repairs at magnification of x1.6

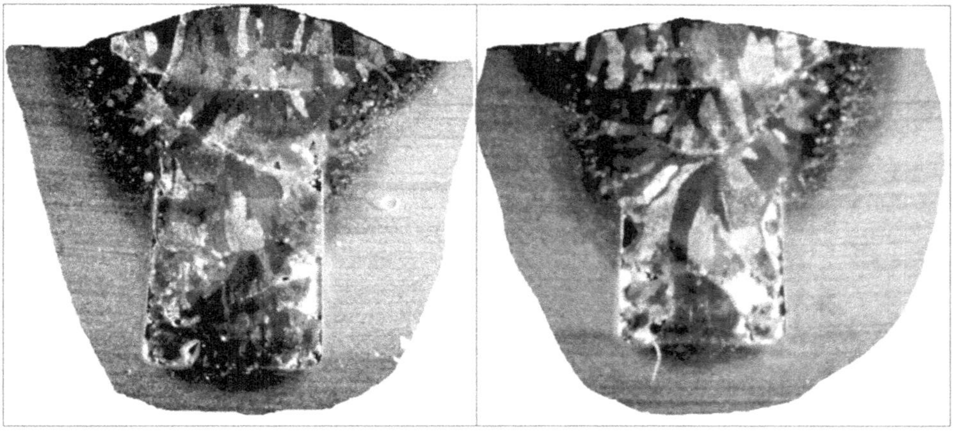

fusion to the extent of exposing the original WED machined rectangular grooves. This was a confirmation of lack of vertical sidewall laser irradiation on the vertical walls of the narrow grooves. Also evident in the optical images was lack of interlayer fusion, which clearly exposed boundaries between the deposited layers. The single remelt run given to S4 and S5 was observed on the macrographs to have slightly eliminated lack of sidewall fusion at the few top layers of the deposited material. This indicated the potential of controlled laser remelt technique to successfully repair narrow rectangular cracks in the Ti substrates.

LAM Successes in Narrow Slots Crack Repair Applications in Ti Alloy Substrates

From the experimental observations made during the preliminary work together with the results reflected on the optical images, further careful selection of parameters was made out of which a multi-track laser remelt crack repair technique was developed using the controlled laser re-melt matrix. The technique was carried out at controlled parameters, mainly laser power, focal length, spot size, powder feed rate and scanning speed. The repaired substrates were characterized through OM and SEM. Table 3 shows the controlled matrix presented with its summary of the parameters used for the repair process. Each set of parameters had 4 replicates, thus yielding a total of 20 welds that were made out of the whole matrix.

Table 3. Controlled laser re-melt matrix

Sample Label	Deposition	Re-Heat	Laser Spot Diameter (mm)	Powder Delivery (rpm)	Gas Flow Rate (l/min)	Bead Width (mm)	Bead Height (mm)	Re-heat Runs	Laser Scanning Speed (m/min)	Deposition Tracks	Number of Weld Runs
	Laser Deposition Power (kW)										
2.5 mm Crack: Focal Distance: 179 mm											
T1	2.2	-	1.3	4	10	1.75	0.60	-	2	10	5
	-	2.2	1.3	-	10	-	-	5	2	-	
T2	1.5	-	1.3	4	10	1.9	0.78	-	2	8	5
	-	2.8	1.3	-	10	-	-	2	2	-	
T3	1.5	-	1.3	3	10	1.8	0.71	-	0.5	9	5
	-	1.8	1.3	-	10	-	-	5	0.5	-	
3.5 mm Crack: Focal Distance: 197 mm											
T4	1.5	-	2.2	3	10	2.7	0.69	-	0.5	9	5
	-	1.8	2.2	-	10	-	-	5	0.5	-	
T5	1.5	-	2.2	4	10	2.9	0.8	-	0.5	8	5
	-	2.8	2.2	-	10	-	-	1	0.5	-	

There were not many changes effected to the laser powers in the new matrix except for the controlled laser remelt runs, reduced laser spot size, increased powder feed rate and the reduced scanning speed. Geometrical studies of the deposited bead width for the 2.5 mm crack, showed the need to reduce the laser spot size to 1.5 mm to achieve a deposit bead that could be deposited within the 2.5 mm groove without powder impedance challenges. This was arrived at after a series of experimental deposition trials on slot-free plate surfaces. The resulting bead widths were measured for every corresponding spot size and it was observed that the 1.3 mm spot size was the most suitable as it yielded bead widths ranging from 2.25 to 2.4 mm which could fit well within the 2.5 mm groove for samples T1, T2 and T3, whose sets of parameters are shown in Table 3. Key changes made on the controlled laser re-melt matrix included the increase in powder feed rate and a significant reduction of the laser scanning speed.

Bead height was also measured to ascertain the total number of deposition layers required and consequently to complete the design of the remelt process. T1 had a bead height of approximately 0.60 mm made by a single track. This arrived at a design of 10 tracks with laser remelt or reheat treatments being introduced after every 2 tracks until completion and this amounted to a total of 5 laser remelt or reheat treatments. A single deposition track on sample T2 produced a bead height of approximately 0.80 mm, from which 8 tracks were made to fill up the groove. Laser remelt or reheat treatments were introduced after every 4 tracks until completion and this amounted to a total of 2 remelt treatments. Sample T3 had a bead height of approximately 0.70 mm made from a single deposition track, from which 9 tracks were made to fill up the groove and remelt treatments were introduced after every 2 tracks until completion and this amounted to a total of 5 laser remelt or reheat treatments. It was observed that every re-melt treatment resulted in higher heat input, which collapsed the impeded deposits and produced denser deposits which showed improved intralayer, interlayer and sidewall fusion. The remelt treatments were made by varying laser power and the number of remelt runs to check their effects on defect reduction. The scanning speed was reduced to allow more time for the laser beam to melt the powder and the substrate. Samples T4 and T5 had 3.5 mm cracks and were repaired at a laser spot size of 2.2 mm. T4 yielded a bead width of 2.8 mm and a height of 0.70 at powder feed rate of 3 rpm and a scanning speed of 0.5 m/min for a single track. A total of 9 tracks and 5 re-melt tracks were made. T5 yielded a bead width of 3.0 mm and a height of 0.8 mm, at a powder feed rate of 4 rpm and a scanning speed of 0.5 m/min for a single track. A total of 8 tracks and 1 top surface re-melt track were made.

Metallographic sample preparation was performed in the same way as in the preliminary investigations. T2 and T5 macrographs (Figures 9 and 10) show presence of lack of sidewall and interlayer fusion at the bottom where the WED machined rectangular grooves are still visible. The defects became more pronounced from the middle portions towards the bottom layers. This suggested the positive effects of the remelt technique that was used for these samples applied on the top layers but could not have significant effect to the subsequent layers.

Figure 9. T2 sample micrographs, taken at 5x magnification

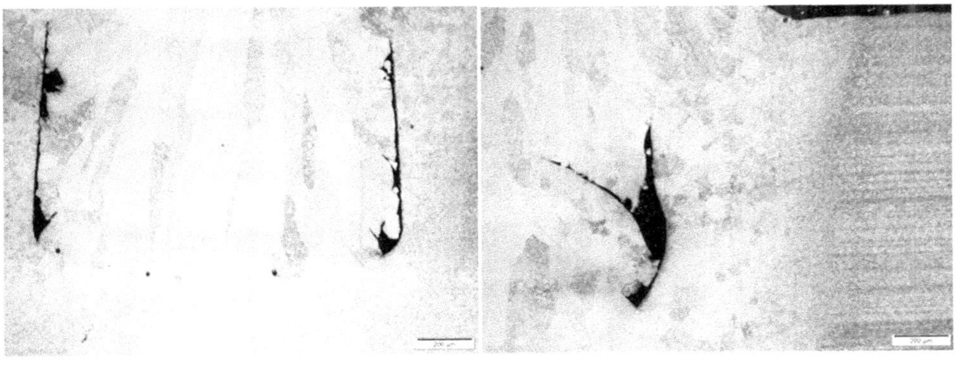

Figure 10. T5 sample micrographs, taken at 5x and 50x magnification

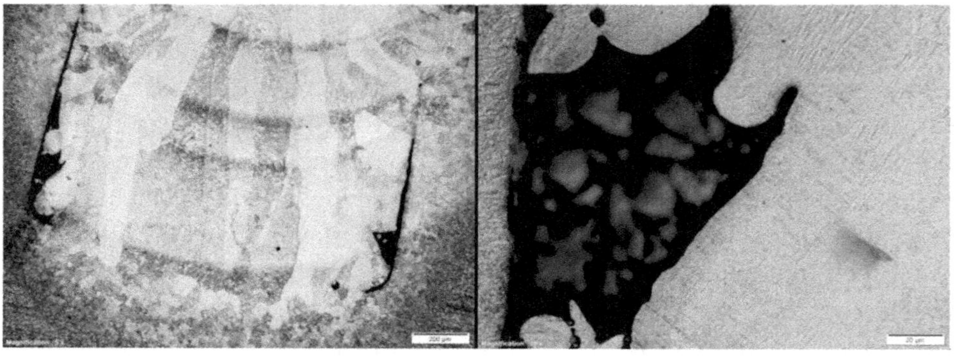

The optical images of the T1, T3 and T4 repaired substrates are shown in Figures 11 through to 13 where they are observed to show densely fused repaired substrates with dendritic macroscopic bands that sink in the direction of the heat flow, which is a sign of superior mechanical attributes. They show defect free clads dominated by fine martensitic, Widmanstätten basket weave structures, fine acicular grains and thick rectangular prior beta grains running normal to the plate. These are common features of well-conducted high-energy processes and in this case effects of the remelt treatments.

The defect free samples T1, T3 and T4 were further subjected to SEM analysis to ascertain absence of defects at higher resolution. The samples were analyzed using SEM at beam electron energy of 20 kV and at 500x and 1000x magnifications. The selected spectra represented the top central,

Figure 11. T1 sample micrographs, taken at 5x and 10x magnification

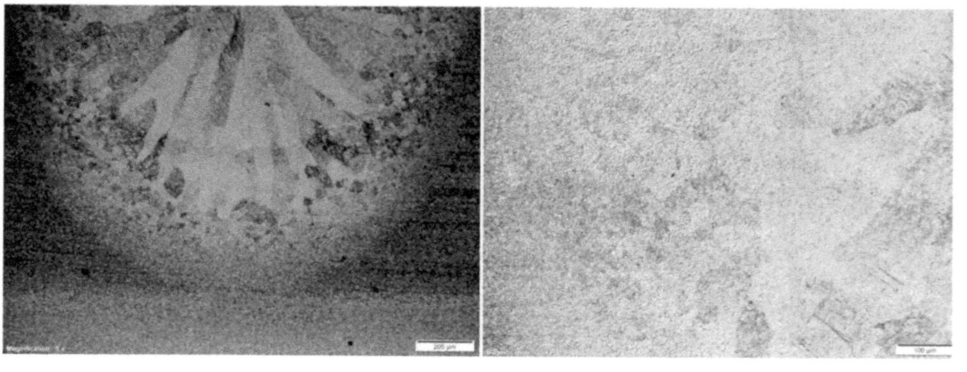

Figure 12. T3 sample micrographs, taken at 10x and 20x magnification

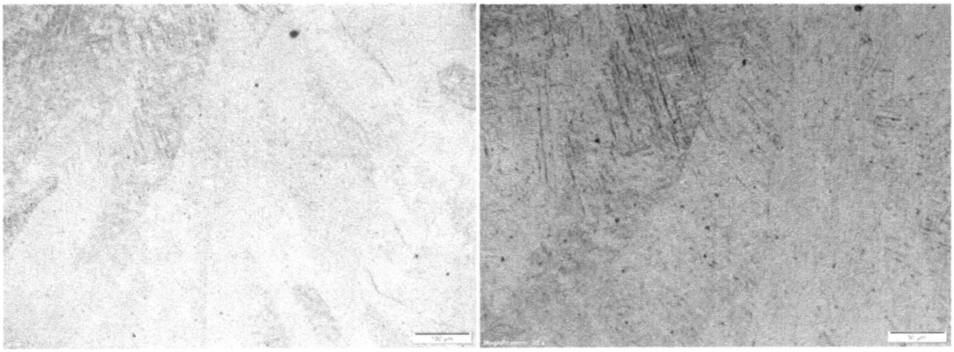

Figure 13. T4 sample micrographs, taken at 20x and 50x magnification

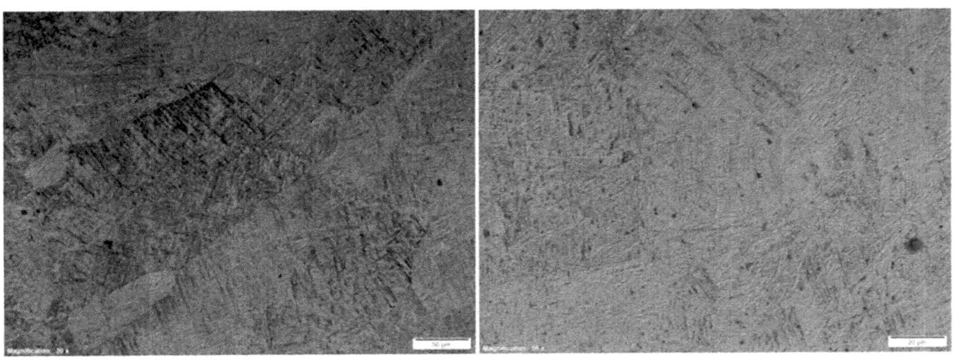

middle, left and right side and the bottom regions of the welds. The SEM images of T1, T3 and T4 are shown in Figures 14-16 respectively.

From the SEM micrographs, it can be seen that the microstructure of T1 and T3 look very similar. They are generally both characterized by α+β phases with the domes (top central regions) dominated by prominent acicular α-plates, basket weave α+β and martensitic α+β structure. Acicular α-plates were also present in the bottom section of T3. Finer basket weaves α+β phases, martensitic α+β, and Widmanstätten structures can be observed in the other sections of both samples. The fusion zone of T3 is characterized by columnar prior β grains, which continued down to the bottom section of the sample. SEM analysis could not pick any defects for both T1 and T3. The finer and homogeneous structure on the rest of the sections of the two samples suggest good mechanical properties.

Figure 14. T1 SEM images

SEM images of defect free sample T4 (Figure 16) are dominated by β phases. For this sample, excellent tensile strength and hardness are both expected to be achieved through the noted β phase distribution as suggested by previous studies (Heng, Qingbin, & Yaole, 2015). The SEM images show complete

Figure 15. T3 SEM images

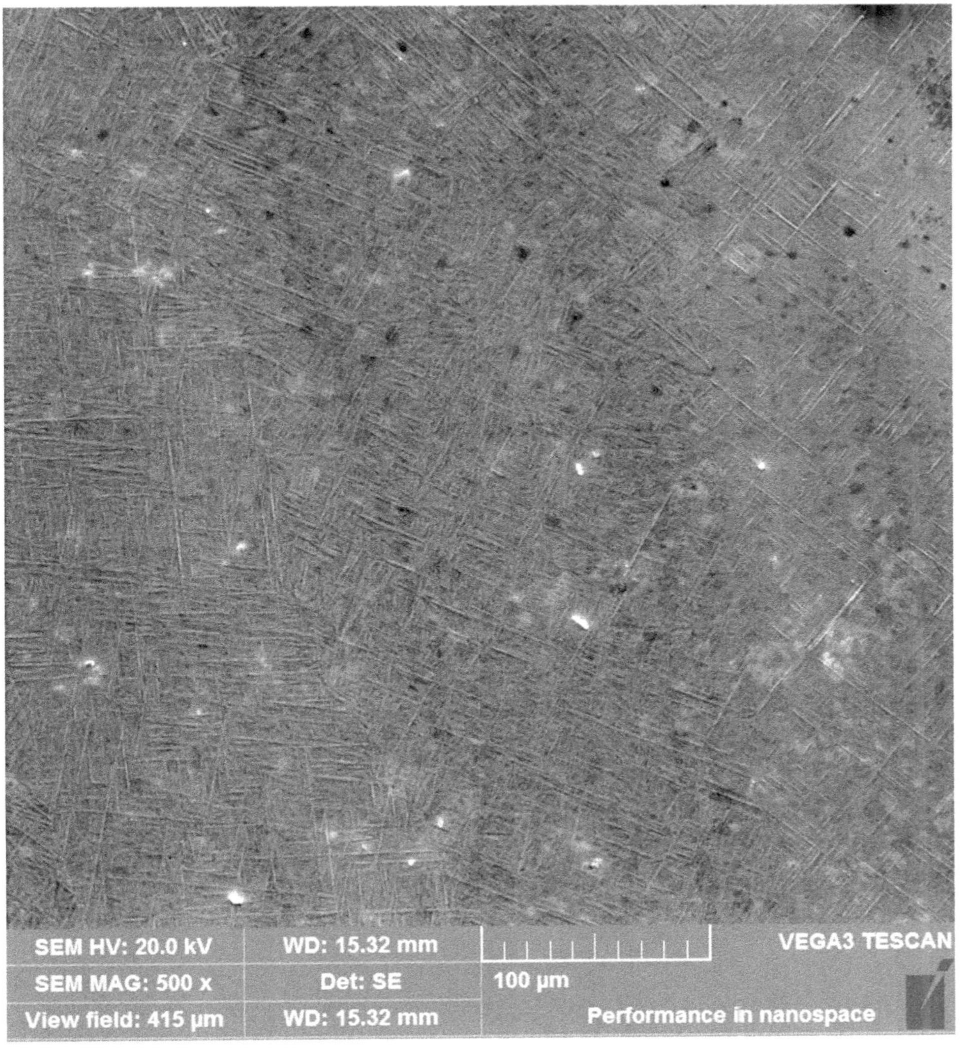

fusion of the deposited bands into the substrate. It can be noted from the T4 SEM images that it is even difficult to distinguish between the deposited layers and the substrate as they all fused perfectly and are so homogeneous with finer basket weave and acicular alpha plates.

Figure 16. T4 SEM images

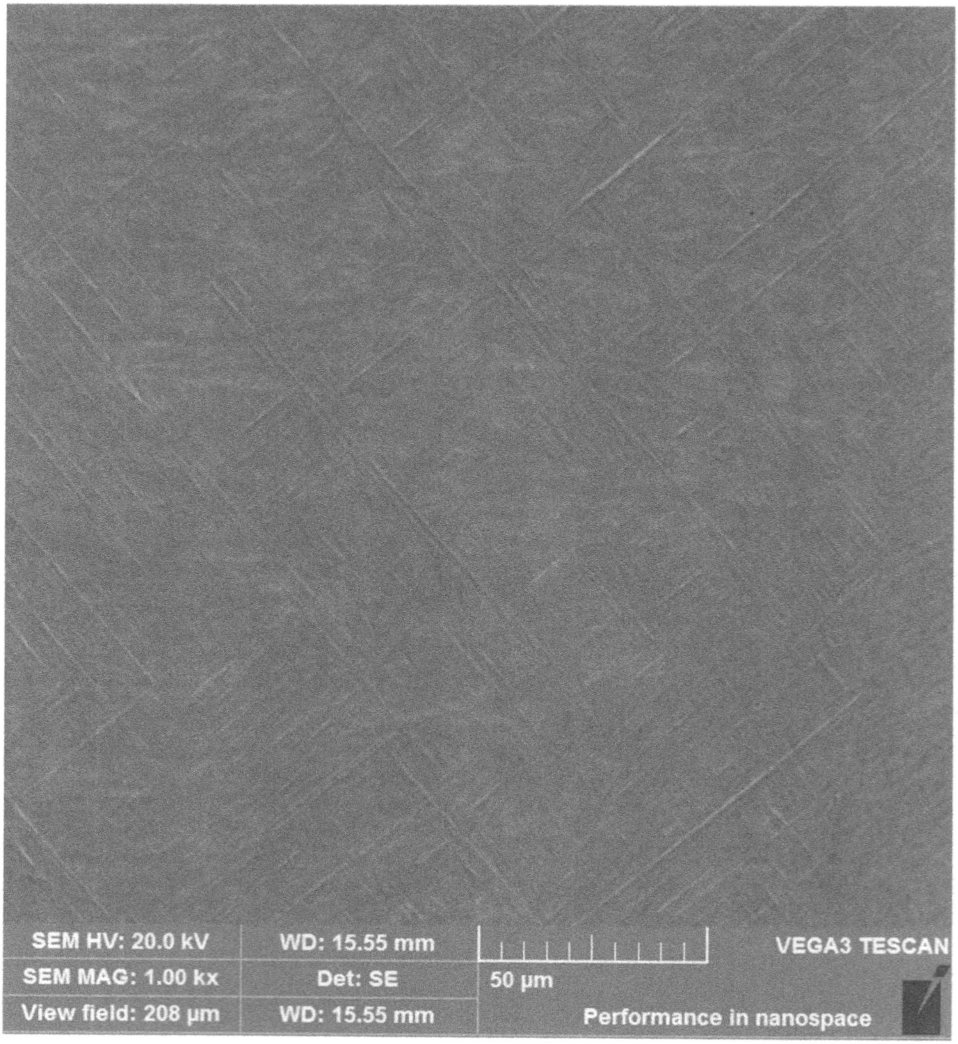

FUTURE DIRECTION OR IMPROVEMENTS

Future studies should look at, but not be limited to, mechanical testing (tensile, Charpy impact and microhardness) and chemical testing (energy dispersive spectroscopy, corrosion and wear) on the repaired substrates. There is also a need to perform simulations to validate the parameters that experimentally produced defect free T1, T3 and T4 samples. It is further suggested that the repair technique used in this study be extended to in-service components for further validation.

CONCLUSION

The developed laser remelt or reheat crack repair technique successfully repaired narrow rectangular cracks in Ti alloy components. It eliminated the challenges of groove inaccessibility, powder impedance, irregular powder delivery, lack of sidewall laser irradiation, lack of intralayer and interlayer fusion, entrapped unmelted powders, unmelted powders and porosity defects. The repaired substrates were examined for defects using OM and SEM analyses and were found to be defect-free with homogeneous clads. From the conducted work, the following conclusions can be drawn as key behind the successful repair of narrow rectangular cracks in Ti alloy components:

- Laser spot size-to-slot width ratio, of not more than 0.6 improved powder delivery and groove accessibility, since it produced beads which could fit within the narrow rectangular grooves.
- Low laser scanning speed allowed better thermal interaction time between the laser beam and both the powder and the substrate, thus eliminating unmelted powders.
- Laser re-heat or re-melt treatments enhanced heat input and melting on both the deposited bands and the vertical sidewalls, collapsing the impeded and irregular depositions, dissipating heat in all directions, thereby eliminating lack of vertical sidewalls laser irradiation, lack of intralayer and interlayer fusion, homogeneously fusing the deposited material with the substrate.
- Repaired slots became more homogeneous and defect-free with increased laser re-heat or re-melt treatments at 1 re-heat or re-melt treatment per every 2 deposition runs.
- Best results were achieved using laser deposition power of 1.5 kW and laser re-heat or remelt power of 1.8 kW, at laser scanning speed of 0.5 m/min, and 5 remelt runs, each made after 2 deposited runs.

REFERENCES

Ahsan, M. N. (2011). *Modelling and analysis of laser direct metal deposition of Ti-6Al-4V alloy* (PhD thesis). Manchester, UK: The University of Manchester.

Ahuja, B., Schaubb, A., Karga, M., Lechnerb, M., & Merkleinb, M. (2014). Developing LBM process parameters for Ti-6Al-4V thin wall structures and determining the corresponding mechanical characteristics. *8th International Conference on Photonic Technologies LANE 2014*. 10.1016/j.phpro.2014.08.102

Akinlabi, E. T. (2015). *Laser deposition of Titanium Carbide on Titanium alloy grade 5*. University of Johannesburg.

ASTM International. (n.d.). *The Global Leader in Additive Manufacturing Standards*. ASTM International. Retrieved November 17, 2018, from https://www.astm.org/ABOUT/OverviewsforWeb2014/Additive-Manufacturing.pdf

Baloyi, N. M., Popoola, A. P., & Pityana, S. L. (2014). Laser coating of Zirconium and ZrO2 composites on Ti6Al4V forbiomedical applications. *South African Journal of Industrial Engineering*, *25*(1), 62–70. doi:10.7166/25-1-661

Barr, C., Sun, S. D., Easton, M., Orchowski, N., Matthews, N., & Brandt, M. (2018). Influence of macrosegregation on solidification cracking in laser clad ultra-high strength steels. *Surface and Coatings Technology*, *340*, 126–136. doi:10.1016/j.surfcoat.2018.02.052

Callister, W. D. (2007). Classification of Materials. In Materials Science and Enmgineering - An introduction (7th ed.; pp. 5-8, 38-44, 80-105, 207-245). John Wiley and Sons Inc.

Chen, Y., Zhang, K., Huang, J., Hosseini, S. R., & Li, Z. (2016). Characterization of heat affected zone liquation cracking in laser additive manufacturing of Inconel 718. *Materials & Design*, *90*, 586–594. doi:10.1016/j.matdes.2015.10.155

Cottam, R. (2012, December 12). *INTECH Open Science Open Minds*. Retrieved April 6, 2015, from www.intechopen.com/...corrosion.../laser-materials-processing-for-impro

Dexter, R., Anami, K., Albretcht, P., Brakke, B., Bucak, O., Connor, R., . . . Fisher, J. (2013). Manual for Repair and Retrofit of fatigue Cracks in Steel Bridges. New York: U.S. Department of Transportation Federal Highway Administration (FHWA).

European Powder Metallurgy Association. (2015). *Introduction to Additive Manufacturing Technology*. European Powder Metallurgy Association. Retrieved November 18, 2018, from http://www.epma.com/am

Farayibi, P. K., Abioye, T. E., Murray, J. W., Kinnell, P. K., & Clare, A. T. (2015). Surface improvement of laser clad Ti–6Al–4V using plain waterjet and pulsed electron beam irradiation. *Journal of Materials Processing Technology*, *218*, 1–11. doi:10.1016/j.jmatprotec.2014.11.035

Graf, B., Gumenyuka, A., & Rethmeiera, M. (2012). Laser metal deposition as repair technology for stainless steel and titanium alloys. *Physics Procedia*, *39*(39), 376–381. doi:10.1016/j.phpro.2012.10.051

Hart, P. (1999). Weld Metal Hydrogen Cracking in Pipeline Girth Welds. *1st International Conference*.

Hashemite-University. (n.d.). *The Hashemite University NDT Centre: Defectology*. Retrieved November 14, 2018, from https://eis.hu.edu.jo/ACUploads/10526/Defectology.pdf

Heng, Z., Qingbin, L., & Yaole, X. (2015). The microstructure and texture analysis of Ti-6Al-4V alloy through linear friction welding. *International Conference on Materials, Environmental and Biological Engineering (MEBE 2015)*.

Klahn, C., Leutenecker, B., & Meboldt, M. (2014). Design for Additive Manufacturing – Supporting the substitution of components in series products. *24th CIRP Design Conference*. 10.1016/j.procir.2014.03.145

Kobryn, P. A., & Semiatin, S. L. (2002). *Mechanical Properties of Laser-Deposited Ti-6AI-4V*. Wright-Patterson Air Force Base.

Kumar, P. (2009). Crack Detection through Non-Destructive Testing. In *Elements of Fracture Mechanics*. McGraw-Hill.

Locknstitch-Inc. (n.d.). *Aluminum Crack Repair Without Welding - Lock-N-Stitch*. Retrieved November 17, 2018, from http://www.locknstitch.com/RepairExamples.htm

Mahamood, R. M., Akinlabi, E. T., Shukla, M., & Pityana, S. (2013). *Proceedings of the International MultiConference of Engineers and Computer Scientists 2013* (*Vol. 2*). Hong Kong: Academic Press.

Mahamood, R. M., Akinlabi, E. T., Shukla, M., & Pityana, S. (2013, March 28). Scanning velocity influence on microstructure, microhardness and wear resistance performance of laser deposited Ti6Al4V/TiC composite. *Materials & Design*, *50*, 656–666. doi:10.1016/j.matdes.2013.03.049

Mahamood, R. M., Akinlabi, E. T., Shukla, M., & Pityana, S. (2014). *Proceedings of the International MultiConference of Engineers and Computer Scientists 2014* (*Vol. 2*). Hong Kong: Academic Press.

Majumdar, J. D. (2011). Laser Gas Alloying of Ti-6Al-4V. *Physics Procedia 12. LiM*, *12*, 472–477.

Mann, A. (2011). Forensic Engineering: Cracks in Steel Structures. *Proceedings - Institution of Civil Engineers*, *164*(FE1), 15–23.

Marazani, T., Madyira, D. M., & Akinlabi, E. T. (2017). Repair of Cracks in Metals. *Procedia Manufacturing*, *8*, 673–679. doi:10.1016/j.promfg.2017.02.086

Pinkerton, A. J., Wang, W., & Li, L. (2008). *Component repair using laser direct metal deposition. In Engineering Manufacture* (Vol. 222, pp. 827–836). Academic Press.

Pityana, S., Mahamood, R. M., Akinlabi, E. T., & Shukla, M. (2013). Gas Flow Rate and Powder Flow Rate Effect on Properties of Laser Metal Deposited Ti6Al4V. *Proceedings of the International MultiConference of Engineers and Computer Scientists 2013* (*Vol 2*). Hong Kong: Academic Press.

Rottwinkel, B., Nölke, C., Kaierle, S., & Wesling, V. (2014). Crack repair of single crystal turbine blades using laser cladding technology. *Procedia CIRP 22 (2014): 3rd International Conference on Through-life Engineering Services: Session: Recent Progress in Jet-Engine Regeneration.*

Sinha, N. (n.d.). *Additive Manufacturing*. Retrieved November 18, 2018, from http://home.iitk.ac.in/~nsinha/Additive_Manufacturing%20I.pdf

Struers. (2015). *The Professional Metallographer Short Version*. Struers.

Stucker, B. (2015). *Additive Manufacturing Technologies*. Louisville, KY: University of Louisville. Retrieved November 12, 2018, from http://www.uta.fi/yky/en/studies/disciplines/northamericanstudies/fulbright/Stucker%20America%20in%20Living%20Color%20presentation.pdf

Sun, G. F., Shen, X. T., Wang, Z. D., Zhang, M. J., Yao, S., Zhou, R., & Ni, Z. (2019). Laser metal deposition as a repair technology 316 stainless steel: Influence of feeding powder compositions on microstructure and mechanical properties. *Optics & Laser Technology*, *109*, 71–83. doi:10.1016/j.optlastec.2018.07.051

Taylor, B., & Weidmann, E. (2015). *Application Notes: Metallographic Preparation of Titanium*. Struers.

Titanium Company. (2014, February 18). *Titanium alloy guide*. Retrieved April 6, 2015, from https://www.scribd.com/doc/207688992/Titanium-Alloy-Guide

Young, W. C., Budynas, R. G., & Sadegh, A. M. (2012). Fatigue and Fracture. In *Roark's Formulas for Stress and Strain* (8th ed.). McGraw-Hill.

Yu, J., Choi, Y., Shim, D., & Park, S. (2018). Repairing cast part using laser assisted metal-layer deposition and its mechanical properties. *Optics & Laser Technology*, *106*, 87–93. doi:10.1016/j.optlastec.2018.04.007

Yu, J., Rombouts, M., Maes, G., & Motmans, F. (2012). Material properties of Ti6Al4V parts produced by laser metal deposition. *Physics Procedia*, *39*, 416–424. doi:10.1016/j.phpro.2012.10.056

Zghair, Y. A., & Lachmayer, R. (2017). Additive repair design approach: Case study to repair aluminium base components. In *21st International Conference On Engineering Design, ICED17* (pp. 141-150). Design to X.

Chapter 5
3D Printing Analysis by Powder Bed Printer (PBP) of a Thoracic Aorta Under Simufact Additive

Hacene Ameddah
University of Batna 2, Algeria

Hammoudi Mazouz
University of Batna 2, Algeria

ABSTRACT

In recent decades, vascular surgery has seen the arrival of endovascular techniques for the treatment of vascular diseases such as aortic diseases (aneurysms, dissections, and atherosclerosis). The 3D printing process by addition of material gives an effector of choice to the digital chain, opening the way to the manufacture of shapes and complex geometries, impossible to achieve before with conventional methods. This chapter focuses on the bio-design study of the thoracic aorta in adults. A bio-design protocol was established based on medical imaging, extraction of the shape, and finally, the 3D modeling of the aorta; secondly, a bio-printing method based on 3D printing that could serve as regenerative medicine has been proposed. A simulation of the bio-printing process was carried out under the software Simufact Additive whose purpose is to predict the distortion and residual stress of the printed model. The binder injection printing technique in a Powder Bed Printer (PBP) bed is used. The results obtained are very acceptable compared with the results of the error elements found.

DOI: 10.4018/978-1-5225-9167-2.ch005

INTRODUCTION

Rapid prototyping was introduced in the early 1980s and applied by the manufacturing industry to design components for various products including automotive, maritime and aerospace (Laschinger et al., 1998; Dekker DL et al., 1974). For these industrial applications, rapid prototyping has been utilized to assess the ease of future product assembly and evaluate the feasibility of developing newly designed products prior to mass production (Olivieri L, 2013). In medicine, 3D printing from radiological images to replicate anatomical structures was initially used in orthopedic and plastic surgery (Laschinger et al., 1998; Estevez ME, 2010). The software was later adapted to accommodate CT and CMR datasets for rapid prototyping of cardiovascular structures. More recently, high-resolution cardiac imaging has ushered in an era where rapid prototyping or 3D printing of congenital heart disease is more feasible (Greil GF, 2007). 3D printed cardiac models can enhance the management of patients by improving interventional and surgical planning and perhaps lead to individualized device deployment targeting specific cardiac defects (Hoyek et al., 2009; Guillot A, 2007). Typically, high-resolution cross-sectional CT and CMR are used as the source datasets to derive whole heart 3D printed models (Jacobs S, 2008; Olivieri L, 2014). 3D printing derived from 3D echocardiographic imaging is also feasible and accurately reflects cardiac morphology, albeit focusing on one part of the anatomy (Samuel BP, 2015; Olivieri LJ, 2014). The integration of multiple imaging modalities for hybrid 3D printing is an additional technique which can be used when one modality is insufficient to give a complete picture of the pathology (Kurup HKN, 2015; Gosnell J, 2016). 3D printed models have been extremely appealing especially for preoperative planning of a variety of cases. Models have been printed on one hand to help doctors fully understand anatomical details and spatial relationship between structures, therefore contributing to improved knowledge and training for certain treatments (Farooqi KM, 2015; Guillot A, 2007), and on the other to more effective communication with patients and their families (Hoyek et al., 2009).

This chapter describe the fundamentals of patient specific modeling for cardiovascular application, including 3D printed and computational analyses. The chapter focuses mostly on cases of thoracic aorta disease (TAD). Particularly focused on the bio-design study of the thoracic aorta in adults. A bio-design protocol was established based on medical imaging. A bio-printing method based on 3D printing that could serve as regenerative medicine has been proposed. A simulation of the bio-printing process was carried out under the software Simufact Additive whose purpose is to predict the distortion and residual stress of the printed model.

RAPID PROTOTYPING IN BIOCONCEPTION

Rapid prototyping broadly indicates the fabrication of a three-dimensional (3D) model from a computer-aided design (CAD), traditionally built layer by layer according to the 3D input (Laoui, T, 2003). Rapid prototyping has also been indicated as solid free-form, computer-automated or layer manufacturing (Rengier, F, 2008). The development of this technique in the clinical world has been rendered possible by the concomitant advances in all its three fundamental steps:

1. Medical imaging (data acquisition),
2. Image processing (image segmentation and reconstruction by means of appropriate software),
3. Rapid prototyping itself (3D printing).

These steps are visually summarized in (Figure 1).

From left to right: data acquisition in this case with magnetic resonance (MR) imaging, image processing, 3D volume reconstruction with appropriate software and final 3D model printed in a transparent resin. The above example (aortic arch of a pediatric patient (Rengier, F, 2010).

The development of biological and anatomical models was one of the first and most important factors that helped introduce rapid prototyping

Figure 1. Stage of rapid prototyping in a clinical setting
Source: (Rengier, F, 2010)

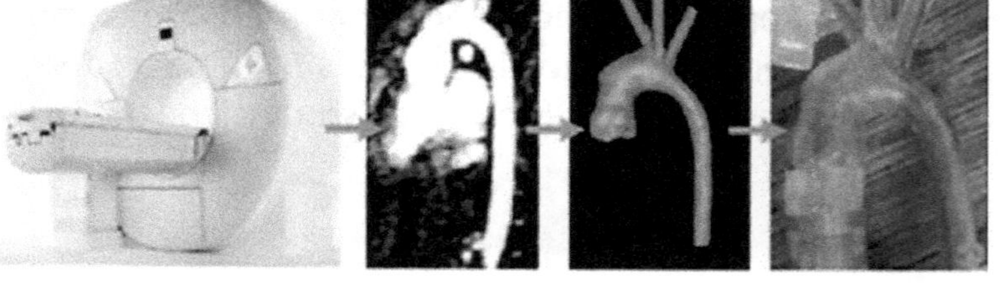

technologies into the biomedical field. These models are usually customized, according to the patient or field of study, to assist pathology diagnosis and subsequent decisions regarding surgery, pharmaceutical treatment or organs replacements. In many cases, having physical prototypes that reproduce the morphology of patients' internal organs is particularly useful for considering later actions, as the prototypes often provide information that is more valuable and easier to interpret than images produced by conventional medical imaging technologies. The use of anatomical models or the implementation of bio printing with replicas of organs and tissues produced using rapid prototyping and manufacturing technologies has made the tests more realistic and improved the final results (Andres Diaz Lantada, 2012). The implications of such anatomical models in diagnostic tasks are remarkable, as discussed in Section results and discussion.

Image acquisition is the most important step in the process of creating a virtual model to be used to print a physical model. A significant determinant in patient selection for 3D printing is the availability of high-quality images. Currently, the imaging modalities used to derive 3D printed models include cardiac CT, CMR, and both 3D TEE and TTE. Each imaging modality has different strengths and weaknesses that impact the quality and accuracy of the 3D printed model (Kurup HKN, 2015). The visualization of extracardiac anatomy and "blood pool" imaging is enhanced by CT (Goitein O, 2014).

PATIENT SELECTION AND IMAGE ACQUISITION

Aortic pathologies are numerous, presenting manifestations are varied, and aortic diseases present to many clinical services, including primary physicians, emergency department physicians, cardiologists, cardiac surgeons, vascular surgeons, echocardiographers, radiologists, computed tomography (CT) and magnetic resonance (MR) imaging (MRI) imagers, and intensivists. Many aortic diseases manifest emergently and are potentially catastrophic unless suspected and detected promptly and accurately. Optimal management of these conditions depends on the reported findings from a handful of imaging modalities, including echocardiography, CT, MRI, and to a lesser extent invasive aortography.

Computed Tomography (CT)

Computed Tomography (CT) is currently the most widely employed technique for the study of the thoracic aorta. Fast scanning achieved with the new wide area detector, low artifact sensibility due to fast velocity tube rotation and 24-hour availability in the emergency rooms are the main advantages of CT usage in the medical practice (Brenner DJ, 2007; Di Cesare E, 2012). The new generation CTs show sensitivities up to 100% and specificities of 98-99%, allowing the possibility to evaluate the entire aorta including lumen and wall, the possible thrombotic apposition and the peri-aortic area. Identification of anatomic variants is also possible (Figure 2), as well as distinction among acute syndromes.

The acquisitions require quite short times and nowadays they are almost universally available (Olivieri L, 2013; Estevez ME, 2010). Sixteen and wider row detectors provide isotropic pixels, mandatory for the ineludible longitudinal reconstruction. Nevertheless, in the thoracic area the ECG-gated technique is strongly suggested to avoid the evidence of false positive flap due to high pulsatility of the root and ascending aorta. Thanks to the ECG-gated technique, it is possible to perform the kinetic evaluation of the aortic leaflets and is useful to assess the valve function. Unfortunately, the ECG-gated technique increases the acquisition time as well as the breath-hold required.

Figure 2. CT showing the anomaly of the right subclavian artery (arrow) with an evident retroesophageal course (also called lusory artery)
Source: (Brenner DJ, 2007)

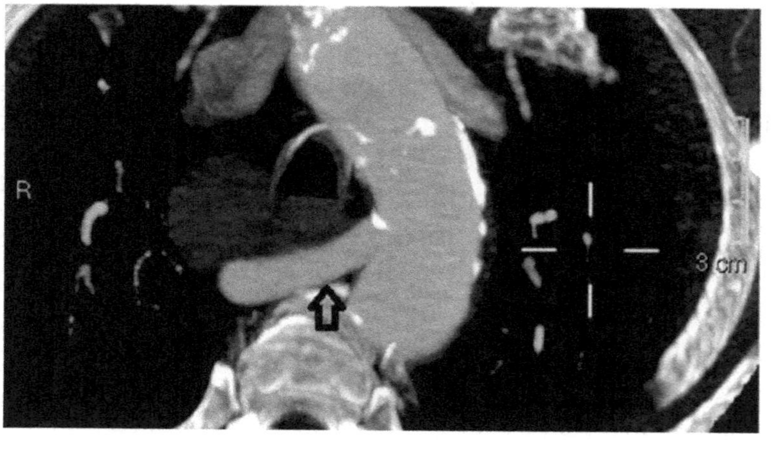

Magnetic Resonance Imaging (MRI)

Magnetic resonance imaging has great potential in the study of the thoracic aorta. However, compared to CT, acquisition times remain longer and movement artifact susceptibility higher (Di Cesare E, 2013; Di Cesare E, 2001).

Different techniques are presently used in the thoracic aorta studies. Both ECG gated and ungated sequences are employed. The employment of contrast media still represents a controversial issue.

The most common acquisition modality is the breath hold 3D Fast SPGR ungated sequences obtained after paramagnetic contrast medium injection. This is a quite fast acquisition obtained using a spoiled gradient echo sequence with multiple k-space line simultaneous acquisition. Shorter Repetition time and Echo time are employed to nullify stationary tissues and enhance the intravascular contrast medium injection. These acquisitions are generally able to reduce the acquisition time to about 10-14 seconds with a sub millimeter voxel resolution. Considerations can be made for large aortic aneurysms, aortic dissection and when compensating for errors in bolus timing (Figure 3). Time resolved techniques may also be applied in the evaluation of the thoracic aorta. This modality also requires contrast medium injection and is acquired without cardiac gating.

AORTIC MEASUREMENTS

A crucial point in the evaluation of both CT and MRI aortic acquisition is the modality in which the measurements are taken. Many radiologists are accustomed to taking measurements on the axial planes. Diameter measurements taken from the axial planes are inherently incorrect unless the aorta being measured is perfectly aligned in the cross-section on the image (22). For this reason, suggest to taking measurements on the longitudinal reformatted imaging. Standard measurements should be obtained at the aortic valve (Table 1), at the maximal diameter of the aortic root, at the sino-tubular junction, at the middle ascending tract, at the arc, between the anonymous trunk and common left carotid artery and in the descending tract posterior to the left atrium. The signed levels are useful to repeat measurements at the same level in subsequent controls. Nevertheless, additional measurements are mandatory if enlargement is evident in any other level. Dimensional evaluation should be obtained on two longitudinal planes and the final diameters should

Figure 3. Aortic root ectasia is evident on 3D FSPGR MRI sequences obtained after gadolinium injection
Source: (Di Cesare E, 2013)

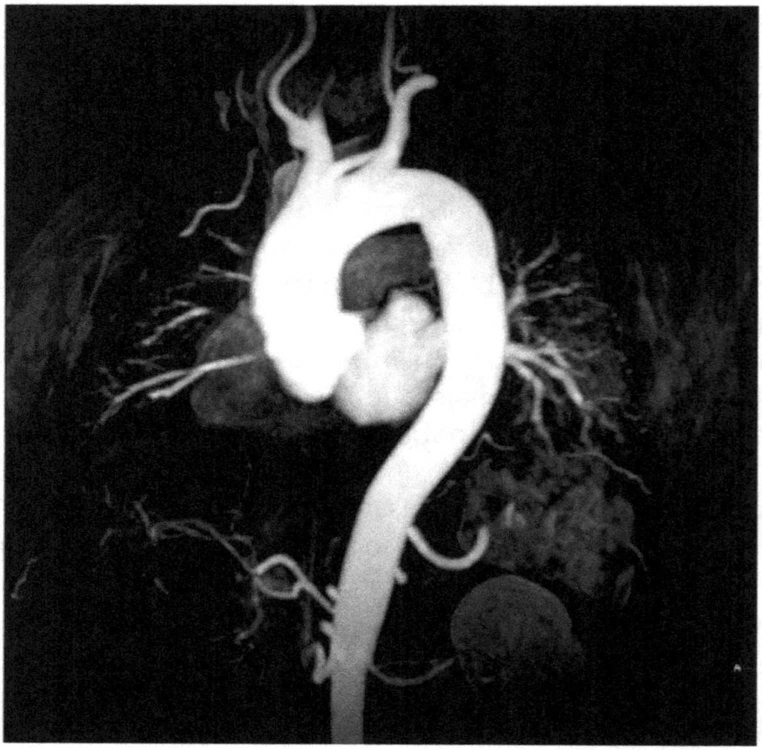

Table 1. The normal values of the thoracic aorta

Site	Normal Values (mm)
Aortic Root	≤ 39
Sinotubular Junction	≤ 30
Ascending Aorta	≤ 37
Aortic arch	≤ 30
Descending Aorta	≤ 25

(Ernesto Di Cesare, 2016)

be obtained by algebraic mean. Moreover, in the analysis of the root, the optimal size measurement should be obtained on images parallel to the root on the so-called short axis view (Ernesto Di Cesare, 2016).

REVERSE ENGINEERING INTERFACE

The reverse engineering interface approach uses a 3D voxel model as the starting point created from the region grow process (figure 4). The 3D voxel model is converted to point cloud data form and are loaded into the reverse engineering software. The points are then used to create triangular facets to form a surface model. The faceted model is further refined and enhanced to reduce file sizes and unwanted features. The freeform surfaces of NURBS patches are used to fit upon the outer shape of the model.

INJECTION PRINTING OF BINDER IN PBP POWDER BED (POWDER BINDER PRINTERS)

3DP or PBP technology was invented by the MIT whose license belongs today to 3D system. Like most 3D printers, the machine is made up of two tanks: a supply tank and a production tank. The temperature within these chambers is controlled by the printer throughout the production, so that the binder injected into the powder has a constant polymerization rate (Figure 5). A first layer of powder with a height of 3mm is deposited at the bottom of the building chamber to facilitate the recovery of the object. The printer,

Figure 4. Process definition to arrive at a CAD model from CT/MRI data
Source: (Hacene. Ameddah, 2010)

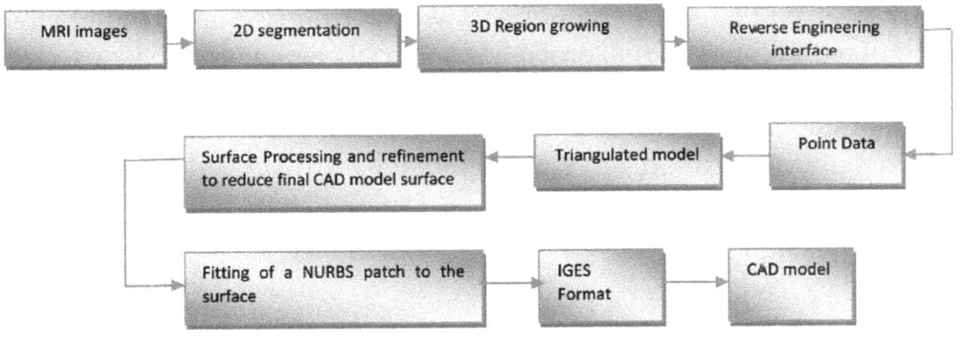

Figure 5. Principle of binder injection printing in PBP powder bed
(Niolas Loucachvsky, 2017)

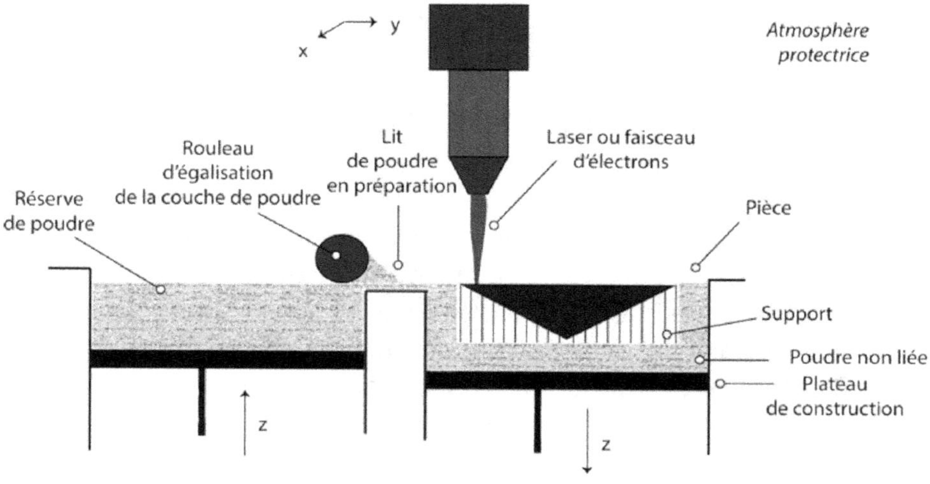

using a supply roller deposits a thin film of about 0.1 mm of powder, then print heads provided with nozzles will come "glue" to the glue the previously deposited powder. This glue may or may not be colored. This protocol makes it possible to produce the first layer of the object. Subsequently, a new layer is deposited and the protocol is repeated until the desired object is obtained. In binder injection printing in a powder bed, the binder can be colored unrestricted with a multitude of colors (Niolas Loucachvsky, 2017).

Additive Manufacturing

AM is a manufacturing method based on additive incremental layer-by-layer manufacturing and divers from conventional subtractive methods such as cutting or casting material. With the use of computer design models, this method can rapidly create components with high precision and without the need of many conventional processes and tools. Since AM can be used to create complex and lightweight components that are created with less waste material and decreased energy usage, interest for using AM has increased within many application areas (Msc simufact©, 2018).

Simulation With MSC Simufact

The software tool used in this study, MSC Simufact, has an application called additive manufacturing, which is adapted to better understand the AM process. The software tool simulates parts of the product development cycle. Some of the design parameters that the software tool examines are the material selection, laser power, powder characteristics and hatch angles (US DOE, 2015). Therefore, comparing how the stress state and distortions in the software tool corresponds to the actual stress state in the AM component will be the main investigative part of this study. The aim of this study is to investigate whether a good prediction of residual stresses and distortions can be performed in AM components, using the software tool MSC Simufact. During this study, thoracic aorta will be manufactured using Binder injection printing in PBP powder bed to compare the residual stresses with the simulation software and to generate the inherent strains.

Simufact Additive

Simufact Additive allows users to quickly predict the breakdown and stress of the parts produced by metal powder bad technologies during production and at the end of the production chain. This predictive ability addresses the main points in the production of additives. The deterioration can in any case be excluded from unusable tolerances or only after a costly treatment such as the desired shape of the mill.

Unwanted high residual stresses can lead to distortion of the part during production or undesired additional deterioration in the subsequent process chain, for example when cutting the part from the production base plate. This allows the user to virtually simulate and optimize factors such as setting parameters and materials, removing direction and support, creating routing, and supporting building displacement (MSC Additive Manufacturing, 2017).

The Simufact Additive software program uses the finite element method to accurately estimate the distortions and residual stresses on the part. One of the most important considerations from the finite element analysis studies is whether the element size used and hence the number of elements is sufficient for that analysis. If the element network is good enough, we can say that the results of the analysis are acceptable, assuming all other inputs of the model is correct. Finite element density is an important metric used to control the accuracy of the analysis (element type and shape affect the accuracy of the analysis).

Assuming there is no singularity region in the model, a high-density network structure will produce results with high accuracy (L. S. Bertol, 2010). However, if the network of elements in the model is very busy, a large amount of computer memory and long running times will be required. This disadvantage is frequently encountered for multiple iteration conditions, especially for non-linear and transient analyzes. One of the main importance of numerical study is voxel mesh benefits compared to other mesh types. Developing model design used and no prototypes needed. The dimensions of the finishing elements used and an objects due to the variability of shapes the geometry can be fully represented. With different materials and geometric properties objects can be examined. Problems related to causal relation, general connected with the stiffness matrix generalized forces and displacements finishing elements. This feature of the method and it makes it possible to simplify.

The boundary conditions can be applied easily. But there is disadvantage like modeling assumptions, difficulty of model connection designs, estimation of component interactions difficulty and damping is usually neglected. The voxel mesh is essential for it to be able to cope with the volume volumetrically and the manufacturing process will be necessary.

In this chapter, using the software Simufact Additive, a simulation of the bio-printing process was carried out, whose purpose is to predict the distortion and residual stress of the printed model.

RESULTS AND DISCUSSION

In the figures, the color scale under the effective stress of the simulated part is the same as in (Figure 6). The color scale shows the parts in the simulated section that have the maximum and minimum values of the effective stress values. The amount of effective stretching in the red zone is greater. Blue is the least in the regions, yellow and green are the middle in the regions. There are distortions of the aorta part. The distortions usually concentrate on the parts and the connections (shown in red color). The external parts are

Figure 6. Results of simulation with MSC Simufact Additive

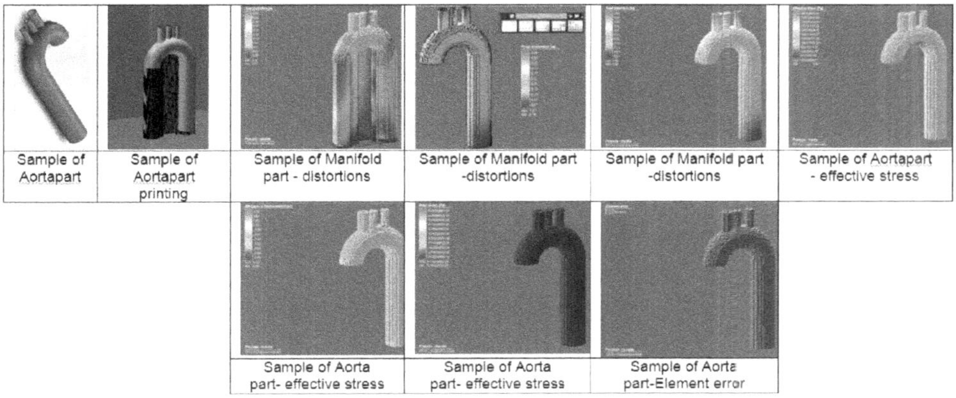

less in the blue, yellow and green regions. The aim of the work within this information is to reduce the distortion results by changing the voxel meshes element size of the part used reaching a conclusion by reducing distortions. After results of simulation with MSC Simufact Additive, Sample of Aorta part was obtained with no element error

The aim of using Simufact Additive software program is to save time and cost. We can see the difference between analytical and numerical values by simulating a specimen once on this numerical and then producing it with layered manufacturing. Computer aided engineering (CAE) and Simufact Engineering is the most powerful tools recently used to provide cost-effective results. It takes time and expense to do physical experiments for each design revision. Design engineers can instead do numeric work to reduce the number of prototypes. This saves considerable effort, money and time. However, no matter how good the performance of the simulation software is, it is unlikely that the numerical analysis will provide 100% accuracy. Mesh generation can be defined as a process of dividing a physical description range into smaller definition ranges (elements). The aim here is to facilitate the solution of a differential equation. The correctness of the approximation of the results obtained by the finite element method due to the finite element depends on the element type and number of elements used in the network.

CONCLUSION

Simultaneous 3D design and component analysis through 3D simulation will reach a greater scope and precision in the coming years, while the need for in-service predictability is at its peak. The tools used for engineering, simulation, production preparation and production with 3D printers are grouped together in a single connected system, which removes the risk of losing the possibility of error-prone data conversion and the information content associated. Generally, very close results can be obtained in the simulation programs. Simufact Additive will provide faster and more accurate as much as possible results with the addition of new features in the first and later versions of this field. Simulation tools will supply and provide ease of production methods. The important thing is to be able to provide the necessary support to a new production method. By using this program which is used by the big companies in the world, which is preferred by the big companies in the Middle Europe, it is possible to increase the added value and provide more support to the digital industry fields by using this simulation program with the progressive versions. This study has been done to raise awareness about this issue.

The final analysis shows that the voxel mesh is modified and the surface mesh is kept constant so that the increase in element size will reduce the amount of distortion for each calculated analysis. Time parameter will decrease with element sizes increase.

REFERENCES

Bertol, L. S., Júnior, W. K., Silva, F. P., & Aumund-Kopp, C. (2010). Medical design: Direct Metal Laser Sintering of Ti–6Al–4V. Materials & Design, 31(8), 3982–3988. doi:10.1016/j.matdes.2010.02.050

Brenner, D. J., & Hall, E. J. (2007). Computed tomography—An increasing source of radiation exposure [Review]. The New England Journal of Medicine, 357(22), 2277–2284. doi:10.1056/NEJMra072149 PMID:18046031

Di Cesare, E. (2001). MRI of the cardiomyopathies. Academic Press.

Di Cesare, E., Cademartiri, F., Carbone, I., Carriero, A., Centonze, M., De Cobelli, F., ... Sardanelli, F. (2013). Natale. Clinical indications for the use of cardiac MRI. By the SIRM Study Group on Cardiac Imaging. La Radiologia Medica, 118(5), 752–798. doi:10.100711547-012-0899-2 PMID:23184241

Di Cesare, E., Carbone, I., Carriero, A., Centonze, M., De Cobelli, F., De Rosa, R., ... Cademartiri, F. (2012). Clinical indications for cardiac computed tomography. From the working group of the Cardiac Radiology Section of the Italian Society of Medical Radiology (SIRM). La Radiologia Medica, 117(6), 901–938. doi:10.100711547-012-0814-x PMID:22466874

Di Cesare, E., Splendiani, A., Barile, A., Squillaci, E., Di Cesare, A., Brunese, L., & Masciocchi, C. (2016). Carlo Masciocchi CT and MR imaging of the thoracic aorta. Open Medicine: a Peer-Reviewed, Independent, Open-Access Journal, 11(1), 143–151. doi:10.1515/med-2016-0028 PMID:28352783

Estevez, M. E., Lindgren, K. A., & Bergethon, P. R. (2010). A novel three-dimensional tool for teaching human neuroanatomy. Anatomical Sciences Education, 3(6), 309–317. doi:10.1002/ase.186 PMID:20939033

Farooqi, K. M., & Sengupta, P. P. (2015). Echocardiography and three-dimensional printing: Sound ideas to touch a heart. Journal of the American Society of Echocardiography, 28(4), 398–403. doi:10.1016/j.echo.2015.02.005 PMID:25839152

Goitein, O., Salem, Y., & Jacobson, J. (2014). The role of cardiac computed tomography in infants with congenital heart disease. The Israel Medical Association Journal, 16(3), 147–152. PMID:24761701

Gosnell, J., Pietila, T., Samuel, B. P., Kurup, H. K. N., Haw, M. P., & Vettukattil, J. J. (2016). Integration of computed tomography and three-dimensional echocardiography for hybrid three-dimensional printing in congenital heart disease. Journal of Digital Imaging, 29(6), 665–669. doi:10.100710278-016-9879-8 PMID:27072399

Greil, G. F., Wolf, I., Kuettner, A., Fenchel, M., Miller, S., Martirosian, P., ... Sieverding, L. (2007). Stereolithographic reproduction of complex cardiac morphology based on high resolution imaging. Clinical Research in Cardiology; Official Journal of the German Cardiac Society, 96(3), 176–185. doi:10.100700392-007-0482-3 PMID:17225916

Guillot, A., Champely, S., Batier, C., Thiriet, P., & Collet, C. (2007). Relationship between spatial abilities, mental rotation, and functional anatomy learning. Advances in Health Sciences Education: Theory and Practice, 12(4), 491–507. doi:10.100710459-006-9021-7 PMID:16847728

Hacene, A., & Mekki, A. (2010). Bio-CAD Reverse Engineering of free-form surfaces by planar contours. Computer-Aided Design and Applications, 7(S1). doi:10.1080/16864360.2010.10738809

Hoyek, N., Collect, C., & Rastello, O. (2009). Enhancement of mental rotation abilities and its effect on anatomy learning. Teaching and Learning in Medicine, 21(3), 201–206. doi:10.1080/10401330903014178 PMID:20183339

Jacobs, S., Grunert, R., & Mohr, F.W. (2008). 3D-Imaging of cardiac structures using 3D heart models for planning in heart surgery: a preliminary study. Interact CardioVasc Thorac Surg., 7(1), 6–9. doi:10.1510/icvts.2007.156588.

Kurup, H. K. N., Samuel, B. P., & Vettukattil, J. J. (2015). Hybrid 3D printing: A game-changer in personalized cardiac medicine? Expert Review of Cardiovascular Therapy, 13(12), 1281–1284. doi:10.1586/14779072.2015.1100076 PMID:26465262

Lantada & Morgado. (2012). Rapid Prototyping for Biomedical Engineering: Current Capabilities and Challenges. Annual Review of Biomedical Engineering, 14. doi:10.1146/annurev-bioeng-071811-150112 PMID:22524389

Laoui, T., & Shaik, S. K. (2003). Rapid prototyping techniques used to produce medical models/implants. In Proceedings of the 4th national conference on rapid and virtual prototyping and applications (pp. 23-32). Buckinghamshire, UK: Chilterns University College.

Laschinger, J. C., Vannier, M. W., & Gutierrez, E. (1974). Preoperative three-dimensional reconstruction of the heart and great vessels in patients with congenital heart disease. Computers and Biomedical Research, an International Journal, 7(6), 544–553. doi:10.1016/0010-4809(74)90031-7 PMID:4457270

Loucachvsky, N. (2017). Impression 3D: Application actuelle en odontologie et perspectives (Thèse du doctorat). Université de Nantes.

MSC Additive Manufacturing, Computer Program, Simufact Volume VII. (2017). Retrieved from http://www.mscsoftware.com/product/simufact-additive

Olivieri, L., Krieger, A., Chen, M. Y., Kim, P., & Kanter, J. P. (2014). 3D heart model guides complex stent angioplasty of pulmonary venous baffle obstruction in a mustard repair of D-TGA. International Journal of Cardiology, 172(2), e297–e298. doi:10.1016/j.ijcard.2013.12.192 PMID:24447757

Olivieri, L. J., Krieger, A., Loke, Y. H., Nath, D. S., Kim, P. C. W., & Sable, C. A. (2015). Three-dimensional printing of intracardiac defects from threedimensional echocardiographic images: Feasibility and relative accuracy. Journal of the American Society of Echocardiography, 28(4), 392–397. doi:10.1016/j.echo.2014.12.016 PMID:25660668

Rengier, F., Mehndiratta, A., von Tengg-Kobligk, H., Zechmann, C. M., Unterhinninghofen, R., Kauczor, H. U., & Giesel, F. L. (2010). 3D printing based on imaging data: review of medical applications. International Journal of Computer Assisted Radiology and Surgery, 5(4), 335-341.

Rengier, F., von Tengg-Kobligk, H., Zechmann, C. M., Kauczor, H. U., & Giesel, F. L. (2008). Beyond the eye – Medical applications of 3D rapid prototyping objects. European Medical Imaging Review, 1, 76-80.

Samuel, B. P., Pinto, C., Pietila, T., & Vettukattil, J. J. (2015). Ultrasound-derived three-dimensional printing in congenital heart disease. Journal of Digital Imaging, 28(4), 459–461. doi:10.100710278-014-9761-5 PMID:25537458

US DOE. (2012). Additive manufacturing: Pursuing the promise. US Department of Energy.

Section 3
Design and Analysis

Chapter 6
Design of Prosthetic Heart Valve and Application of Additive Manufacturing

Dheeman Bhuyan
https://orcid.org/0000-0001-9772-0028
National Institute of Technology Meghalaya, India

ABSTRACT

Heart valve prostheses are well known and can be classified in two major types or categories: biological and mechanical. Biological valves (i.e., Homografts and Heterografts) make use of animal tissue as the valving mechanism whereas mechanical valves make use of balls, disks, and other mechanical valving mechanism. Mechanical valves carry considerable risk and require lifelong medication. The design of these valves is usually done on a "one size fits all" basis, with only the diameter changing depending on the model being produced. The author seeks to present an application of additive manufacturing in the design process for mechanical valves. This is expected to provide patients with customized prostheses to match their physiology and reduce the risk associated with the implantation.

DOI: 10.4018/978-1-5225-9167-2.ch006

INTRODUCTION

Heart valve prostheses are well known in the art and can be classified in two major types or categories. One type employs tissue valves of animal origin in its valve mechanism, known as a heterograft or homograft according to the source of the valve. The other type of heart valve prosthesis utilizes a ball, a disc, valve leaflets or other mechanical valving devices to regulate the direction of blood flow through the prosthesis. The latter type of prosthesis is usually known as "mechanical" heart valve prosthesis. Valve implants pose the risk of thromboembolic complications, forcing the patient to require chronic anticoagulation. This chapter explores a design for a mechanical heart valve which is expected to mimic the physiological heart valve in functioning and haemodynamic performance. The proposed valve showed better open-close characteristics as compared to existing designs. This should effectively reduce the complications arising from the implant, especially regurgitation and thrombosis. The design is made such that it can be manufactured using layered manufacturing techniques or additive manufacturing.

Layered manufacturing technology describes a range of techniques where 3-dimensional objects are constructed from a laminated form. There is a range of methods by which this can be achieved but all rely on the same fundamental set of processes. First the object to be fabricated must be described in terms of an accurate 3-dimensional design representation. This must then be reformatted to describe the object in terms of a number of slices with finite thickness. This "slice information" is then used to fabricate the appropriate number of slices from the desired material.

Finally, these slices are assembled to form the solid object. In practice, a number of the current technologies combine the slice fabrication and assembly processes by using a previously deposited slice as a template for the deposition of subsequent slices.

The current interest in layered manufacturing has its roots in the technology of Rapid Prototyping. Rapid prototyping (RP) develops to fill a need in manufacturing industry to develop representative or functional prototypes of objects normally manufactured in large quantities by tooled processes. RP is a process in which a part is produced using layer-by-layer deposition of material. To reduce the product development time and the cost of manufacturing, rapid prototyping technology has emerged, offering the potential of dramatically changing manufacturing processes. Models and prototypes can be manufactured with RP technology not only for visualization purposes but also to build functional parts. It is an important technology as it has potential to reduce the manufacturing lead time of the product up to 30–50% even when the relative part complexity is very high (Kai & Fai, 1997).

BACKGROUND

Four heart valves, two on either side of the heart, ensure that each muscle contraction produces efficient, unidirectional flow. On the right side of the heart, the tricuspid and pulmonary valves regulate the flow of blood that is returned from the body to the lungs for oxygenation, whereas on the left side, the mitral and aortic valves control the flow of oxygenated blood to the body (Hillis et al., 1995).

One of the main afflictions of the cardiovascular system is heart valve disease, which is generally caused by diseases such as rheumatic fever congenital birth defects or ageing. In the Indian context, rheumatic diseases are the most common factor leading to valvular disease. Such heart valve disease compromises the functionality of the valve by restricting the motion of the valve leaflets or by damaging its supporting structure (Dasi et al., 2009). This leads to either valve stenosis (calcification of the leaflets associated with narrowing of the valve, resulting in greater resistance to blood flow and a greater cross – valvular pressure drop) or regurgitation (failure of the valve to close completely), both eventually leading to valve failure (Dasi et al., 2009).

The need for prosthetic heart valves had been felt for a long time but seemed impossible before 1952 when Dr Charles Hufnagel clinically introduced a ball valve that he placed into the descending thoracic aorta for treatment of aortic valvular insufficiency (DeWall et al., 2000). Development of the Starr-Edwards heart valve marked a new era in the treatment of valvular heart disease. Until the development of the Starr-Edwards valve, there were no published reports of patients who had lived longer than 3 months with a prosthetic valve in the mitral position (Matthews, 1998). Caged ball valves have a high tendency to forming blood clots, so the patient would have a high degree of anti-coagulation therapy.

Caged ball valves eventually gave way to tilting disk valves. The Bjork – Shiley Delrin Valve was made clinically available in 1969. The BSD valve provided a low-profile, quiet prosthesis with excellent hemodynamics (Weiting

Figure 1. Tilting disc valves of the 1970s (a) Bjork–Shiley Delrin valve, (b) Bjork–Shiley standard, (c) Lillehei–Kaster, (d) Medtronic–Hall, (e) Zorin and (f) Omniscience
Source: (Lefrak & Starr, 1979)

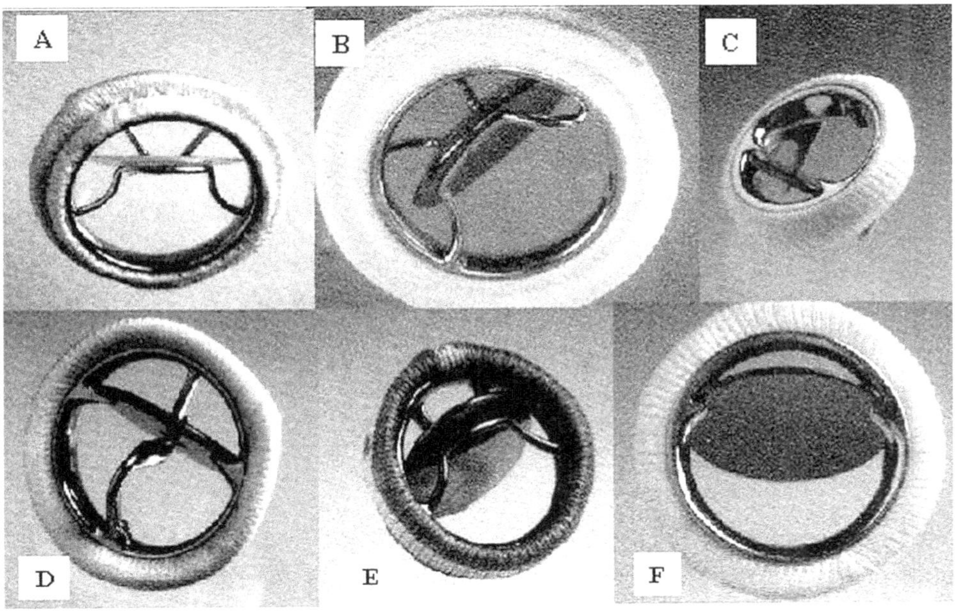

1996). The purpose in creating the titling-disc valve was to restore the central blood flow that was lost with the ball valve design. The Medtronic–Hall valve, first introduced in 1977 remains one of the more popular tilting disk valves today.

Leaflet designs for valves were experimented on by Albert Starr and Lowell Edwards. The Starr –Edwards leaflet valve consisted of 2 silicone-rubber leaflets that were hinged on a central crossbar made of solid Teflon; it included a Teflon cloth margin for fixation. The leaflet valves were plagued by thrombus formation. Thrombus would originate at the suture line and grow by direct extension onto the leaflets. In most cases, the valve became totally occluded after only 2 or 3 days (Lefrak & Starr, 1979)

The bileaflet heart valve design was introduced in 1979. Bileaflet heart valves consist of two semicircular leaflets that rotate about struts attached to the valve housing, and while they take care of some of the issues that exist in the other models, bileaflets are vulnerable to backflow and so they

Figure 2. Starr-Edwards leaflet valve
Source: (Oregon Health and Science University, n.d.)

cannot be considered as ideal. Bileaflet valves do, however, provide much more natural blood flow than caged-ball or tilting-disc implants. In recent times, the St. Jude Medical bileaflet valve became one of the more popular mechanical valves.

Modern replacement valves are divided broadly into three categories:

- **Mechanical:** These have a virtually zero primary failure rate but require anticoagulation and are usually used for the relatively younger patient. Modern bileaflet mechanical valves are broadly classed as open pivot (e.g. ATS valve) or closed pivot (St Jude Medical, OnX, Sorin). The major differentiating factor various bileaflet designs apart from the pivoting is the composition and purity of the pyrolytic carbon, the shape and opening angle of the leaflets, the design of the pivots, the size and shape of the housing and the design of the sewing ring (Chambers, 2014).
- **Biological:** Biological replacements have limited durability but do not require anticoagulation and are usually used for the relatively elderly. Homografts (human valve) were first introduced in 1956. However, the most frequently implanted biological replacement valves are the stented xenografts (animal tissue) and these were first introduced in 1965. The

stent is a plastic or wire structure covered in fabric with the cusps of the valve usually placed inside and a sewing ring attached outside. The valve cusps usually consist of pericardium (e.g. Edwards Perimount, Sorin Mitroflow) or a porcine aortic valve (e.g. St Jude Epic, Medtronic Hancock). Modern xenografts differ in anticalcification treatment and fixation pressure. A different class, the stentless heterograft valve consists of a preparation of porcine aorta or shaped pericardium and was introduced in the hope of better haemodynamic function and fewer complications. However, these expectations have not generally been fulfilled and the longer bypass times and increased complexity of the surgery means that stentless valves are not now commonly implanted (Chambers, 2014).

- **Transcatheter:** Transcatheter valves are used for patients in whom conventional surgery is not technically feasible or who have significant co-morbidities. These consist of biological tissue mounted within an expandable stent. A prototype was first implanted in 2002 and the first two commercially available designs received their CE mark in 2007. The Edwards SAPIEN has pericardial tissue mounted within a balloon-expandable stent which is introduced either via the femoral artery or a small incision transapically or via the aorta or subclavian artery. The Medtronic Core Valve has a nitinol self-expanding stent for transfemoral insertion. A new valve, the Portico, with bovine pericardial leaflets within a self-expanding nitinol stent has more recently been introduced and many more designs are being developed or introduced (Chambers, 2014).

Various RP processes have been developed and used in the past decade. The better known among these processes are Stereolithography (SL), Selective Laser Sintering (SLS), Fused Deposition Modeling (FDM) and Laminated Object Manufacturing (LOM).

- **Stereolithography:** Stereolithography (SL) is a popular RP process in which intricate parts of a plastic monomer are directly built by photo polymerization process with the model constructed using a computer-aided design (CAD) package [2]. The process of SL involves; modeling of part with a CAD package to generate 3D solid model; conversion of 3D solid model into standardized triangular language (STL) file format to create volumetric mesh and creation of support structure; slicing of STL format of 3D solid model to provide a series of cross-sectional layers; exporting the sliced model to the computer of stereolithography apparatus (SLA); building the support structure and the part layer by layer (from bottom to top) over a vat of specially designed liquid resin with a helium–cadmium or argon laser, which traces the outline of the planar sections and solidify the resin in SLA; removal of support structure to get the green part; post-curing of green part to undergo final polymerization in a post-curing apparatus (PCA), which is either a controlled furnace or an ultraviolet oven. The dies made through SL process are subjected to high tension due to high injection pressure. Strength is crucial in case of rapid tooling since the parts have to withstand pressures during the test of fitment and when they are used as a die for injection moulding.

- **Selective Laser Sintering:** Selective laser sintering (SLS) is an additive manufacturing technique that uses a high power laser (for example, a carbon dioxide laser) to fuse small particles of plastic, metal (Direct Metal Laser Sintering), ceramic, or glass powders into a mass representing a desired 3-dimensional object. The laser selectively fuses powdered material by scanning cross-sections generated from a 3-D digital description of the part (for example from a CAD file or scan data) on the surface of a powder bed. After each cross-section is scanned, the powder bed is lowered by one layer thickness, a new layer of material is applied on top, and the process is repeated until the part is completed. Compared to other methods of additive manufacturing, SLS can produce parts from a relatively wide range of commercially available powder materials. These include polymers such as nylon, (neat, glass-filled or with other fillers) or polystyrene, metals including steel, titanium, alloy mixtures, and composites and green sand. The physical process can be full melting, partial melting, or liquid-phase sintering. In many cases large numbers of parts can be packed within the powder bed, allowing very high productivity. Additionally, the selective laser sintering (SLS) process makes parts that have greater stability than SL and selective laser sintering (SLS) parts generally do not lose their shape or post cure over time.
- **Fused Deposition Modeling:** Fused deposition modeling (FDM) is an additive manufacturing technology commonly used for modeling, prototyping, and production applications. The technology was

developed by S. Scott Crump in the late 1980s and was commercialized in 1990. FDM works on an "additive" principle by laying down material in layers. A plastic filament or metal wire is unwound from a coil and supplies material to an extrusion nozzle which can turn on and off the flow. The nozzle is heated to melt the material and can be moved in both horizontal and vertical directions by a numerically controlled mechanism, directly controlled by a computer-aided manufacturing (CAM) software package. The model or part is produced by extruding small beads of thermoplastic material to form layers as the material hardens immediately after extrusion from the nozzle. The FDM process constructs three-dimensional objects directly from CAD data. A temperature-controlled head extrudes thermoplastic material layer by layer. The FDM process starts with importing an STL file of a model into the preprocessing software. This model is oriented and mathematically sliced into horizontal layers varying from +/- 0.127 to 0.254 mm thickness. A support structure is created where needed, based on the part's position and geometry. After reviewing the path data and generating the tool paths, the data is downloaded to the FDM machine. Several materials are available, with different trade-offs between strength and temperature properties like Acrylonitrile Butadiene Styrene (ABS) polymer, Polycarbonates, Polycaprolactone, Polyphenylsulfones and Waxes. A "water-soluble" material can be used for making temporary supports while manufacturing is in progress, this soluble support material is quickly dissolved with specialized mechanical agitation equipment utilizing a precisely heated sodium hydroxide solution.

- **Laminated Object Manufacturing:** The first commercial Laminated Object Manufacturing (LOM) system was shipped in 1991. LOM was developed by Helisys of Torrance, CA. The main components of the system are a feed mechanism that advances a sheet over a build platform, a heated roller to apply pressure to bond the sheet to the layer below, and a laser to cut the outline of the part in each sheet layer. Parts are produced by stacking, bonding, and cutting layers of adhesive-coated sheet material on top of the previous one. A laser cuts the outline of the part into each layer. After each cut is completed, the platform lowers by a depth equal to the sheet thickness (typically 0.002 to 0.020 in), and another sheet is advanced on top of the previously deposited layers. The platform then rises slightly and the heated roller applies pressure to bond the new layer. The laser cuts the outline and the process is repeated until the part is completed. After a layer is cut, the extra material remains in place to support the part during build. The excess material supports overhangs and other weak areas of the part during fabrication. The cross-hatching facilitates removal of the excess material. Once completed, the part has a wood-like texture composed of the material layers. Moisture can be absorbed by the paper, which tends to expand and compromise the dimensional stability. Therefore, most models are sealed with paint or lacquer to block moisture increase.

DESIGN AND MANUFACTURE OF THE VALVE

Design and Modelling

The material of the valve needed to fulfil two criteria – biotolerance and hyperelasticity. The valve needed to open and close within the pressure gradients generated during systole and diastole. However, biotolerance was a major factor driving the material selection process. Based on these, Silicone was shortlisted for the valve material.

This design uses a tricuspid design with each of the cusps/leaflets being concavo-convex such that the diastolic pressure completely closes the valve much like in the physiological valve. The design allows the systolic pressure to open the valve while preventing prolapse of the same.

The structure of the valve includes three cusps or flaps, affixed on the annulus. These are affixed to the body at the root. During forward flow, the cusps deform outward and open the valve. When the pressure in the chamber drops, the flaps fill with blood and collapse sealing the chamber off, preventing backflow.

While the method and apparatus of the valve is applicable for implantation in patients with any cardiopulmonary condition, for the sake of practical demonstration, a process of systole and diastole at normal haemodynamic conditions has been described. However, the present design is not limited to normal haemodynamic conditions.

The blood pressure in the human cardiopulmonary system varies with time for the same individual. However, for most individuals, the two normal clinically significant pressures involved are 120mm of Hg in systole and

80mm of Hg in diastole. Although the position of the valve in use will vary, depending on whether it is replaces an aortic valve or a mitral valve, it is described in the upright position indicated by Figure 3 for ease of illustration. The valve in the figure is ready for insertion into a heart.

The valve is formed from two coaxial rings and flaps. The outer ring is a Polytetrafluoroethylene (PTFE) ring is provided for fixation. The inner ring is the body of the valve and forms the inlet as well as the outlet for the valve.

The flaps, shown in Figure 4, are made of the same material as the valve body. The surfaces are curved to form cusps which will fill up with blood and collapse. Three exactly identical flaps are arranged in the valve such that on collapsing, the flaps completely close of the valve.

During systole, the contraction of the ventricles will force blood to push the flaps open with the pressure being applied on the convex face. The material

Figure 3. Valve ready for implantation

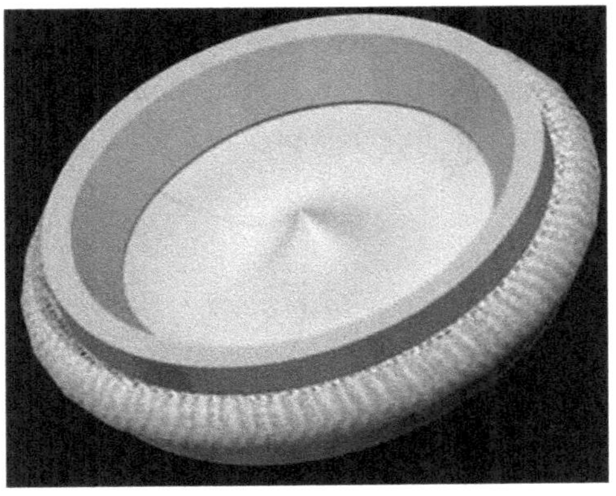

Figure 4. Shape of flap/cusp

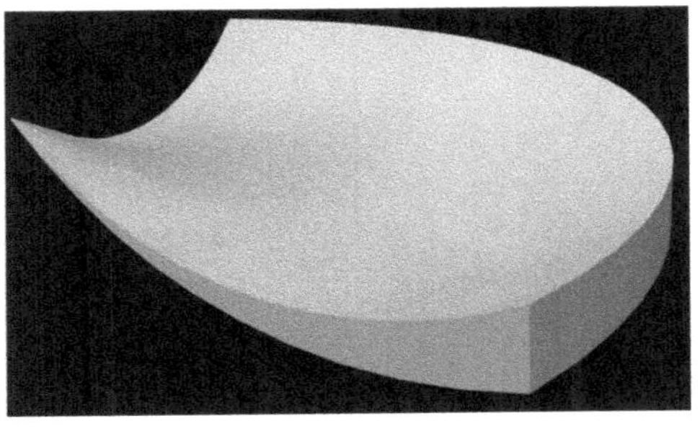

being hyperelastic, will deform and open the valve. The blood flows through the orifice thus created.

During diastole, the pressure in the ventricles drops. This causes the flow of blood to be reversed. When this happens, the cusp formed by the flaps, fill out with blood. The material being hyperelastic, the individual flaps will collapse into the ventricles. However, the arrangement of the flaps, as shown on Figure 5, is such that when they collapse, they engage each other, thus sealing off the orifice and preventing regurgitation of blood.

PTC Creo was used to generate 3D models of the valve components and for the assembly

Selection of Appropriate Technique for Present Study

It has been mentioned earlier that the differences in the layering methods have significant effects on achievable accuracy, surface finish, manufacturing time and building cost. As a result, considerations for the selection of suitable orientation that relates to above mentioned criteria vary with different RP processes. The ability to evaluate and determine the best part building orientation for different rapid prototyping (RP) processes is important for building a satisfactory part/prototype. It is also an essential step towards the identification of the most suitable RP process for a given RP application.

Figure 5. Arrangement of flap/cusp with respect to the valve body

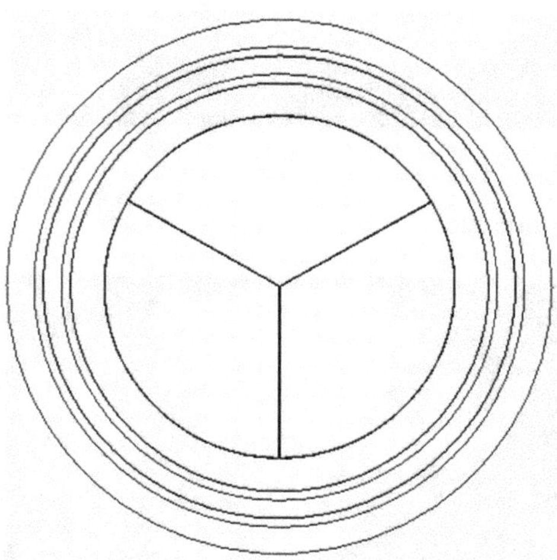

The dynamic mechanical properties of products are significantly dependent upon the type of materials and processing techniques. In practice, polymer materials exhibit more than one relaxation region or so-called transition over a wide range of temperature during dynamic thermomechanical analysis.

SIMULATION AND SOLUTIONS

Finite element analysis was conducted on the present designs using ANSYS 16.2. Static structural analysis of the design was done on a model having 10mm diameter. The deformation of the valving mechanism i.e. the flaps in case forward flow was always greater than 2mm at the tips of the cusps under 140 mmHg pressure. On reversing the direction of pressure to simulate diastole, the maximum deflection was 0.4mm at the tips of the cusps. However, it was observed that the valve did not prolapse and remained closed under a pressure of 100 mmHg. This indicated better regurgitant characteristics as compared to other valve designs.

The calculation of cardiac valvular orifices is done using the Gorlin Formula, the final form of which is:

$$A = \frac{CO / \left(DFP\ or\ SEP\right)\left(HR\right)}{44.3C\sqrt{\Delta P}}$$

where A is the orifice area, CO is cardiac output (cm^3/minute), DFP is diastolic filling period (seconds/beat), SEP is the systolic ejection period (seconds/beat), HR is heart rate (beats/minute), C is an empirical constant and P is the pressure gradient (Carabello & Grossman, 2006).

FUTURE RESEARCH DIRECTIONS

While mechanical prosthetics are far from ideal, a concerted effort can be made to improve the functioning of the same with flow patterns and haeodynamic characteristics of the physiological valves as the benchmark.

CONCLUSION

Despite the widespread use of heart valve prostheses neither mechanical nor bioprosthetic heart valves are free from complications. The overall complications associated with prosthetic heart valves can be divided into six main categories: structural valvular deterioration, non-structural dysfunction, valve thrombosis, embolism, bleeding and endocarditis (Grunkemeier & Anderson, 1998)

These complications are believed to be associated with non-physiological blood flow patterns in the vicinity of heart valves. The potential of abnormal flow patterns promoting blood cell damage has long been recognized, because they may initiate thrombus formation.

The most common causes being: (i) imposition forces on cell elements (regions of high shear stress cause tearing of the blood elements, thus leading to haemolysis and platelet activation); and (ii) changing frequency of contact (recirculation and flow stagnation regions increase the contact time between blood elements, in particular activated platelets, thereby promoting thrombus formation). In addition, these abnormal flow patterns may induce leaflet calcification and tearing in tissue and polymeric valves by creating elevated regions of shear in the immediate vicinity of the leaflet surfaces (Dasi et al., 2009).

The present design shows promise in reducing these factors if not complete elimination of the same. This can however only be verified by further testing and clinical trials.

Also, with the deformation patterns shown during diastole simulations showing reduced regurgitation, studies need to be conducted to verify them.

With the introduction of additive manufacturing into the overall process, the lead time in the manufacture of the valves will come down. Also, the availability of modern imaging processes in the field of medical sciences guarantees that the valves can be customized as per the requirements of the patient. Both these factors come together to ensure that the risk to the patient undergoing the procedure is minimized. Also, the requirement for anti-coagulants will be reduced as the valve will closely resemble the physiological valve being replaced.

REFERENCES

Carabello, B. A., & Grossman, W. (2006). Calculation of stenotic valve orifice area. In D. S. Baim (Ed.), *Grossman's Cardiac Catheterization, Angiography, and Intervention* (7th ed.; pp. 173–183). Philadelphia: Lippincott Williams & Wilkins.

Chambers, J. (2014, October). Prosthetic heart valves. *International Journal of Clinical Practice*, *68*(10), 1227–1230. doi:10.1111/ijcp.12309 PMID:24423099

Dasi, L. P., Simon, H. A., Sucosky, P., & Yoganathan, A. P. (2009, February). Fluid mechanics of artificial heart valves. *Clinical and Experimental Pharmacology & Physiology*, *36*(2), 225–237. doi:10.1111/j.1440-1681.2008.05099.x PMID:19220329

DeWall, R. A., Qasim, N., & Carr, L. (2000, May). Evolution of mechanical heart valves. *The Annals of Thoracic Surgery*, *69*(5), 1612–1621. doi:10.1016/S0003-4975(00)01231-5 PMID:10881865

Grunkemeier, G. L., & Anderson, W. N. (1998). Clinical evaluation and analysis of heart valve substitutes. *The Journal of Heart Valve Disease*, (7): 163–169. PMID:9587856

Hillis, L. D., Lange, R. A., Winniford, M. D., & Page, R. L. (1995). *Manual of Clinical Problems in Cardiology*. Little, Brown and Company.

Kai, C. C., & Fai, L. K. (1997). Rapid Prototyping: Principles and Applications in Manufacturing. Singapore: John Willey & Sons (Asia) Pte Ltd.

Lefrak, E. A., & Starr, A. (1979). Starr-Edwards ball valve. In *Cardiac valve prostheses* (pp. 67–117). New York: Appleton-Century-Crofts.

Matthews, A. M. (1998). The development of the Starr-Edwards heart valve. *Texas Heart Institute Journal*, *25*(4), 282–293. PMID:9885105

Oregon Health and Science University. (n.d.). Retrieved March 2016 from http://www.ohsu.edu/xd/education/library/about/collections/historical-collections-archives/exhibits/miles-lowell-edwards.cfm

Wieting, D. W. (1996, August). The Björk-Shiley Delrin tilting disc heart valve: Historical perspective, design and need for scientific analyses after 25 years. *The Journal of Heart Valve Disease*, *5*(2S), 157–168. PMID:8905516

Chapter 7
Designing Thin 2.5D Parts Optimized for Fused Deposition Modeling

James I. Novak
https://orcid.org/0000-0003-4082-4322
Deakin University, Australia

Mark Zer-Ern Liu
University of Technology Sydney, Australia

Jennifer Loy
Deakin University, Australia

ABSTRACT

This chapter builds new knowledge for design engineers adopting fused deposition modeling (FDM) technology as an end manufacturing process, rather than simply as a prototyping process. Based on research into 2.5D printing and its use in real-world additive manufacturing situations, a study featuring 111 test pieces across the range of 0.4-4.0mm in thickness were analyzed in increments of 0.1mm to understand how these attributes affect the quality and print time of the parts and isolate specific dimensions which are optimized for the FDM process. The results revealed optimized zones where the outer wall, inner wall/s, and/or infill are produced as continuous extrusions significantly faster to print than thicknesses falling outside of optimized zones. As a result, a quick reference graph and several equations are presented based on fundamental FDM principles, allowing design engineers to implement optimized wall dimensions in computer-aided design (CAD) rather than leaving print optimization to technicians and manufacturers in the final process parameters.

DOI: 10.4018/978-1-5225-9167-2.ch007

INTRODUCTION

As a range of technologies, additive manufacturing (AM), also known as 3D printing, has been the subject of research for several decades. A considerable body of knowledge has been built on the topic across disciplines and there are many publications focusing on processes, materials and applications. Research-based guidelines have been developed to provide designers and engineers with core principles to consider when adopting AM processes; for example Gibson, Rosen and Stucker's (2015) 'Additive Manufacturing Technologies' textbook, as well as books by Lipson and Kurman (2013) and Redwood, Schöffer and Garret (2017). However, as AM shifts from a predominantly prototyping technology, known as rapid prototyping (RP), towards an end-use manufacturing technology (Campbell, Bourell, & Gibson, 2012; Gibson et al., 2015), further guidance is needed for designers to understand how to design specifically for end-use production appropriate to the specific AM technology. Designers need to be aware of the constraints and opportunities of individual AM processes, just as they would when designing for traditional manufacturing technologies. For example, designing for injection molding requires a thorough understanding of draft angles, part-lines and appropriate wall thicknesses, which can vary between injection molding machines and individual molds. These constraints influence the design decisions made throughout product development, and ultimately impact the final form and function. Likewise, there are constraints when designing for additive manufacturing.

Design for additive manufacturing (DfAM) is emerging as an interdisciplinary field of research to address these constraints, helping design engineers effectively adopt AM through the development of more specific methodologies and discourse. This chapter builds on recent DfAM guidelines (Kumke, Watschke, & Vietor, 2016; Pradel, Zhu, Bibb, & Moultrie, 2018; Thompson et al., 2016), focusing specifically on research into the relationship between computer-aided design (CAD) geometry and stereolithography (STL, also known as Standard Triangulation Language) files for part thicknesses in the range of 0.4-4.0mm. Furthermore, it identifies the relationship between such thin geometry and the quality and speed of 3D printing, providing design engineers with specific settings to optimize a design for fused deposition modeling (FDM) as the end manufacturing process. Thin test pieces in 0.1mm increments are analyzed using Cura software from Ultimaker, alongside three printed *wall thicknesses* (also called the *shell*) related to nozzle diameter, and graphed alongside three STL export settings (fine, medium and coarse). The vast data set is presented in a visual quick-reference graph with optimal dimensions for FDM printing with the most common Ø0.4mm nozzle highlighted, allowing design engineers to implement settings for maximum printing speed and accuracy, or calculate them using the provided equations for other nozzle diameters. This is particularly important when part designs may only consist of a small number of layers, often described as a 2.5D print (Galbally & Satta, 2016; Zhu, Dancu, & Zhao, 2016), with an increasing range of projects being manufactured using 2.5D printing. The value of this research and experimental study is that it allows designers to significantly improve the final manufacturability of a thin part design, prior to a technician or manufacturer modifying process parameters which are often outside the control of the designer.

BACKGROUND

Fused deposition modeling (FDM) technology, also known as fused filament fabrication (FFF), is part of the "material extrusion" category of technologies defined within the ISO/ASTM 52900 standards for additive manufacturing ((ISO), 2015). Polymer filament is directed through a heated print head where it reaches a semi-viscous state, and through an electro-mechanically applied force extruded from a nozzle. The molten polymer is selectively dispensed to form a horizontal cross-section of the part being printed, before the next layer is printed on top in a repeating layer-by-layer process until the final object is formed. The FDM method of 3D printing has become the most mainstream 3D printing technology after expiry of key patents several years ago, and a rapid decline in hardware costs (Gibson et al., 2015; Quinlan, Hasan, Jaddou, & Hart, 2017). Despite proliferation of these 3D printers, designing specifically for FDM additive manufacturing remains challenging (Seepersad, 2014), requiring designers to learn a new set of rules which may change as hardware and software rapidly improve, with the industry experiencing exponential growth and improvement likened to Moore's Law (Benson, Triulzi, & Magee, 2018; Greenfield, 2017; B. Krassenstein, 2014).

Recent research has begun to define the specific details of FDM technology that will affect the final outcome; for example Vasilescu and Groza (2017) measured the roughness of flat FDM parts after varying settings such as infill density, print temperature and layer height, presenting a range of settings that improves surface finish, a quality which is particularly useful when FDM is used for end-use parts. Similarly, Huang and Singamneni (2015) assessed the stair-stepped layer effect of FDM for curved surfaces, proposing a new "Curved Layer Adaptive Slicing (CLAS)" process to improve the surface finish and print speed by varying layer height throughout the printing process. Such an adaptive layer process of printing has recently been integrated into mainstream slicing software Cura (Ultimaker, The Netherlands), with a recent case study showing the 3D printing of a bottle could be achieved 10% faster using the new setting (Jani, 2018). Materials research by Coogan and Kazmer (2017) analyzed the tensile properties of Acrylonitrile Butadiene Styrene (ABS) plastic after being printed with FDM in different conditions, recommending that higher nozzle temperatures than those used in traditional extrusion and injection molding be used in order to improve layer adhesion, amongst a variety of other optimal settings. However, these studies, and many like them, are related to process parameters modified after a design is complete and ready for production or prototyping, and do not provide designers with knowledge to help optimize a products functional characteristics or manufacturability early in the design process.

Specific guidelines to design for FDM are more elusive and often combined within generic DfAM guidelines such as consolidating parts, increasing complexity or customizing parts for different needs or user-fit requirements (Gibson et al., 2015; Lipson & Kurman, 2013; Petrick & Simpson, 2013). More tangible guidelines have been collated by Adam and Zimmer (2014) who compared laser sintering, laser melting and FDM across a variety of geometries and tolerances, resulting in a visual design rule catalog which includes specific design requirements for FDM such as an unsupported overhang length ≤1.8mm or a gap between parts ≥0.4mm. However, these tests were performed on a Fortus 400mc (Stratasys, United States of America) using Ultem material, which is a commercial quality FDM printer, and may not translate well to more ubiquitous desktop and open source 3D printers. Similarly, online 3D printing bureaus like i.Materialise and Shapeways provide designers with technical guidelines for FDM printing, for example i.Materialise (2018) recommends a minimum wall thickness of 1.0mm for FDM using ABS material, or 1.2mm for larger parts, while Shapeways (2018) recommends 1.0mm as a minimum for its FDM process with Polylactic Acid (PLA) material. However, these recommendations are framed within a commercial context where failed 3D prints cost time and money to the bureau, and do not provide designers with true minimum values possible with FDM, or an understanding of how wall thickness is calculated and how it might be adjusted depending on the particular printer being used. Additionally, these bureaus use commercial quality 3D printers and slicing software, which may transfer to some manufacturing situations where similar machines are being utilized, but is less likely to be applicable in lower volume production facilities and small entrepreneurial centers using lower-end machines and open source software.

A particular assumption within guidelines and broader AM discourse is that parts being printed are three-dimensional, often featuring complex geometries with well-known examples including topology optimized aerospace parts (Fendrick, 2016), a customized lattice bicycle frame (Novak, 2015) and titanium hip implants featuring bone-like textured surfaces (Wyatt, 2015). Given the increasing distribution of 3D printing technology, designers, engineers, makers, hobbyists and educators are applying FDM printing to a vast range of problems, some of which require relatively simple geometries that can be designed with rudimentary CAD modeling skills. A single extrude, or collection of extrudes that are combined together, may be all that's required of a particular geometry for 3D printing; this is often described as a 2.5D print since the geometry can be defined using a single 2D sketch with height perpendicular to the sketch axis. Designers may choose to use 3D printing in this way to take advantage of novel materials, minimize material waste (compared with laser cutting for example), or because of access to affordable 3D printers over other technologies.

Performing an extrude operation in CAD is typically the first tool taught during 3D modeling workshops and education (Novak, 2018), whether this is in free entry-level software like Tinkercad (Autodesk, San Rafael, USA), or high-end commercial software like Solidworks (Dassault Systèmes, Vélizy-Villacoublay, France). Designers create a two-dimensional sketch made up of line-work, such as a square, and then extrude this geometry to form a three-dimensional shape, such as a cube. This produces the first piece of digital "material" which can be further manipulated into a product for 3D printing through numerous cutting, extruding, patterning and assembly operations. Alternatively, a 2.5D object such as a name tag or keyring (common products designed and printed during introductory 3D printing workshops (Novak, 2018)), or more complex repeating patterns, may be built from one or several 2D sketches, some examples of which are shown in Figure 1.

Figure 1. Examples of 2.5D prints produced with FDM 3D printers

Recent research has begun to examine 2.5D printing opportunities and properties. For example, a study quantifying the effects of varied geometry on print time and material use found that "an 'optimal', 'basic' or 'simple' geometry for FDM exists; and that shape may be different from a shape considered 'optimal' for conventional manufacturing technologies" (Pradel, Zhu, Bibb, & Moultrie, 2017). Through application of such research, designers may directly shorten manufacturing times and material use as they would during design of components for traditional manufacture. More recently, research has optimized 2.5D geometry for strength properties using topology optimization methods (Kandemir, Dogan, & Yaman, 2018), resulting in a variety of structures that are easily 3D printed on FDM machines without the need for support material. Studies such as these are improving knowledge of 2.5D structures; however, there is a notable lack of discourse on the topic with most research focused on more 3D structures and geometries.

Whilst the academic discourse on this topic is still under development, designers are exploiting 2.5D printing in industry. For example, Danit Peleg is a fashion designer who made headlines in 2015 for 3D printing an entire fashion collection on desktop FDM 3D printers (Boruslawski, 2015). Peleg's pieces are made in smaller panels and assembled together, with most of her original collection consisting of 2.5D patterns that form a textile-like structure suitable for creating full-size wearable garments. Peleg now offers an online customization platform on her website for people to create their own jacket which utilizes 2.5D patterns similar to those shown in the second example in Figure 1. The letterpress industry has also experimented with 2.5D printing to create custom fonts which are functional within a letterpress machine, where more complex 3D forms are unnecessary. For example, a studio called A2-Type produced a font named *A23D* for London's New North Press studio (Steven, 2014), updating the centuries old technology into the twenty-first century. 2.5D prints are also common for gears, brackets, logos and lithophanes (etched photograph which can be clearly seen when held up to a light). These examples require different knowledge to design and manufacture using FDM compared with more complex 3D forms typically discussed within AM discourse, especially since they may have minimal height, require no support material, or be very thin in the case of Peleg's fashion to allow flexibility. These examples show how designers are engaged in their own experimentation and development of 2.5D knowledge, signaling a need for new guidelines and critical research to support ongoing development of the field.

EXPERIMENTAL STUDY OF THIN PART PRINTING

In order to better quantify and understand FDM printing of thin 2.5D geometry, a study was conducted to find optimized settings for designers to implement in their practice, and develop new knowledge about the relationship STL export settings and wall thickness settings play on the final printed part. The goal was to provide a holistic understanding of the factors affecting thin part printing based on fundamental FDM principles, rather than producing guidelines specific to a single machine or piece of CAD software. The study consisted of four phases which will be discussed separately:

1. **Parametric Design:** Creation of a 3D CAD part which could be varied in thickness and test both straight wall and curved wall features.
2. **STL Export:** Conversion of each CAD file to a STL file using three different resolution settings to understand whether triangulation will affect the quality of thin geometry.
3. **3D Printing Simulation:** Use of Cura to collect data about each file, followed by analysis of results.
4. **Case Study:** Based on more complex geometry similar to Peleg's jacket design, data was collected to quantify how settings affect print results and print times in a real-world manufacturing scenario.

Parametric Design

The part used for testing in this study was designed in Solidworks using the *straight slot* geometry (also known as a *stadium*), with external dimensions shown in Figure 2. Solidworks is a parametric CAD tool, meaning the geometry is linked by relationships within the 2D and 3D elements of the part, providing high accuracy and the ability to rapidly modify the design history of the part to create new variations. Unlike a mesh modeling CAD

Figure 2. Dimensions of test piece in millimeters

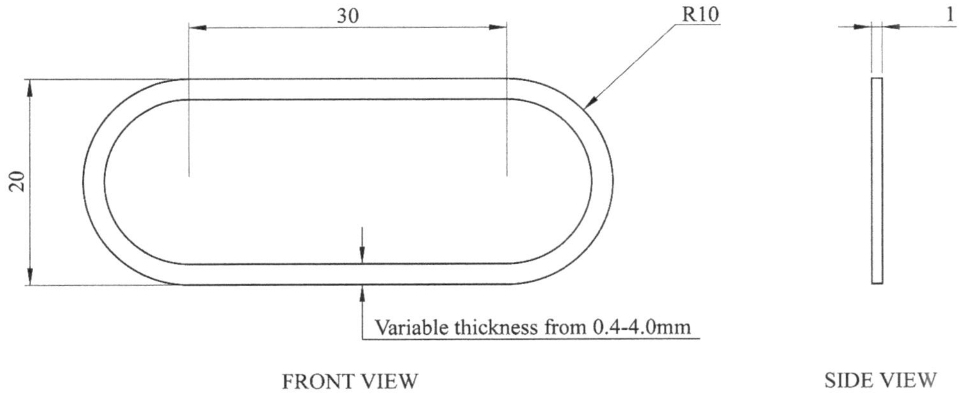

program, the geometry in Solidworks is dimensionally exact, much like vector-based 2D graphics software. The external dimensions and extrusion distance of 1mm shown in Figure 2 remained constant for all test pieces, with the thickness of the geometry varied towards the inside of the part, starting at 0.4mm (to match the most common diameter of the extrusion nozzle of desktop FDM printers) up to 4.0mm in 0.1mm increments. The design of the part provided both straight sections and curves to assess the impact of STL conversion and printing.

STL Export

Despite the recent introduction of the more accurate Additive Manufacturing File (AMF) format ((ISO), 2016), current 3D printers predominantly rely on the STL file format, which turns a dimensionally accurate CAD model into an approximated mesh constructed of triangular planar surfaces. This can result in geometric tolerance errors (Gibson et al., 2015; Zha & Anand, 2015), with numerous studies exploring novel methods of improving STL quality (Wu & Cheung, 2006; Zha & Anand, 2015). In order to build new knowledge about the affect of STL settings on thin parts, each of the 37 CAD files resulting from Figure 2 were exported to a STL file using three different settings chosen to represent fine, medium and coarse resolution, defined in Table 1. The *deviation* setting relates to the whole-part tessellation of the mesh, with lower numbers resulting in a higher number of mesh triangles and greater resolution, which is particularly important for curved geometry.

Table 1. STL export settings from Solidworks

Resolution	Deviation (mm)	Angle (degrees)	Mesh Triangles
Fine	0.01	3	976
Medium	0.02	10	376-416
Coarse	0.06	30	224-256

The *angle* relates to smaller detail tessellation, and similarly increases quality with smaller angles. As quality increases across both measures, the number of mesh triangles, and correspondingly the file size, also increases. However, for this simple test piece file size was not an important factor with all files in the range of 12-48KB. The total result was 111 STL files for testing, made up of three varied STL resolutions for each of the 37 different thicknesses in the range of 0.4-4.0mm.

3D Print Settings

3D printers typically require a separate piece of software to control the slicing of the STL file external to the original CAD program. Slicing takes the 3D geometry of the STL file and slices it into discrete layers suitable to print one layer at a time, normally in the range of 0.1-0.5mm thick for desktop FDM printers. This is then output as G-code, the machine language used by many 3D printers and other computer-numerically controlled (CNC) processes. Numerous slicing programs exist from free to paid, with some being proprietary and tied to a particular brand or printer, while others are more universal and allow for files to be sliced for a range of different 3D printers. For this study a freely available universal program called Cura (version 3.1.0) from 3D printer manufacturer Ultimaker was selected for a number of reasons:

1. Being freely available, it is widely used by designers, engineers and hobbyists for FDM printing. Therefore the results of this research will be readily replicable, and applicable to a broad range of designers.
2. Cura features a view mode called *Layer View* to visualize and simulate the 3D print process, providing data relevant for this study without the need to observe each of the 111 files print and record information about how the printer produced the parts. Simulation is a common practice in AM research, allowing consistent data to be collected without the unpredictable influence of hardware, and without the time delays required

to physically produce a broad range of test pieces (Kim, Zhao, & Zhao, 2016; Pradel et al., 2017). Different colors are used to separate the outer wall (red), inner wall (green) and infill (yellow) portions of printing for each layer, as clarified in Figure 3.

Within Cura hundreds of settings can be modified to accommodate different materials and print outcomes, however, for this testing the default *fine* settings for ABS were chosen, with the primary settings summarized in Table 2. These settings were identical for all simulations, with the only setting changed for testing being the *wall thickness* in order to understand the effect of increasing the number of perimeter walls for thin parts. The wall thickness is a dimension relative to the diameter of the print nozzle, with 0.4mm representing a single wall thickness of a common 0.4mm nozzle, and any remaining part thickness filled with infill. A 0.8mm wall thickness represents two walls (outer wall + inner wall), and 1.2mm represents three walls (outer wall + 2 inner walls). This is visually explained in Figure 3. The hypothesis of this study was that certain dimensions would make optimum use of wall thickness to produce a part without any infill, whereas other dimensions would take longer to produce as they require zigzagging infill structures.

Images of each simulation were captured and data recorded in a spreadsheet regarding the quality of the outer wall, inner wall, and the infill. Each of these elements was recorded separately in one of four states:

1. **Missing:** No material is extruded.
2. **Failure:** When sections of the wall or infill are disjointed and not solid/ continuous.

Figure 3. 0.8mm wall thickness example = 0.4mm outer wall (red) + 0.4mm inner wall (green) with remaining volume filled with infill (yellow)

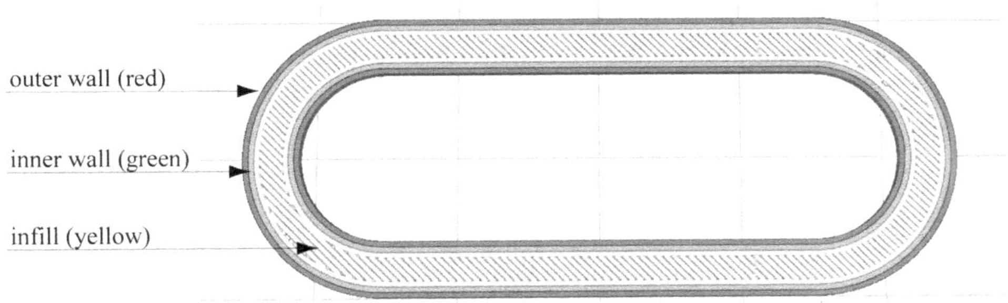

Table 2. Primary settings used for testing based on the default "fine" setting within Cura

Consistent Settings	
Layer Height	0.1mm
Nozzle Diameter	0.4mm
Line Width	0.4mm
Infill Density	100%
Infill Pattern	Lines
Printing Temperature	230°C
Bed Temperature	80°C
Filament Diameter	1.75mm
Flow	100%
Print Speed	60mm/s
Build Plate Adhesion Type	None
Variable Settings	
Wall Thickness	0.4mm, 0.8mm, 1.2mm

3. **Messy:** When the sections of the wall or infill appear joined, but may be printed in a haphazard (not continuous) fashion or made up of both infill and wall components.
4. **Solid:** The wall or infill structure is continuous and solid.

These states do not necessarily coincide with whether the physical print would fail, for example Figure 4 shows a solid outer wall but failed infill, which would most likely result in a successful 3D print, although the part would be structurally weaker than a solid infill. Furthermore, each of the 111 STL files was assessed at three orientations on the simulated print plate: 0°, 45° and 90° as shown in Figure 5. This was to identify the effect of print orientation on results, which has been shown to significantly affect part strength in mechanical testing (Zelený, Safka, & Elkina, 2014), and play a role in the roughness of the visible surfaces (Delfs, Tows, & Schmid, 2016).

Results and Discussion

Across the 3 print orientations in Figure 5, all results were identical for the 111 STL files except for one occurrence at 1.2 mm thick, using the 0.8mm wall thickness setting and fine STL resolution. With these settings, the 0°

Figure 4. 0.9mm thick test piece simulated with 0.4mm wall thickness showing a solid outer wall and failed infill

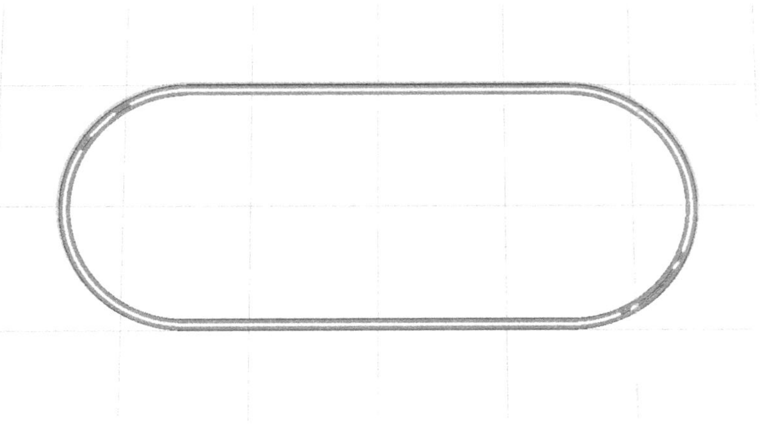

Figure 5. Test pieces oriented at 0° (left), 45° (centre) and 90° (right)

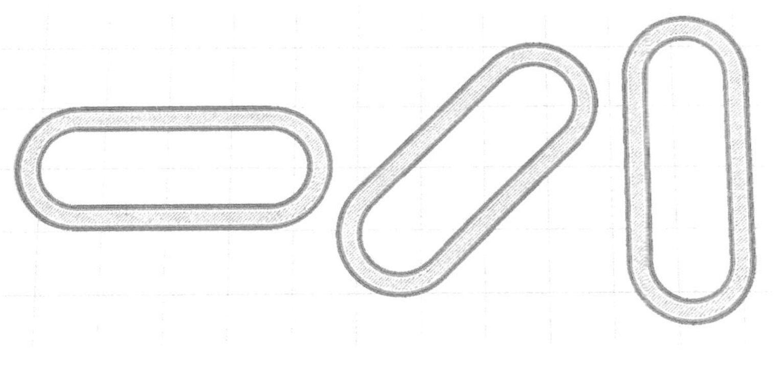

and 90° orientations presented a single outer wall and messy infill pattern, while for the 45° orientation a mix of an inner wall structure and messy infill pattern was observed within the outer wall boundary. This is shown in Figure 6. This would be unlikely to affect the ability to 3D print this part, with the overall part at the 45° orientation appearing 100% filled just like the 0° and 90° orientations. However, the 45° orientation may have an increased chance of being dislodged from the build plate as the extruder maneuvers between inner wall and infill sections during printing, which may reduce the visual quality and mechanical strength of the printed result. With all other test parts exhibiting identical wall and infill characteristics at their three orientations,

Figure 6. Results of 1.2mm thick test piece with 0.8mm wall thickness setting in Cura at 90° and 45° orientations

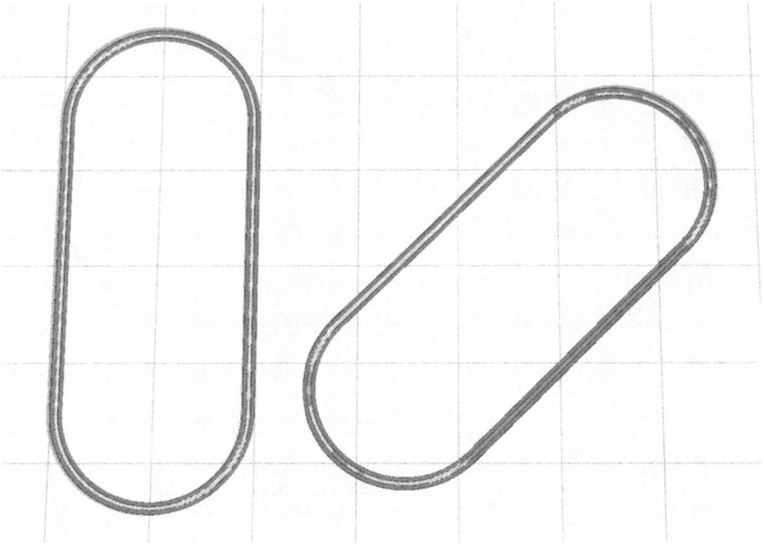

the overall results of this study presented in Figure 7 are simplified to the 0° orientation.

As shown in the overall results of Figure 7, the outer wall failed with a test piece thickness of 0.4mm across all STL resolutions and print settings, meaning that despite the matching 0.4mm extruder nozzle diameter, designing to this minimum thickness in CAD will not succeed in producing a printed object even with a fine STL export setting. The minimum part thickness to successfully produce a result was found to be 0.5mm at any STL resolution. The simulated results of printing at 0.4mm and 0.5mm can be seen in Figure 8, which shows all transitional characteristics of the test pieces using a 0.4mm wall thickness and fine STL resolution. Through these transitions it is possible to see what is deemed a failed infill (1.0mm) with gaps in the structure, compared to a messy infill (1.1mm) which is complete but somewhat haphazard, or a solid infill (1.2mm and beyond). Additionally, the dashed boxes in Figure 7 identify what are termed *optimized wall zones*, an example of which can be seen in the 0.5mm setting in Figure 8. These zones correspond with continuous concentric lines as a perimeter, without the need for any infill. A similar structure occurs once the wall thickness set in Cura is achieved and infill is produced as a continuous perimeter matching the

Figure 7. Overall results of testing at 0° print orientation

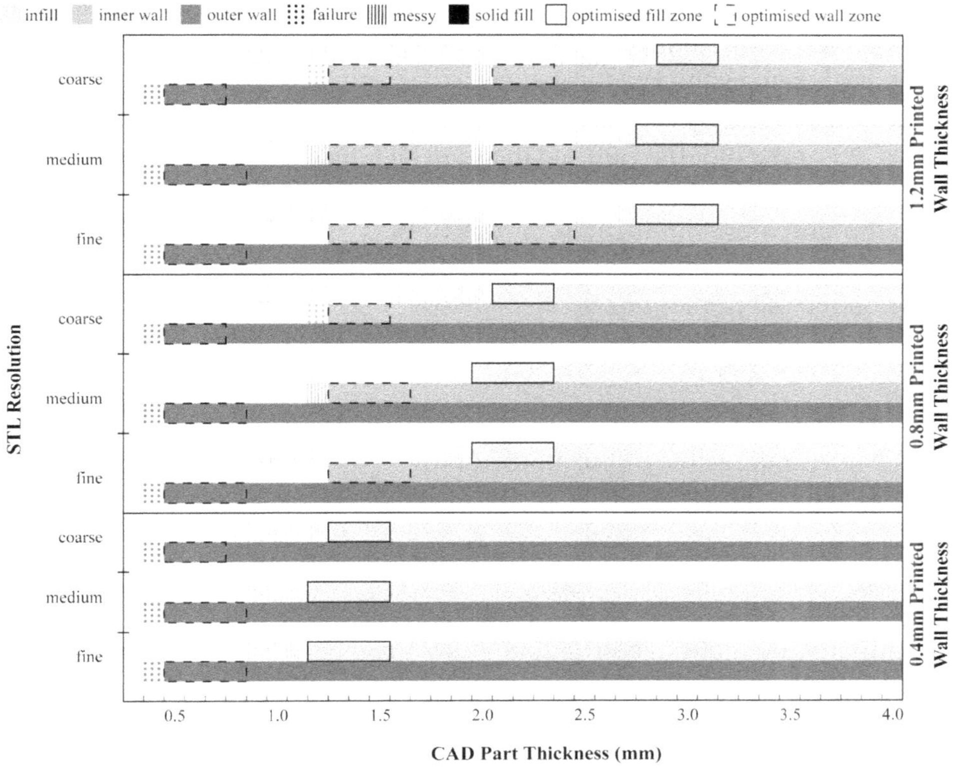

Figure 8. Key characteristics of test pieces using a 0.4mm wall thickness and fine STL resolution (yellow = infill, red = outer wall)

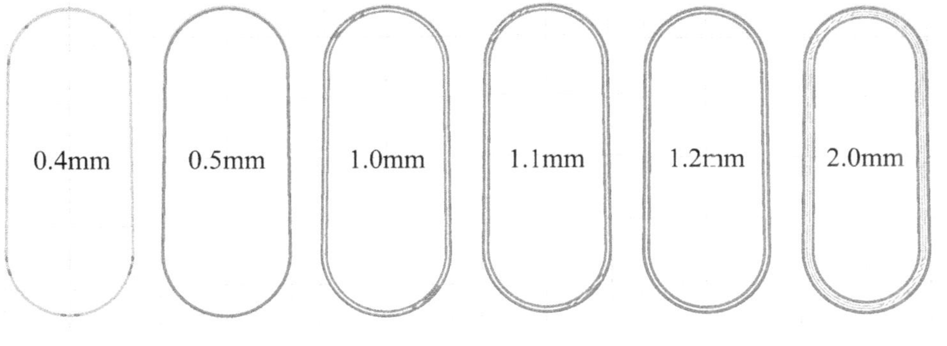

outer wall, an example of which can be seen in the 1.2mm setting in Figure 8. This is called an *optimized fill zone*, and occurs for several fractions of a millimeter prior to the more common zigzag structure emerging as part geometry becomes thicker.

From the regularly repeating occurrence of optimized wall and fill zones shown in Figure 7 for the fine and medium STL resolutions (which would be the most common STL settings to maintain tolerances to the original CAD part), the narrowest optimized wall zone (OWZ_{min}), which corresponds to the minimum printable part thickness, can be calculated using a simple equation based on the 3D printer nozzle diameter ($Ø_N$). This is labeled Equation 1:

$$OWZ_{min} = \varnothing N + 0.1$$

The maximum value of this initial optimized wall zone (OWZ_{max}) can be calculated with Equation 2:

$$OWZ_{max} = 2\varnothing N$$

Similarly, the optimized fill zone (OFZ) can be calculated using the 3D printer nozzle diameter ($Ø_N$) and the intended wall thickness for printing (W). The equation (Equation 3) from this study to calculate minimum part thickness fitting this optimized fill zone (OFZ_{min}) can be summarized as:

$$OFZ_{min} = \varnothing N\left(\left(\frac{2W}{\varnothing N}\right)+1\right)$$

The equation (Equation 4) to calculate the maximum part thickness fitting this optimized fill zone (OFZ_{max}) can be summarized as:

$$OFZ_{max} = \varnothing N\left(\left(\frac{2W}{\varnothing N}\right)+1\right)+\varnothing N$$

When designing in CAD to fit within the range of OWZ_{min} to OWZ_{max}, or OFZ_{min} to OFZ_{max}, these equations allow designers to create parts that will be solid and printed with maximum speed. For example, Table 3 compares

Table 3. Cura simulation of 1.5mm versus 1.6mm fine resolution parts using 0.4mm wall thickness

Test piece thickness	Quantity	Print Time	Time per part
1.5mm	20	56 mins	168 seconds
1.6mm	20	69 mins	207 seconds
Total difference		**13 mins**	**39 seconds**

the printing of 20 test pieces designed at 1.5mm thick (within the OFZ) and 1.6mm thick (solid fill but outside the OFZ) using a 0.4mm wall thickness. The total print time is 13 minutes quicker for the 1.5mm thick part, which works out to be 39 seconds quicker per individual part. This is a time saving of 19% despite only being 0.1mm different in size. If this were a larger piece, or being printed in significant quantities, designing a part to be 1.5mm thick would significantly improve the speed to produce the parts, reducing costs over the long term and allowing higher throughput of the machine. In some circumstances, predominantly for rapid prototyping applications, reducing a wall thickness by 0.1mm may simply be impossible due to tolerances or interfacing with other parts of an assembly; however, as a DfAM consideration, it is most efficient to design within this optimized fill zone. Table 4 and Table 5 show similar differences at the 0.8mm and 1.2mm wall thickness settings, with time savings of 14% and 13% respectively.

Table 4. Cura simulation of 2.3mm versus 2.4mm fine resolution parts using 0.8mm wall thickness

Test piece thickness	Quantity	Print Time	Time per part
2.3mm	20	75 mins	225 seconds
2.4mm	20	87 mins	261 seconds
Total difference		**12 mins**	**36 seconds**

Table 5. Cura simulation of 3.1mm versus 3.2mm fine resolution parts using 1.2mm wall thickness

Test piece thickness	Quantity	Print Time	Time per part
3.1mm	20	88 mins	264 seconds
3.2mm	20	101 mins	303 seconds
Total difference		**13 mins**	**39 seconds**

From the results in Figure 7 another pattern that emerges as wall thickness increases to 0.8mm (2 shells) and 1.2mm (3 shells) is the appearance of infill structures prior to an inner wall structure being used. Across all fine and medium resolution tests, infill structure begins at 0.9mm part thickness, while for coarse resolution it begins earlier at 0.8mm part thickness. As infill emerges it improves in quality from a fail structure to a messy structure, before becoming a solid structure either directly, or in the form of becoming a new inner wall. This will repeat with regularity until the maximum wall thickness is reached, in which case the infill structure has a 0.4mm range where it transitions from a fail to a solid structure for fine and medium resolution STL files, or 0.6mm in the case of a coarse resolution STL file. Following on from Figure 8, Figure 9 shows all the transitional characteristics of the test pieces after 1.2mm using a 0.8mm wall thickness and fine STL resolution as inner wall structures becoming evident. The 1.6mm value in Figure 9 also exemplifies a print within the *optimized wall zone*, which occurs as the inner wall structure integrates into the printed part and forms a continuous perimeter before the part thickens and requires infill alongside the inner wall. Optimized wall zones and optimized fill zones are identical in their ability to reduce print time. This means that when printing using a 1.2mm wall thickness setting, there are four optimized zones which occur between 0.5-0.8mm (initial optimized wall zone which can be calculated using Equation 1-2), 1.3-1.6mm (optimized wall zone), 2.1-2.4mm (optimized wall zone) and 2.8-3.1mm (optimized fill zone which can be calculated with Equation 3-4) as shown in Figure 7.

Figure 9. Key characteristics of test pieces using a 0.8mm wall thickness and fine STL resolution (yellow = infill, green = inner wall, red = outer wall)

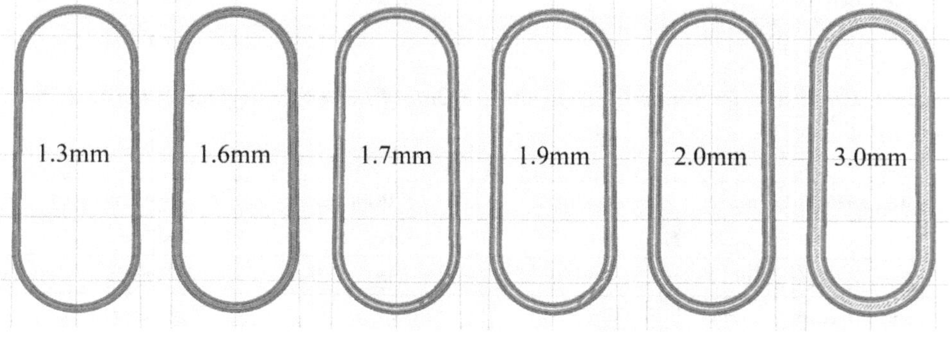

Fine and medium resolution STL results are identical for all tests in Figure 7; however, the results are slightly different for the coarse STL setting, with the tolerances of the STL being more relaxed from the original CAD file and creating areas where the printer must add extra material, particularly as curved surfaces become more faceted. These STL settings are extreme and it is unlikely designers would use such low quality files; however, the results show how optimized fill zones become smaller with low resolution STL files, and there is an increased likelihood of failed or messy infill structures with these zones occurring across broader part thicknesses.

Designing at the extreme range of thin wall sections <1mm, below the recommendations from commercial 3D printing bureaus, test pieces are made entirely from outer wall structure with two passes of the nozzle. The settings used for this study did not allow the nozzle to create a wall thickness larger than its diameter, and as a result, 0.5mm requires the nozzle to make two passes to build up material, just as it does for 0.8mm (although this can be difficult to see in the images). The simulations within Cura from 0.5-1.0mm would result in successful prints, however, within the scope of this study physical prints have not been produced to determine how reliably they can be printed at this fine scale. This is an area for future research and would vary based on the specific hardware of each FDM printer and the rheology of the filament (Cicala et al., 2018).

Case Study: Mesostructure

To validate the results of this study a more complex test piece based on a mesostructure was created as shown in Figure 10. Designers such as Peleg (Boruslawski, 2015), Bastian (E. Krassenstein, 2014) and Novak (2016), as well as researchers such as Li, Chen, Hoe and Yin (2016) have explored mesostructure geometry using desktop FDM technology for its flexibility and complex aesthetic. The geometry also features a selection of straight and curved sections like the original straight slot test piece, providing a range of challenges within a single part. A number of variations in thickness were created and exported as STL files like the original test piece, although only fine resolution settings were used (see Table 1) and a 0.8mm wall thickness. The resultant print times estimated in Cura can be seen in Figure 11.

What is clear from these test results is that the mesostructure is fastest to print within the range of a 1.3-1.6mm part thickness, aligning with the data captured in Figure 7. Despite 1.1mm and 1.2mm thick parts requiring less

Figure 10. Dimensions of mesostructure in millimeters

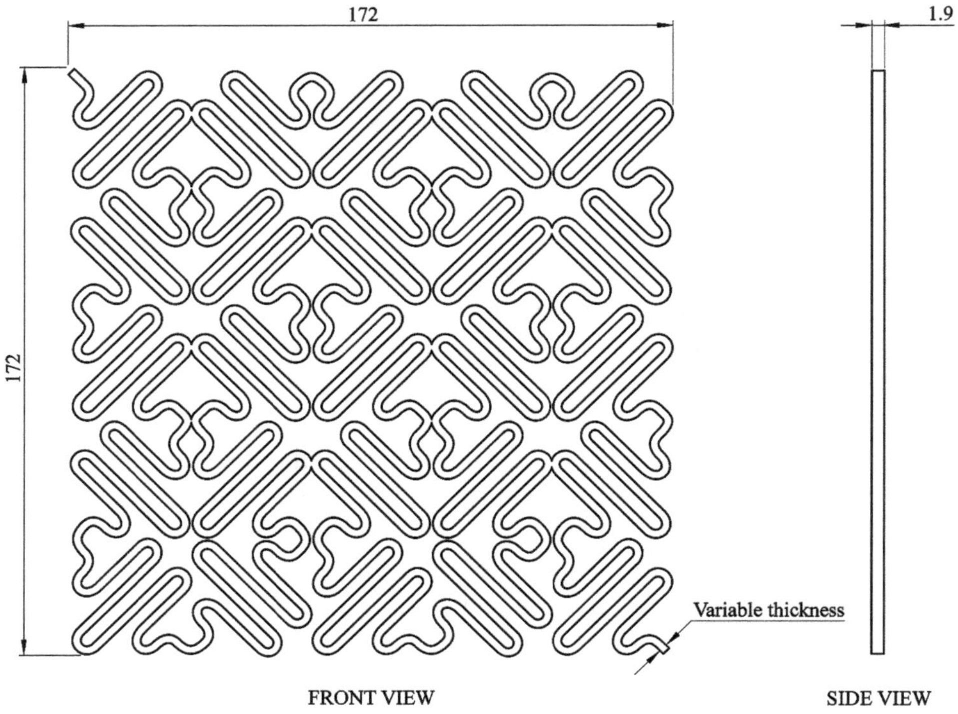

Figure 11. Cura simulated print times for Mesostructure using 0.8mm wall thickness setting

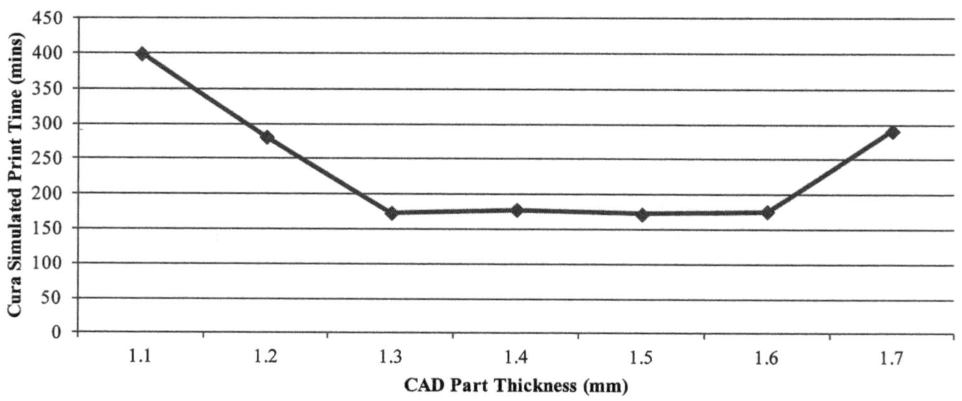

material, it is considerably faster to print parts in the optimized zone, with 1.3mm being 57% faster to 3D print than 1.1mm, and 38% faster than 1.2mm. These significant time variations of several hours for a single part are purely related to fractional variations in design geometry, with no modification to process parameters. Multiplied over the number of components in the fashion of Peleg or another similar product, this may result in days worth of time saved by understanding how to design specifically for the FDM process using optimized thin-wall features. Figure 12 compares a detailed section of the mesostructure at 1.2mm and 1.3mm, with 1.2mm exhibiting a messy combination of both inner wall and infill structures which takes time for the extruder to move between, while 1.3mm features three concentric wall structures which are most efficient to 3D print due to their continuity.

It is unclear whether designers such as Peleg 3D model parts with an understanding of these optimized zones; according to a recent article (Krassenstein, 2015), each A4 sized section of a Peleg dress takes twenty hours to print, with each complete dress taking approximately four hundred hours. If the pieces of each dress have not been designed with an understanding of these optimized zones, there may be an opportunity to significantly reduce the time required to produce each dress, saving days of time without any change to the 3D printer hardware or software. If time is factored into retail cost

Figure 12. Detail of mesostructure at 1.2mm thickness (left) compared with 1.3mm (right)

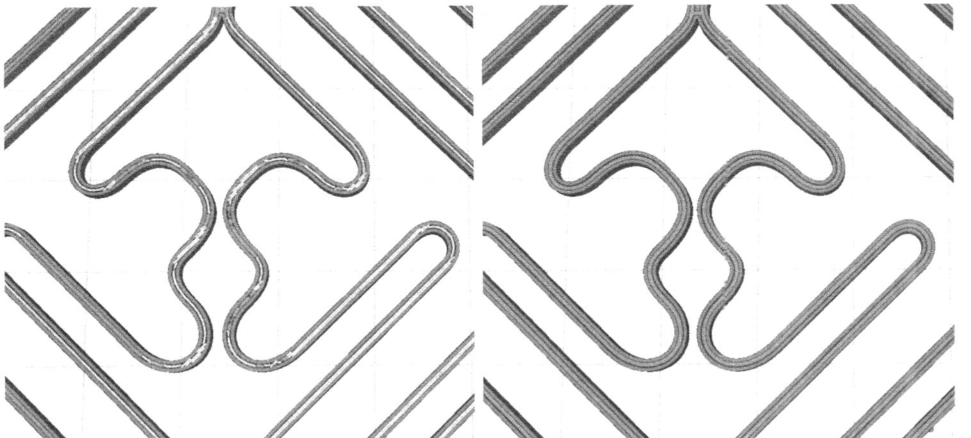

equations, this may also result in a significant reduction in cost for customers. These considerations align with the research of Pradel et al. (2017), and are important aspects to the DfAM approach.

LESSONS AND FUTURE RESEARCH DIRECTIONS

While 3D printing thin wall sections presents numerous challenges, and the results from simulations are not always linked to whether a part will successfully 3D print, the data from this study suggests there are specific ranges of dimensions, described as *optimized wall zones* and *optimized fill zones*, which result in thin yet solid parts, printed with maximum machine efficiency. For designers and engineers, these dimensions can be used to specifically design for the FDM print process, optimizing the design before any process parameters are modified at the final stage of production. What is significant about Equations 1-4 is that they can be used to calculate the initial optimized wall zones and optimized fill zones for any FDM 3D printer, since the wall thickness setting in slicing software directly corresponds to nozzle diameter, and the equations are based on this fundamental data. However, future testing is needed to confirm this hypothesis, and to more broadly understand how more commercial varieties of FDM technology, such as Fortus machines from Stratasys, relate to these optimized zones given they work with their own proprietary software.

Future research will also assess the magnitude of these factors across different nozzles diameters, particularly as large area FDM machines with build volumes measuring in meters emerge. Examples include the 'BigRep ONE' (BigRep, Berlin, Germany) with a print volume of 1.005 × 1.005 × 1.005m (Cartesian style printer), the 'Delta WASP 3MT' (Wasp, Massa Lombarda, Italy) with a print volume of Ø1 × 1.2m (delta style printer), and the 'Tractus3D T3500' (Tractus3D, Ammerzoden, The Netherlands) with a print volume of Ø1 × 2m (delta style printer). Such printers feature nozzle diameters larger than 1mm, meaning a significant amount of material is being continuously extruded, and potential time savings for large object prints may be measured in days rather than hours. Optimization in these situations becomes critical to productivity. Preliminary experiments on a BigRep ONE using a

0.6mm diameter nozzle have proven successful, as shown in Figure 13, with a part thickness of 0.8mm fitting within the calculated initial optimized wall zone of 0.7-1.2mm (Equation 1-2), requiring no infill or inner wall structures. This will form part of a follow up study with a range of nozzle diameters available including 1mm and 2mm on this printer.

While 3D printing bureaus like i.Materialise (2018) and Shapeways (2018) recommend a minimum wall thickness of 1.0mm, the data from this study suggests 1.0mm is not an efficient thickness in terms of printing speed on a FDM printer with 0.4mm nozzle diameter. While these bureaus use more sophisticated slicing software than Cura, they may also use a smaller nozzle diameter which shifts the optimized zones in favor of 1.0mm, driven by the need for fast and accurate prints within a commercial context. However, this is speculative. For design engineers, this study highlights why it is necessary to understand the end FDM production technology during design, and use this knowledge to design specifically to produce the desired outcome, particularly for thin parts where there is minimal room for error. An advantage of programs like Cura is the ability to visualize print paths prior to printing, and iteratively develop a design to make it more efficient for additive manufacture. While Cura provides useful visualization tools and time estimates, not all software provides such a print preview, with "plug-n-play" programs like UPStudio

Figure 13. Preliminary experiment 3D printing 0.8mm thick geometry on a BigRep ONE with a 0.6mm nozzle

(Tiertime, Beijing, China) for popular 3D printers such as the UP Plus 2 and UP Box providing limited settings and visualizations to inform a designer; therefore the data in this study is a valuable resource, with Figure 7 a quick reference guide for any FDM 3D printer with a 0.4mm diameter nozzle. Furthermore, Equations 1-4 offer a quick method to calculate optimal part thicknesses for 3D printers with a different nozzle diameter.

This study utilized simulations to reveal patterns in FDM settings, however, as mentioned throughout the discussion, the appearance of messy and even failing infill or inner wall structures does not necessarily mean the corresponding physical print will fail. Gaps in extrusion may reduce functional strength properties, yet parts will most likely be successfully printed as long as the outer wall is consistent, particularly in 2.5D parts which have minimal geometric complexity requiring consistent layer infill in order to support layers on top. Future research should consider the relationship between the simulations from this study and the actual 3D printed outcomes, particularly those less than 1mm thick which are most vulnerable to being dislodged from the print plate during haphazard extruder movements or gaps in infill.

CONCLUSION

Using common fused deposition modeling printer settings, this study has revealed important relationships between CAD geometry, STL resolution and print settings that are critical when 3D printing thin geometries. New data was collected from a study examining 111 STL files in the range of 0.4-4.0mm in part thickness. The results reveal that specific ranges of dimensions, described as *optimized fill zones* and *optimized wall zones*, result in thin yet solid parts, printed with maximum machine efficiency, most notably a faster print time compared to parts falling outside of these optimized zones by only a fraction of a millimeter. The appearance of these zones follows a regular repeating pattern as wall thickness settings increase, with the initial optimized wall zone able to be calculated using Equation 1 (minimum) and Equation 2 (maximum) using only the machine nozzle diameter as the input data. A second optimized zone, called the optimized fill zone, can be further calculated using Equation 3 and Equation 4 by inputting the intended nozzle diameter and wall thickness setting of a FDM machine.

The principal recommendation for design engineers from this study is to apply these optimized dimensions early in the design process when FDM is used as the final manufacturing method. The visual quick reference graph generated for common 0.4mm nozzles, or Equations 1-4, will allow these zones to be calculated and applied with clear understanding of how geometrically accurate CAD files will be translated through the STL conversion process and affect print quality and speed. Time savings in the order of 13-19% were found in the basic test pieces by removing 0.1mm of part thickness to fit within an optimized zone, while a time saving of 38% was recorded for a more complex mesostructure part by counter-intuitively adding 0.1mm of thickness to fit within an optimized zone. The second recommendation for design engineers is to use medium to high resolution STL conversion settings when exporting files from CAD software in order to maximize the range of part thicknesses that are optimized for FDM printing. This is particularly important for thin parts and geometries that are 2.5D in nature, with low resolutions causing greater discontinuity in wall and infill details, and a smaller range of optimized wall and infill dimensions. These specific guidelines for the FDM production of thin geometries will help designers implement optimization protocols during design development, just as they would for other manufacturing technologies, rather than relying on print technicians and manufacturers to modify process parameters outside of the designer's control in order to optimize production. Designing for additive manufacture requires new knowledge and guidelines such as these to help expedite adoption of the technology within industry, and train designers to be fluent in the processes at work between digital design and final production.

REFERENCES

Adam, G. A. O., & Zimmer, D. (2014). Design for Additive Manufacturing—Element transitions and aggregated structures. *CIRP Journal of Manufacturing Science and Technology*, *7*(1), 20–28. doi:10.1016/j.cirpj.2013.10.001

Benson, C. L., Triulzi, G., & Magee, C. L. (2018). Is There a Moore's Law for 3D Printing? *3D Printing and Additive Manufacturing, 5*(1), 53-62. doi:10.1089/3dp.2017.0041

Boruslawski, P. (2015). *Danit Peleg 3D Prints Entire Graduate Fashion Collection at Home*. Retrieved from https://www.designboom.com/technology/danit-peleg-3d-prints-fashion-collection07-27-2015/

Campbell, I., Bourell, D., & Gibson, I. (2012). Additive Manufacturing: Rapid Prototyping Comes of Age. *Rapid Prototyping Journal*, *18*(4), 255–258. doi:10.1108/13552541211231563

Cicala, G., Giordano, D., Tosto, C., Filippone, G., Recca, A., & Blanco, I. (2018). Polylactide (PLA) Filaments a Biobased Solution for Additive Manufacturing: Correlating Rheology and Thermomechanical Properties with Printing Quality. *Materials (Basel)*, *11*(7), 1191. doi:10.3390/ma11071191 PMID:29997365

Coogan, T. J., & Kazmer, D. O. (2017). Bond and Part Strength in Fused Deposition Modeling. *Rapid Prototyping Journal*, *23*(2), 414–422. doi:10.1108/RPJ-03-2016-0050

Delfs, P., Tows, M., & Schmid, H. J. (2016). Optimized Build Orientation of Additive Manufactured Parts for Improved Surface Quality and Build Time. *Additive Manufacturing*, *12*, 314–320. doi:10.1016/j.addma.2016.06.003

Fendrick, J. (2016). 3D Printing is Passing the Aerospace Test. *Manufacturing Engineering*, 37.

Galbally, J., & Satta, R. (2016). Three-dimensional and two-and-a-half-dimensional face recognition spoofing using three-dimensional printed models. *IET Biometrics*, *5*(2), 83–91. doi:10.1049/iet-bmt.2014.0075

Gibson, I., Rosen, D., & Stucker, B. (2015). *Additive Manufacturing Technologies: 3D Printing, Rapid Prototyping, and Direct Digital Manufacturing* (2nd ed.). New York: Springer. doi:10.1007/978-1-4939-2113-3

Greenfield, A. (2017). *Radical Technologies: The Design of Everyday Life*. London: Verso.

Huang, B., & Singamneni, S. B. (2015). Curved Layer Adaptive Slicing (CLAS) for Fused Deposition Modelling. *Rapid Prototyping Journal*, *21*(4), 354–367. doi:10.1108/RPJ-06-2013-0059

(ISO) International Organization for Standardization. (2015). Additive manufacturing - General principles. ISO - International Organization for Standardization.

(ISO) International Organization for Standardization. (2016). Specification for Additive Manufacturing File Format (AMF) Version 1.2. *International Organization for Standardization*. ISO.

Jani, M. (2018). *Print detailed objects faster using adaptive layers in Ultimaker Cura*. Retrieved from https://ultimaker.com/en/blog/52520-print-detailed-objects-faster-using-adaptive-layers-in-ultimaker-cura

Kandemir, V., Dogan, O., & Yaman, U. (2018). Topology optimization of 2.5D parts using the SIMP method with a variable thickness approach. *Procedia Manufacturing*, *17*, 29–36. doi:10.1016/j.promfg.2018.10.009

Kim, H., Zhao, Y., & Zhao, L. (2016). *Process-level modeling and simulation for HP's Multi Jet Fusion 3D printing technology*. Paper presented at the 2016 1st International Workshop on Cyber-Physical Production Systems (CPPS), Vienna, Austria. 10.1109/CPPS.2016.7483916

Krassenstein, B. (2014). *The Moore's Law of 3D Printing… Yes it Does Exist, and Could Have Staggering Implications*. Retrieved from http://3dprint.com/7543/3d-printing-moores-law/

Krassenstein, E. (2014). *Andreas Bastian Creates Incredible Bendable 3D Printed Mesostructured Material*. Retrieved from https://3dprint.com/2739/bastian-mesostructured/

Krassenstein, E. (2015). *Danit Peleg Creates First 3D Printed Fashion Collection Printed Entirely at Home*. Retrieved from https://3dprint.com/83423/danit-peleg-3d-printed-fashion/

Kumke, M., Watschke, H., & Vietor, T. (2016). A new methodological framework for design for additive manufacturing. *Virtual and Physical Prototyping*, *11*(1), 3–19. doi:10.1080/17452759.2016.1139377

Li, Y.-J., Chen, C.-H., Hoe, Z.-Y., & Yin, Z.-X. (2016). Design a Stretchable Elbow Brace by the Use of 3D Printed Mesostructure. In R. Goonetilleke & W. Karwowski (Eds.), *Advances in Physical Ergonomics and Human Factors* (Vol. 489, pp. 739–750). Cham: Springer International Publishing. doi:10.1007/978-3-319-41694-6_71

Lipson, H., & Kurman, M. (2013). *Fabricated: The New World of 3D Printing*. Somerset, NJ: Wiley.

iMaterialise. (2018). *ABS: Design Guide*. Retrieved from https://i.materialise.com/3d-printing-materials/abs/design-guide

Novak, J. I. (2015). *A Study of Bicycle Frame Customisation Through the use of Additive Manufacturing Technology*. Paper presented at the RAPID 2015, Long Beach, CA. Retrieved from https://www.sme.org/globalassets/sme.org/about/awards/a-study-of-bicycle-frame-customization-through-the-use-of-additive-manufacturing-technology.pdf

Novak, J. I. (2016). *Mesostructure*. Retrieved from https://edditiveblog.wordpress.com/2016/03/07/mesostructure/

Novak, J. I. (2018). Re-educating the Educators: Collaborative 3D Printing Education. In I. M. Santos, N. Ali, & S. Areepattamannil (Eds.), *Interdisciplinary and International Perspectives on 3D Printing in Education* (pp. 28–49). Hershey, PA: IGI Global.

Petrick, I. J., & Simpson, T. W. (2013). 3D Printing Disrupts Manufacturing. *Research Technology Management*, *56*(6), 12–16. doi:10.5437/08956308X5606193

Pradel, P., Zhu, Z., Bibb, R., & Moultrie, J. (2018). A framework for mapping design for additive manufacturing knowledge for industrial and product design. *Journal of Engineering Design*, *29*(6), 291–326. doi:10.1080/09544828.2018.1483011

Pradel, P., Zhu, Z., Bibb, R. J., & Moultrie, J. (2017). *Complexity Is Not For Free: The Impact of Component Complexity on Additive Manufacturing Build Time*. Paper presented at the Rapid Design, Prototyping & Manufacturing (RDPM2017), Newcastle, UK.

Quinlan, H. E., Hasan, T., Jaddou, J., & Hart, A. J. (2017). Industrial and Consumer Uses of Additive Manufacturing: A Discussion of Capabilities, Trajectories, and Challenges. *Journal of Industrial Ecology*, *21*(S1), S15–S20. doi:10.1111/jiec.12609

Redwood, B., Schöffer, F., & Garret, B. (2017). *The 3D Printing Handbook*. 3D Hubs.

Seepersad, C. C. (2014). Challenges and Opportunities in Design for Additive Manufacturing. *3D Printing and Additive Manufacturing, 1*(1), 10-13. doi:10.1089/3dp.2013.0006

Shapeways. (2018). *PLA Material Information*. Retrieved from https://www.shapeways.com/materials/pla

Steven, R. (2014). *A2 & New North Press' 3D-printed letterpress font*. Retrieved from https://www.creativereview.co.uk/a2-new-north-press-3d-printed-letterpress-font/

Thompson, M. K., Moroni, G., Vaneker, T., Fadel, G., Campbell, R. I., Gibson, I., ... Martina, F. (2016). Design for Additive Manufacturing: Trends, Opportunities, Considerations, and Constraints. *CIRP Annals*, *65*(2), 737–760. doi:10.1016/j.cirp.2016.05.004

Vasilescu, M. D., & Groza, I. V. (2017). Influence of Technological Parameters on the Roughness and Dimension of Flat Parts Generated by FDM 3D Printing. *Revista de Tehnologii Neconventionale*, *21*(3), 18–23.

Wu, T., & Cheung, E. H. M. (2006). Enhanced STL. *International Journal of Advanced Manufacturing Technology*, *29*(11), 1143–1150. doi:10.100700170-005-0001-5

Wyatt, M. C. (2015). Custom 3D-printed acetabular implants in hip surgery - innovative breakthrough or expensive bespoke upgrade? *Hip International*, *25*(4), 375–379. doi:10.5301/hipint.5000294 PMID:26351112

Zelený, P., Safka, J., & Elkina, I. (2014). The Mechanical Characteristics of 3D Printed Parts According to the Build Orientation. *Applied Mechanics and Materials, 474*, 381-386. Retrieved from www.scientific.net/AMM.474.381

Zha, W., & Anand, S. (2015). Geometric Approaches to Input File Modification for Part Quality Improvement in Additive Manufacturing. *Journal of Manufacturing Processes*, *20*, 465–477. doi:10.1016/j.jmapro.2015.06.021

Zhu, K., Dancu, A., & Zhao, S. (2016). *FusePrint: A DIY 2.5D Printing Technique Embracing Everyday Artifacts.* Paper presented at the 2016 ACM Conference on Designing Interactive Systems, Brisbane, QLD, Australia. 10.1145/2901790.2901792

ADDITIONAL READING

Booth, J., Alperovich, J., Chawla, P., Ma, J., Reid, T., & Ramani, K. (2017). The Design for Additive Manufacturing Worksheet. *Journal of Mechanical Design*, *139*(10), 100904. doi:10.1115/1.4037251

Oropallo, W., & Piegl, L. A. (2016). Ten Challenges in 3D Printing. *Engineering with Computers*, *32*(1), 135–148. doi:10.100700366-015-0407-0

Pradel, P., Zhu, Z., Bibb, R., & Moultrie, J. (2018). Investigation of design for additive manufacturing in professional design practice. *Journal of Engineering Design*, *29*(4-5), 165–200. doi:10.1080/09544828.2018.1454589

Rosen, D. W. (2007). *Design for Additive Manufacturing: A Method to Explore Unexplored Regions of the Design Space*. Paper presented at the 18th Annual SFF Symposium.

Yang, S., & Zhao, Y. F. (2015). Additive manufacturing-enabled design theory and methodology: A critical review. *International Journal of Advanced Manufacturing Technology*, *80*(1), 327–342. doi:10.100700170-015-6994-5

KEY TERMS AND DEFINITIONS

2.5D Printing: The use of 3D printing to produce a relatively simple geometric form which can be described in a single orthogonal drawing and extruded in a single axis.

3D Printing (Additive Manufacturing): A digital fabrication technology that allows the production of an object by adding material layer-by-layer in three dimensions.

Computer-Aided Design (CAD): The use of computer systems to assist in the creation, modification, analysis or optimization of a design in 2D or 3D.

Fused Deposition Modeling (FDM) or Fused Filament Fabrication (FFF): The most common form of extrusion-based 3D printing technology that works similar to a hot glue gun; plastic filament is fed through a heating element, where it softens and is extruded through a small nozzle, which can move in 3D space to deposit the plastic layer-by-layer as it builds up an object.

Infill: Within the perimeter *wall thickness* describing a part, infill is the material used to fill the interior space of a part, and can range from empty (0%) to solid (100%) and gradients in between where infill patterns are used to create solid and hollow zones within a part.

Slicing: The process of converting a three-dimensional STL file into layer-by-layer information that can be 3D printed.

STL File: Originally short for *Stereolithography* file and now often described as a *Standard Triangulation Language* file, this is the native 3D file type exported from CAD software and imported into a slicing program linked to a 3D printer. A STL file is a mesh made up of triangles describing the exterior surface of an object.

Wall Thickness: Within the context of slicing, this is the thickness of the perimeter of a part, directly proportional to the nozzle diameter e.g. a wall thickness of 0.8mm requires two passes with a 0.4mm nozzle, or a single pass with a 0.8mm nozzle. It may also be called the *shell* thickness.

Chapter 8
Additive Manufacturing:
Laser Metal Deposition and Effect of Preheating on Properties of Deposited Ti-4822-4 Alloy

Kamardeen Olajide Abdulrahman
University of Johannesburg, South Africa

Esther T. Akinlabi
University of Johannesburg, South Africa

Rasheedat M. Mahamood
University of Ilorin, Nigeria & University of Johannesburg, South Africa

ABSTRACT

Three-dimensional printing has evolved into an advanced laser additive manufacturing (AM) process with capacity of directly producing parts through CAD model. AM technology parts are fabricated through layer by layer build-up additive process. AM technology cuts down material wastage, reduces buy-to-fly ratio, fabricates complex parts, and repairs damaged old functional components. Titanium aluminide alloys fall under the group of intermetallic compounds known for high temperature applications and display of superior physical and mechanical properties, which made them most sort after in the aeronautic, energy, and automobile industries. Laser metal deposition is an AM process used in the repair and fabrication of solid components but sometimes associated with thermal induced stresses which sometimes led to cracks in deposited parts. This chapter looks at some AM processes with more emphasis on laser metal deposition technique, effect of LMD processing parameters, and preheating of substrate on the physical, microstructural, and mechanical properties of components produced through AM process.

DOI: 10.4018/978-1-5225-9167-2.ch008

INTRODUCTION

Additive manufacturing (AM) technique, a process of fabricating component parts from three-dimensional computer aided design (CAD) data has become a modern and advanced manufacturing technique that continues to witness tremendous improvements. Layer upon layer build-up are formed from the computational instructions of CAD to fabricate required parts (Aliakbari, 2012; Herderick, 2011). This method used by AM technology makes it possible to provide an alternative path as compared to conventional subtractive manufacturing path like machining. The AM technique is able to provide solutions to issues regarding global competitiveness facing the manufacturing industries. Some of the advantages of AM include improved product quality, customized product demand, low cost of production and reduce down time.

History of AM can be traced back to 1987 when 3D systems manufactured the first stereolithographic equipment. The main goal of AM is to improve the performance of fabricated parts through production cost, lead time and material usage (Kobryn et al., 2006). Some of the AM techniques use lasers as primary source of energy. These lasers provide fast heating and make it possible to manipulate operation processes. This is because, convection forces created by these lasers produce melt pool having the capacity to improve the diffusion rate and make it possible for powders introduced onto the melt pool to mix (Tlotleng et al., 2016). Common AM techniques available include: electron beam melting, direct laser sintering, easyclad and laser metal deposition. This chapter briefly explains some of these lasers AM techniques, their properties and applications. More emphasis was made on laser metal deposition technique of titanium aluminide, importance of preheating of substrate and effect of laser deposition parameters on the quality of fabricated parts.

LASER ADDITIVE MANUFACTURING TECHNOLOGY

Recently, light amplification by stimulated emission of radiation (LASER) gained more ground in different areas of applications. Laser is produced from light or sometimes referred to as electromagnetic radiation and then amplified (Singh et al., 2012). The ability of laser seen in it far reaching light travelling properties and showing very little divergence made laser highly useful for different applications. The applications of lasers as constantly witnessed in communication and office equipment (such as bar-code scanners, laser printers, laser scanners and video players), medicine for surgery, military (for locating and targeting) and in construction industries (for cutting, melting, boring, etc.). Laser is unique in its ability to concentrate its energy to create high energy intensities and remains almost the same over far distance due to its low divergent property. The high intensity of laser beam makes it possible to melt hard materials like metals within a short time. This useful ability of laser beam is what is employed in laser AM technology, and makes it possible to create melt pool on substrate on which metal powder is deposited.

Additive or layer manufacturing technology uses data from 3D model to build near-net-shape components through layer upon layer build-ups. This technique of manufacturing process is developing because of its ability to reduce energy usage, material wastage and component lead time. It also has the ability of fabricating complex and what seem impossible parts when traditional production methods are to be used (Herderick, 2011). Laser metal deposition process, an AM technology has the ability to produce near-net dense structures with improved mechanical properties (Dinda et al., 2009; Kobryn et al., 2006).The sophisticated ability of AM have made it possible to be used in numerous applications including medicine, biomedical, aerospace, automobile, defense, and energy generation (Abdulrahman et al., 2018).

PROCESSES OR TECHNIQUES OF AM

There are different AM techniques currently being used. The name of each technique was given either by the company that developed the process or the method of process employed. Some of these processes are discussed below.

Laser Beam Melting (LBM) or Selective Laser Melting (SLM)

LBM, sometimes referred to as SLM, is the technique that has the ability to directly produce intricate parts from metal powder via computer aided design (CAD) model data. This is done by slicing the CAD model into tiny layers and transferring the data generated to LBM machine for production. The form or geometric data of each sliced layer are transferred onto the powder bed in an inert environment, where the surface area is scanned and a solid layer piece is then produced. This process (shown in Figure 1) continues until the whole solid part is produced from the sliced model layers. This process produced parts with excellent mechanical properties when compared to that produced by traditional process (Bremen et al., 2012). Research is still on going as to the appropriateness of the process for series production.

Electron beam melting (EBM): The energy needed for melting in this process comes from a high power electron beam. This process usually takes place in a vacuum at high temperature. The electron beam supplied the temperature needed for each layer of component to be achieved. Component produced through this method are mostly free of residual stresses and their microstructure are free of martensitic structures. Svensson et al., (2010) demonstrated the fabrication of gamma titanium aluminide component using the EBM process. 3 KW power was used to melt the metallic powder (-140/+325 powder size). The components produced have fine grain size with minimal internal defects. The EBM schematic set-up is shown in Figure 2.

Laser Metal Deposition (LMD)

The process employs the use of nozzle to directly deposits metal powder onto a desired molten surface created by laser beam, where solidification of the melted metal powder later takes place. The process has proven to be very effective when compared to the method of selective laser melting. One of the usefulness of laser metal deposition technique is it strength to concurrently use multiple material to produce composite and functionally graded materials

Figure 1. Laser beam melting process
Source: (Bremen et al., 2012)

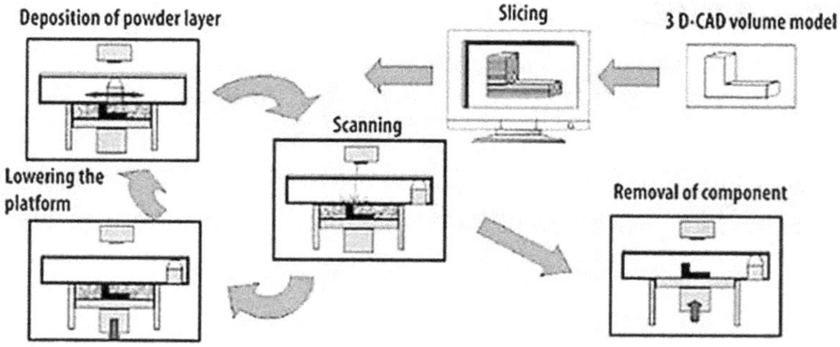

Figure 2. Schematic EBM process set-up
Source: (Svensson et al., 2010)

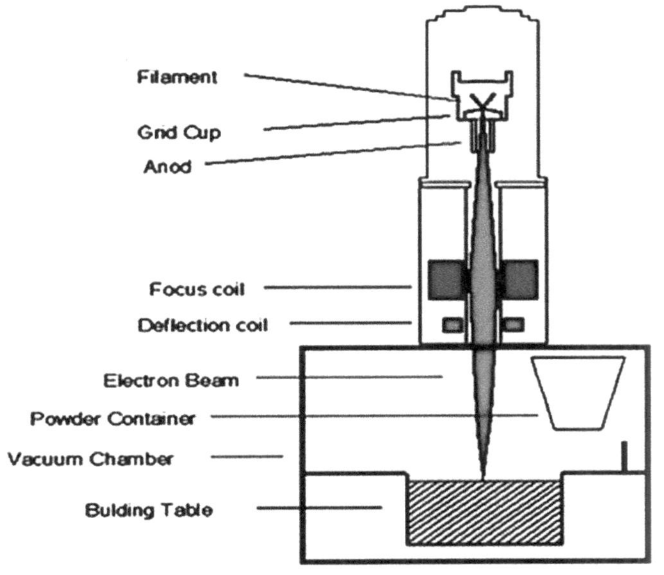

(Balla et al., 2016; Mahamood & Akinlabi, 2017). The technique produces high quality parts. Some of the other qualities of the process include it ability of repair damaged parts, addition of new features on existing parts and deposition control ability. The limitation of the technique lies in the size of parts that can be produced as this mainly depends on the size of the LMD machine. Schematic view of laser deposition process is shown in Figure 3. Difficult to machine materials such as titanium and its alloy can easily be fabricated using AM technologies.

TITANIUM AND ITS ALLOYS

Titanium is an impure oxide that was first found by William Gregor in 1791. Mathew Albert Hunter in 1910 was able to produce titanium metal after processing titanium tetrachloride and sodium. And by 1932, Wilhelm Kroll produced large amount of titanium from titanium tetrachloride and calcium and by 1948, Dupont Company became the first company to commercially produce titanium (Peters et al., 2003b). Titanium is one of the most abundant and one of the lightest metals known to man (Balla et al., 2016). Titanium has

Figure 3. Schematic view of laser metal deposition process
Source: (Akinlabi & Akinlabi, 2016)

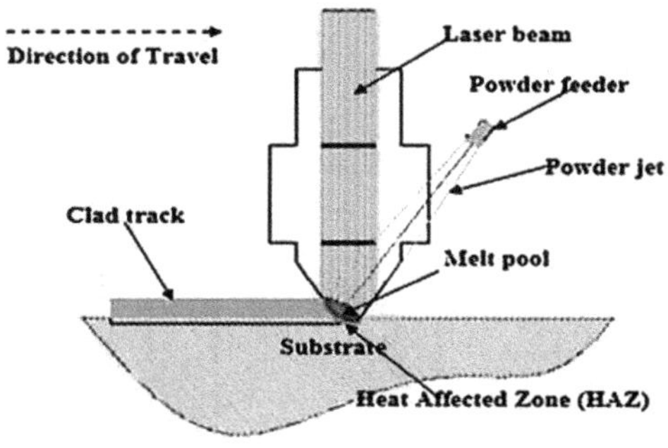

low density, good strength, high resistant to corrosion and good resistance to fracture. At high temperature, alloys of titanium are highly reactive with atmospheric gases such as nitrogen, oxygen and hydrogen (Nurul Amin & Shah Alam, 2012).

Titanium aluminide is an intermetallic compound classified under high temperature structural materials with superior properties which make it highly applicable in automotive, aircraft engines and gas turbines (Lapin, 2009). The unique properties of such intermetallic compound have been known to include: high strength to low weight ratio and good corrosion resistance which is the main reason why they are mostly needed in the aerospace, chemical and medical industries.

Titanium alloying process by adding other elements (such as Mo, Vn, Sn, Nb, Cr, etc.) are done as a way to further improve the properties of the titanium alloy. This action is seen when element such as tin are added to produce Ti-5Al-2.5Sn specifically for high temperature applications. Also molybdenum, a β-stabilizer have been added to titanium alloy to produce Ti-7Al-4Mo (α+β) alloy suitable for high strength applications. Other alloys that have been developed include Ti-13V-11Cr-3Al (β titanium alloy), Ti-6Al-4V (α+β) alloy. Ti-6Al-4V alloy possesses superior properties and despite being so expensive, still remains the most preferable titanium alloy. The high cost of titanium alloy is one of the factors responsible for creating a manufacturing process that will reduce and made rapid prototyping (additive manufacturing) technique a viable alternative rout (Yvonni-Effrosyni, 2014). Silicon has also been introduced (as a creep resistant enhancer) to titanium alloy to form Ti-4Al-4Mo-2Sn-0.5Si (Lutjering & Williams, 2003). Other alloys that have been developed over the time include Ti-6Al-7Nb, Ti-5Al-2.5Fe and Ti-48Al-2Cr-2Nb. It is possible to achieve very high strength of titanium alloys at higher temperatures but the high oxidation behavior at maximum temperature application process makes it difficult to achieve, which is why intense research are still on in the development of other titanium alloys (Peters et al., 2003a).

RESEARCH WORKS ON LASER METAL DEPOSITION (LMD) PROCESS

As earlier discussed, the laser metal deposition (LMD) process is one of the techniques of additive manufacturing that is used to clad, repair, add new feature or produce an entirely new part. The technique uses laser to create melt pool on to which metal powder is deposited. Several researches are being carried out in the deposition process using this technique because of its numerous advantages.

Akinlabi and Akinlabi, (2016) did a research using LMD process to deposit aluminum powder on titanium substrate. The depositions were done at a constant laser power of 1 kW, gas flow rate of 1.5 l/min and scanning speed varied from 0.5 m/min to 3 m/min. The research revealed that at a lower scanning speed, alpha phase grains microstructures were observed. While at higher scanning speed, beta phase grains were noted. It was however noted that the laser-material interaction lead to changes in the geometrical properties (like height, width and heat affected zone) of deposited samples. The research summarized that increase in scanning speed lead to decrease in the geometrical properties and also lead to increase in the microhardness and corrosion rates of deposited samples.

In another research carried out by Yvonni-Effrosyni, (2014), the direct laser metal deposition technique was used to study the effect of deposition parameters on the mechanical properties and microstructure of deposited samples. Ti6Al4V was deposited at 500 W laser power and other samples re-melted at 600 W at scanning speed of 200 and 400 mm/min. The re-melted samples revealed homogenous distribution of dendrites and the samples deposited at 200mm/min has better surface roughness and hardness.

Effect of deposition parameters in the deposition of pure titanium powder using laser net-engineered shaping (LENS) process was studies by Hu et al., (2016). Laser power, scanning speed and powder feed rates were varied for the deposition sixteen layers of for each sample where first layers where taken as substrate on which other layers were deposited. The outcome of the work revealed that the height of deposits increases with respect to increase in laser power, decrease in scanning speed and increase in powder flow rate. The hardness of the deposited samples increases with respect to increase in laser power and decrease in scanning speed. Also it was noted that increase in powder feed rate from 0.5 to 1 rpm resulted to a sharp drop in hardness.

In the work of Yan et al., (2017), functionally graded material (FGM) was produced using LMD process. Ti-48Al-2Cr-2Nb powder was deposited on pure titanium substrate. The deposition process was carried out by varying the incident energy input which is mostly responsible for crack formations in deposited samples if not properly controlled (Balla et al., 2016). The deposition process uses high energy input at the beginning and later reduced to avoid overheating. Scanning speed was kept at 600 mm/min while the laser power was varied. Primary phases of α, α_2 and γ were discovered within the alloy gradient. Basket-weave microstructures were noted at the region where the energy input was higher and lamella structure was seen at the final region where the energy input has been reduced. Fine grain structures were noted at the top due to increase in cooling rate from the top to the bottom of deposits which account for the higher ultimate tensile strength achieved. Hardness slightly increased as the percentage weight of Ti4822 increased in the region and led to material brittleness.

DEPOSITION PARAMETERS IN LMD PROCESS

Deposition parameters or sometime referred to as processing parameters in deposition processes play paramount role in the quality and properties of fabricated parts. Proper understanding of the interaction of between the parameters will ensure the right selection of deposition parameters needed in the deposition processes (Bayode et al., 2018). As such, careful selection and control of deposition parameters is paramount in ensuring that high quality parts are produced. Most commonly employed processing parameters in deposition processes include the spot size of laser, laser power, and laser scanning speed and powder flow rate. These parameters are briefly discussed below.

Spot Size of Laser

The laser spot size plays a critical role in the laser deposition process. This spot size is determined by the laser beam diameter which can be used to influence the laser beam concentration on any particular spot. The relationship between the spot size and beam intensity is inversely proportional. As such, the smaller the spot size, the higher the beam intensity.

Laser Power

Laser power is another vital process parameter that plays an incredible role in deposition process. This is because it is not only responsible for the creation of melt pool but also play important role in the quality and properties of deposited parts as it was clearly demonstrated in literatures (Sharman et al., 2018; Yan et al., 2017; Yvonni-Effrosyni, 2014). Some of the work that have been carried out clearly demonstrated that laser power apart from having the ability to reduce cracks in fabricated parts also play tremendous role in microstructural and mechanical properties of parts. This is because laser power plays a role in laser-material interaction and rate of cooling in the solidification process.

Laser Scanning Speed

Laser scanning speed is another deposition parameter that have been proven to also have a great influence in the deposition process and in determining the quality and property of deposited parts. Laser scanning speed is a parameter used in describing the speed at which the laser moves along a desired path. Just like laser power, laser scanning speed has a relationship with the laser material interaction and rate of cooling in the solidification process. Sobiyi et al., (2017) looked at the influence of laser scanning speed on titanium and titanium carbide metal powders on titanium alloy substrate. The outcome revealed that the heat affected zone (HAZ) increases with increase in scanning speed and that the height of deposits also varied with the change in laser scanning speed. It was also found that the microhardness of samples increases with increase in laser scanning speed. However, the conclusion reached that HAZ increased with increase in scanning speed might not be the case, because increase in scanning speed means the reduction in energy density and will results to reduction in laser material interaction time. Therefore, it is logical to think that there will be reduction in HAZ as the scanning speed increases.

It is however good to note that the laser spot size, laser power and scanning speed do have a relationship that connect them together. This is because the three parameters central around energy input involve in the deposition process. The interaction between these parameters gives what is referred to energy density. This energy density is the total incident energy input per unit area (Mazumder et al., 1999) and calculated with the equation below:

$$Energy\ density\left(E\right) = \frac{P}{VD}$$

where the energy density *(E)* is in J/mm^2, laser power *(P)* in watt (W), scanning speed *(V)* in mm/s and spot size dimeter *(D)* in mm. The energy density therefore becomes a very crucial factor and its influence in deposition process has well been established in literatures (Hu et al., 2016; Tang et al., 2007; Zyl et al., 2016). The influence of energy density was also noted by Liu & Dupont, (2004), that the increase in laser beam incident energy will result in crack frequency decrease in deposited parts. This is because the cooling rate is inversely proportional to laser heat input (Kou, 1987). The cooling rate of any material in deposition process is largely dependent on laser heat input.

Powder Flow Rate

Powder flow rate referred to the amount of powder flowing out to the deposition zone per unit time. It normally carries the unit g/min and sometimes revolution per minute (rpm). The flow ability of any metal powder mainly revolves around its morphology and particle size. Spherical shaped metal powder is believed to be most preferable for laser metal deposition process because of its ability to react with laser beam better (Schade et al., 2014). Some literatures have discussed on the influence of powder flow rate in deposition processes. Powder flow rate have been proven to have an effect on material efficiency and dimensional accuracy of laser deposited parts (Kumar et al., 2014; Schade et al., 2014). Powder flow rate have also been investigated to affect the resistance of deposited parts (Saboori et al., 2017; Shukla et al., 2012). Other property such as surface roughness can also be affected due to powder flow rate as noted by Shah et al., (2010).

ESSENCE OF PREHEATING IN LMD PROCESS

Different researchers continue to stress the importance of preheating especially in the LMD process of titanium aluminide parts. This is because LMD process involves rapid heating and cooling which sometimes resulted to residual stresses in fabricated parts and made them highly susceptible to cracks. To prevent or reduce cracks in LMD processes, careful control of process parameters and preheating are mostly recommended. As such, several works that have been carried out on laser deposition additive manufacturing of titanium aluminide all agreed that it is quite impossible to fabricate parts that are crack-free, unless additional heating system is provided which may be in the form of an induction coil or a heating bed to control the cooling rate (Sharman et al., 2018). In the study carried out by Liu & Dupont, (2004), it was discovered that thermal cracking can only be reduced by increasing incident laser energy through the use of higher laser power and lowering the laser scanning speed. This action only reduces the thermal cracking and to completely eliminate cracking, it was recommended that additional heating system should be provided to preheat the substrate to 450^{0}C before commencing the deposition.

EXPERIMENTAL CASE STUDY OF LMD OF TITANIUM ALUMINIDE (TI-4822-4)

There are currently few studies on the fabrication of titanium aluminide (Ti-48Al-2Cr-2Nb or simply known as Ti-4822-4 according to manufacturer's label). These limited studies currently show limitations in its usage as more still need to be known in terms of its properties and suitability in different applications. Most of the studies are geared to finding optimum processing parameters, studying the emerging properties and fabricating crack-free parts. This is because titanium aluminides are crack sensitive materials (Brueckner et al., 2015).

Studies done by Weishiet et al., (2000) on TiAl revealed that lowering the power density and laser scanning speed only reduced the cracking and did not totally eliminate it and that cracking could only be prevented through preheating of substrate to about 400^{0}C. The work of Srivastava et al., (2000) revealed that the continuous increase in the built height of TiAl strips keeps led to increase in length and frequency of cracks which is often linked to residual stress produced by high thermal gradient. As such, different attempt made researchers to conclude that it is impossible to produce TiAl parts that are crack-free except when an additional heating system is provided in the laser deposition process (Sharman et al., 2018).

In the experimental case study that is about to be discussed, TiAl powder has been deposited on preheated pure titanium substrate (CP-Ti) via laser deposition process called laser engineered net shaping (LENS) technique. Studies on quality of deposits, microstructure and microhardness have been carried out.

MATERIALS AND METHODS

In this study, spherical shaped TiAl powder of particle size range of 45 to 150 μm gotten from Praxair surface technologies USA was deposited on CP-Ti (of size 10 x 10 x 6 mm) substrate. LENS 850R laser machine at Council for Scientific and Industrial Research (CSIR), Pretoria, South Africa have been used for the deposition process. The elemental composition of the metal powder used for the deposition is shown in Table 1. The substrate was first sand blasted and cleaned with acetone to remove the oil deposit and other impurities that may be on it after which the substrate was preheated to 450^0C before the deposition was started using a spot size of 1.4 mm. Deposits of five layers were produced according to parameters highlighted in Table 2 while the powder flow rate was kept constant at 2.8 g/min.

After the depositions, a digital Vernier caliper was used to measure the heights of each deposit, to determine the relationship between deposition parameters and height of deposits. The deposits were later section perpendicularly and prepared using standard metallographic procedure for titanium alloy (Taylor, 2015). A Tescan scanning electron microscope was used to analyze the microstructure of deposited samples. The microhardness of deposits was determined using a Vickers microhardness tester. Microhardness

Table 1. Metal powder elemental composition

Aluminum	Chromium	Niobium	Other elements, Total	Titanium
34	2.6	4.8	<0.10	Balance

Table 2. Deposition parameters

Sample designation	Laser power (KW)	Scanning speed (mm/s)	Energy input (J/mm^2)
A	0.40	3.17	90
B	0.45	3.17	101
C	0.40	2.65	108
D	0.45	2.65	122

measurements were taken by applying a load of 500 g over a dwelling time of 15 s to specimens from top to bottom at 0.5 mm interval.

RESULTS AND DISCUSSION

After the deposition process, the physical examination of the deposits with naked eye, revealed the presence of micro-cracks on deposited sample A (as shown in Figure 4a) and C. After the samples were examined under the scanning electron microscope (SEM), the result revealed that increase in energy input (increase in laser power from sample A to B and reduction in scanning speed from sample C to D) led to tremendous decrease in micro-cracks. Table 3 gives the deposits height and microhardness of deposited samples.

From Table 3, it is clear that increase in laser power or reduction in scanning speed in the deposition process influenced the outcome of heights and microhardness of deposited samples. Increase in laser power form 0.4 W to 0.45 W and reduction in scanning speed from 3.17 mm/s to 2.65 mm/s resulted in a corresponding increase in deposits height. Also, reduction in microhardness was witnessed when the laser power was increased from 0.4 W to 0.45 W at scanning speed of 3.17 mm/s. The relationship between deposition parameter and height and microhardness of deposits can be linked to the laser material interaction and rate of cooling experienced in the deposited samples.

Table 3. Height and microhardness of deposits

Sample designation	Deposit height (mm)	Deposit microhardness (Hv)
A	1.6	565
B	2.1	536
C	2.3	550
D	2.7	560

Figure 4. Shows the (a) Cross sectional image of sample A (b) Phase transition in sample B (c) microstructure of sample A and (d) microstructure of sample B

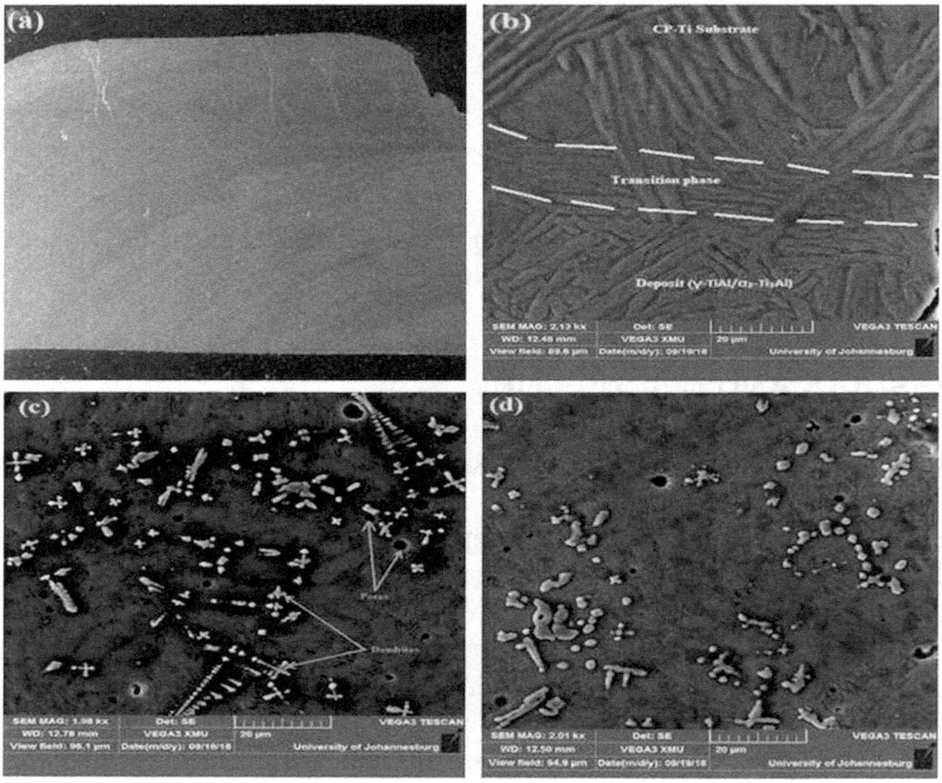

Figure 4b shows a transition phase of sample B characterized by needle-like (columnar) grains and witnessed across all the deposited samples. The transition layer observed between the substrate and the deposit show a good bonding between the materials. The SEM result show that the samples (as shown in Figure 4c and 4d) are characterized with dendrites and pores which reduce as the laser power increases and scanning speed reduces. Considerably, a homogenous microstructure comprising of lamellar γ-TiAl and α_2-Ti_3Al regions were obtainable in the deposits.

CONCLUSION

Laser metal deposition (LMD) is an additive manufacturing technique that has proven to be very effective in the production and repair of parts. However, the parts formed through the technique are prone to cracks to due to residual stresses. Cracking in deposited samples can be reduced or eliminated by careful selection of processing parameters and provision of additional heating system. This chapter has clearly looked at some of the AM processes with more attention on laser metal deposition (LMD), effect of LMD deposition parameters and preheating on the physical, microstructural and microhardness of components produced through the process. The outcome of a case study revealed that increase in input energy and preheating of substrate can hugely reduce cracking. The outcome also revealed that process parameters also have a great influence in the microstructural and mechanical properties of deposited parts.

REFERENCES

Abdulrahman, K. O., Akinlabi, E. T., & Mahamood, R. M. (2018). Manufacturing of aluminium composite materials: A review. In K. Kumar & P. Davim (Eds.), *Hierarchical composite materials: Materials, Manufacturing and Engineering*. Berlin: Witer de Gruyter. doi:10.1515/9783110545104-002

Akinlabi, E. T., & Akinlabi, S. A. (2016). Advanced Coating : Laser Metal Deposition of Aluminium Powder on Titanium Substrate. In *Proceedings of the World Congress on Engineering* (*Vol. 2*). London: Academic Press.

Aliakbari, M. (2012). *Additive Manufacturing: State-of-the-Art, Capabilities, and Sample Applications with Cost Analysis*. KTH. Retrieved from https://www.diva-portal.org/smash/get/diva2:560827/FULLTEXT02.pdf

Balla, V. K., Das, M., Mohammad, A., & Al-Ahmari, A. M. (2016). Additive manufacturing of γ-TiAl: Processing, Microstructure, and Properties. *Advanced Engineering Materials*, *18*(7), 1208–1215. doi:10.1002/adem.201500588

Bayode, A., Akinlabi, E. T., & Pityana, S. (2018). Fabrication of stainless steel-based FGM by laser metal deposition. In K. Kumar & P. Davim (Eds.), *Hierarchical composite materials: Materials, Manufacturing and Engineering*. Berlin: Walter de Gruyter.

Bremen, S., Meiners, W., & Diatlov, A. (2012). Selective Laser Melting A manufacturing technology for the future? *Laser-Technik-Journal*. doi:10.1002/latj.201290018

Brueckner, F., Finaske, T., Willner, R., Seidel, A., Nowotny, S., Leyens, C., & Beyer, E. (2015). Laser additive manufacturing with crack-sensitive materials. *Laser-Technik-Journal*, *12*(2), 28–30. doi:10.1002/latj.201500015

Dinda, G. P., Dasgupta, A. K., & Mazumder, J. (2009). Laser aided direct metal deposition of iconel 625 superalloy: Microstructural evolution and thermal stability. *Materials Science and Engineering A*, *509*(1), 98–104. doi:10.1016/j.msea.2009.01.009

Herderick, E. (2011). Additive Manufacturing of Metals : A Review. *Materials Science and Technology*, (176252), 1413–1425.

Hu, Y., Wang, H., Ning, F., & Cong, W. (2016). Laser Engineered Net Shaping of Commercially Pure Titanium : Effects of Fabricating Variables. *Proceedings of the ASME 2016 International Manufacturing Science and Engineering Conference*. 10.1115/MSEC2016-8812

Kobryn, P. A., Ontko, N. R., Perkins, L. P., & Tiley, J. S. (2006). Additive Manufacturing of Aerospace Alloys for Aircraft Structures. In *Meeting Proceedings RTO-AVT-139* (pp. 3-1-3–14). Neuilly-sur-Seine, France: Academic Press. Retrieved from http://www.rto.nato.int/abstracts.asp

Kou, S. (1987). *Welding Metallurgy*. John Wiley & Sons, Inc.

Kumar, S., & Choudhary, A. K. S., & Rakesh. (2014). Effect of the Process Parameters on Geometrical Characteristics of the Parts in Direct Metal Deposition: A Review. *International Journal of Mechanical Engineering and Technology*, *5*(4), 116–122.

Lapin, J. (2009). TiAl-based alloys: Present status and future. *Hradec Nad Moravici*, 1–12. Retrieved from http://metal2013.tanger.cz/files/proceedings/metal_09/Lists/Papers/077.pdf

Liu, W., & Dupont, J. N. (2004). Fabrication of Carbide-Particle-Reinforced Titanium Aluminide – Matrix Composites by Laser-Engineered Net Shaping. *Metallurgical and Materials Transactions. A, Physical Metallurgy and Materials Science*, *35A*(March), 1133–1140. doi:10.100711661-004-0039-2

Lutjering, G., & Williams, J. C. (2003). *Titanium*. New York: Springer-Verlag. doi:10.1007/978-3-540-71398-2

Mahamood, R. M., & Akinlabi, E. T. (2017). *Functionally graded materials* (C. P. Bergmann, Ed.). Springer Nature. doi:10.1007/978-3-319-53756-6

Mazumder, J., Schifferer, A., & Choi, J. (1999). Direct materials deposition: Designed macro and microstructure. *Materials Research Innovations*, *3*(3), 118–131. doi:10.1007100190050137

Nurul Amin, A. K. M., & Shah Alam, M. (2012). *High Speed Machining of Titanium Alloy, Ti6Al4V: to Achieve Nano Level Surface Roughness*. LAP Lambert Academic Publishing.

Peters, M., Hemptenmacher, J., Kumpfert, J., & Leyens, C. (2003a). Structure and Properties of Titanium Alloys. In C. Leyens & M. Peters (Eds.), *Titanium and Titanium Alloys, Fundamentals and Applications*. Weinheim: WILEY-VCH. doi:10.1002/3527602119.ch1

Peters, M., Hemptenmacher, J., Kumpfert, J., & Leyens, C. (2003b). Structure and Properties of Titanium and Titanium Alloys. In C. Leyens & M. Peters (Eds.), *Titanium and Titanium Alloys, Fundamentals and Applications*. Weinheim: WILEY-VCH. doi:10.1002/3527602119.ch1

Saboori, A., Gallo, D., Biamino, S., Fino, P., & Lombardi, M. (2017). An Overview of Additive Manufacturing of Titanium Components by Directed Energy Deposition: Microstructure and Mechanical Properties. *Applied Sciences*, *7*(9), 883. doi:10.3390/app7090883

Schade, C. T., Murphy, T. F., & Walton, C. (2014). Development of atomized powders for additive manufacturing. *World Congress on Powder Metallurgy and Particulate Materials,* 215–226. Retrieved from https://pdfs.semanticscholar.org/164c/56f1e3a3e525162cc28a65d975e2dd39e357.pdf

Shah, K., Pinkerton, A. J., Salman, A., & Li, L. (2010). Effects of Melt Pool Variables and Process Parameters in Laser Direct Metal Deposition of Aerospace Alloys. *Materials and Manufacturing Processes*, *25*(12), 1372–1380. doi:10.1080/10426914.2010.480999

Sharman, A. R. C., Hughes, J. I., & Ridgway, K. (2018). Characterisation of titanium aluminide components manufactured by laser metal deposition. *Intermetallics*, *93*, 89–92. doi:10.1016/j.intermet.2017.11.013

Shukla, M., Mahamood, R. M., Akinlabi, E. T., & Pityana, S. (2012). Effect of Laser Power and Powder Flow Rate on Properties of Laser Metal Deposited Ti6Al4V. *World Academy of Science and Technology*, *6*, 44–48.

Singh, S. C., Zeng, H., Guo, C., & Cai, W. (2012). Lasers: Fundamentals, Types, and Operations. In S. C. Singh, H. Zeng, C. Guo, & W. Cai (Eds.), *Nanomaterials: Processing and Characterization with Lasers*. Wiley-VCH Verlag GmbH & Co. KGaA. Retrieved from https://application.wiley-vch.de/books/sample/3527327150_c01.pdf

Sobiyi, K., Akinlabi, E., & Akinlabi, S. (2017). The influence of scanning speed on the laser metal deposition of Ti / TiC powders. *Materials Technology*, *51*(2), 345–351.

Srivastava, D. U., Chang, I. T. H., & Loretto, M. H. (2000). The optimisation of processing parameters and characterisation of microstructure of direct laser fabricated TiAl alloy components. *Materials & Design*, *21*(4), 425–433. doi:10.1016/S0261-3069(99)00091-6

Svensson, M., Sabbadini, S., Pelissero, F., Gennaro, P., Filippini, M., Beretta, S., … Badini, C. (2010). Additive Manufacturing of Gamma Titanium Aluminide Parts by Electron Beam Melting (EBM ®). In Titanium 2010, Orlando, FL.

Tang, L., Ruan, J., Landers, R. G., & Liou, F. (2007). *Variable Powder Flow Rate Control in Laser Metal Deposition Processes*. Retrieved from https://sffsymposium.engr.utexas.edu/Manuscripts/2007/2007-03-Tang.pdf

Taylor, B. (2015). *Metallographic preparation of titanium Application Notes*. Struers Ltd. Retrieved from http://cdnstruersproduction.azureedge.net/-/media/Struers-media-library/Materials/Application-reports/Application_Note_Titanium_2015_ENG.pdf

Tlotleng, M., Masina, B., & Pityana, S. (2016). Characteristics of laser In-situ alloyed titanium aluminides coatings. *Procedia Manufacturing*, *7*, 39–45. doi:10.1016/j.promfg.2016.12.013

Weishiet, A., Mordike, B. L., Smarsly, W., & Richter, K.-H. (2000). Laser surface remelting and laser surface gas alloying of an intermetallic TiAl alloy. *Lasers in Engineering*, *10*, 63–81.

Yan, L., Chen, X., Zhang, Y., Newkirk, J. W., & Liou, F. (2017). Fabrication of functionally graded Ti and γ-TiAl by laser metal deposition. *JOM: The Materials. Metals & Materials Society*, *69*(12), 2756–2761. doi:10.100711837-017-2582-5

Yvonni-Effrosyni, D. (2014). *Direct Laser Metal Deposition- [DLMD] of Titanium Matrix Composites : Analysis of Microstructure and Mechanical*. National Technical University of Athens. Retrieved from https://higherlogicdownload.s3.amazonaws.com/SNAME/a09ed13c-b8c0-4897-9e87-eb86f500359b/UploadedImages/DamianidouY_SNAME_ThesisPresentation_23jan2014 (no videos)_(EN).pdf

Zyl, I., Van, & Yadroitsava, I., & Yadroitsev. (2016). Residual stress in ti6al4v objects produced by direct metal laser sintering. *South African Journal of Industrial Engineering*, *27*(December), 134–141.

Section 4
Optimization

Chapter 9
Optimization of Additive Manufacturing for Layer Sticking and Dimensional Accuracy

Syed Faiz Ali
Abdullah Gül University, Turkey

Fasih Munir Malik
Abdullah Gül University, Turkey

Emin Faruk Kececi
Abdullah Gül University, Turkey

Burak Bal
Abdullah Gül University, Turkey

ABSTRACT

When the 3D printing process is considered, there are also other parameters, such as nozzle size, flow rate of material, print-speed, print-bed temperature, cooling rate, and pattern of printing. There are also dependencies that will be addressed in between these parameters; for example, if the printing temperature is increased, it is not clear if the viscosity of the material will increase or decrease. This chapter aims to explain the effect of printing temperature on layer sticking while dimensional accuracy is achieved. Theoretical modelling and experimental testing will be performed to prove the relationship. This type of formulation can be later adapted into a slicer program, so that the program automatically selects some of the printing parameters to achieve desired dimensional accuracy and layer sticking.

DOI: 10.4018/978-1-5225-9167-2.ch009

INTRODUCTION

Additive manufacturing is gaining importance in mechanical design processes since it allows an easy way of prototyping. However, even though 3D printers are becoming more common and easy to use, the relationship in between the printing parameters and the part properties is not clearly known.

In this chapter, the strength of the part is defined as the sticking of the layers to each other. The effect of the printing parameters, namely print speed, print temperature, fan speed and layer height, to the layer sticking is experimentally tested.

3D printing technology has been defined as to have transformative implications (Jong & Bruijn, 2013). In the current literature there are several works on this topic. Some of the hot studies are fabrication procedures in rapid prototyping (Suresh & Narayana, 2017), surface characterization (Boschetto, 2017), laser additive manufacturing (Mahamood & Akinlabi, 2017), laser metal deposition process (Mahamood, 2017) and production of fully functional 3-D printed components (Kocovic, 2017). There are studies on ABS material and in these studies the results of injection molded parts with FDM printed parts were compared. The injection molding showed higher resistance to the impact, hardness, and tensile fractures than the 3D printed parts did. Moreover, their study also includes the 3D printed parts printed at different layer heights. It is also experimentally concluded that the decrease in layer height resulted in parts with higher strength (Pritish et al., 2016). Moreover, the build parameters of low print speed and small layer height resulted in the most resistance to flexural and tensile fractures (Christiyan et al., 2016). This is explained by better layer sticking under low print speed and small layer height. The increase of nozzle head temperature resulted in higher tensile fracture resistance (Behzadnasab & Yousefi, 2016). It is also proved that increasing the layer thickness decreases the tensile strength (Vaezi & Chua, 2011).

The effect of build orientation, layer thickness and feed rate on mechanical properties has been studied in (Chacon et al., 20170 and proven that build orientation also plays a vital role on the mechanical properties of a 3D printed part.

The effect of cooling air speed is studied in (Lee and Liu, 2019) and the dimensional accuracy and the strength of the part is measured. It has been seen that increasing the cooling air speed decreases the dimensional accuracy but improves the part strength.

The effect of layer thickness on the strength of parts made with ABS and manufactured by fused deposition is studied in (Rankouhi et al., 2016). The research shows that decreased layer thickness increases the strength and this information should be taken into account for the manufacturing of the load bearing parts.

Binder jetting additive manufacturing process parameters such as saturation level, power level, drying time and spread speed have also been studied in order to find the effects of these parameters on strength and dimensional accuracy (Miyanaji et al., 2016). For Selective Laser Melting, a type of additive manufacturing method, parameters have been studied in (Hanzl et al., 2015) to measure the effect on the strength of the manufactured part. Laser power, scan speed, layer thickness, overlap rate and building direction were the parameters that have been studied.

This chapter focuses on the optimization of parameters that result in an optimum strength provided by inter-layer sticking of printed layers. An experimental approach is adopted, and obtained results are theoretically analyzed to find the reason for different behaviors exhibited by the specimens. The 3D printing mechanism adopted is Fused Deposition Modeling (FDM). The material used in experiments is PLA with 2.85mm diameter. The CURA slicing software is used and the specimen are printed by an Ultimaker GO2 3D printer. The following part of the chapter explains the experiments and the test results showing the effect of 4 different parameters on inter layer sticking.

LAYER STICKING OF 3D PRINTED PART

The strength of the 3D printed part is influenced by many factors. The most important is the mechanical properties of the material. However, there are other parametric influences that also function in deciding the strength of the part. In FDM 3D printing, the printer prints the part in layer-over-layer structure. Therefore, in a FDM printed part, the strength of the material also depends on the strength of the bonding between layers. This strength depends on the composition of the material and can only be changed at the time of manufacturing. However, the strength rendered by the layer-sticking or layer-bonding in the part can be altered through the different parametric controls provided by the slicing software. In this chapter the strength of the part due to the slicing parameters is experimentally measured. The strength of the printed part offered by the inter-layer sticking will simply be referred to as strength in the rest of the chapter.

As a slicing software, CURA offers a number of parameters to the user to choose from. The general purpose that a parametric section serves and its general effect on the printing is summarized in the Table 1 below.

Table 1 gives a general idea to the reader about the parameters offered by a slicing software. However, in the later part of this chapter specific parameters and their effect on inter layer sticking are explained.

The aim of this chapter is to establish experimentally the responsible parameters that cause a sensitive effect on the strength of the printed part. A 3 Point Bending Test (3PBT) was performed on different specimens to measure the respective force at fracture. Therefore, the experiment was modeled to be conducted on specimens printed with effective parameters. The effective parameters were then tested on their rational extreme values and the optimum in-between value. The parameters that exhibited effective impact were selected - Table 2. Each parameter was tested separately by keeping all the other parameters constant at recommended defined values by the software.

Effective Parameters

The parameters that are tested and have an effect on the printed parts' strength are: print speed, print temperature, fan speed and layer height. Each parameter was tested on 4 different values, and each test of a given value was repeated 3 times. The general overview of the effect of each effective

Table 1. Parameters of custom settings in CURA

Parametric Section	Purpose / Effect
Surface Quality	The layer height, layer thickness, and other parameters affect or may affect the quality of the printed part in terms of surface quality and surface smoothness. Layer height and layer thickness can be changed via the slicing software.
Shell	A shell is the wall's thickness, the outer wall's wipe distance, the outer wall's thickness, top/bottom print pattern, and other parameters that deal with the shell of the part. The shell can be roughly considered as the outer unit holding the inner unit, the infill.
Infill	The infill is the inner unit that fills the outer unit (shell). Infill density (up to 100%) and infill pattern are some prime options offered in this section.
Material	The material flow properties can be changed. The program offers the ability to set the material flow rate and material retraction rate at the initial position.
Speed	The user sets different speeds. The important ones are print speed and travel speed. Print speed can be adjusted from 10 mm/sec to 100 mm/sec, while travel speed can go up to 250 mm/sec.
Cooling	Cooling is achieved by the fan and its speed can be selected at the initial position or for minimum layer time. Fan speed ranges from 0 to 100%.

Table 2. Effective parameters and their general effects on the printed part

Effective Parameter	Effect on Printing
Print speed	This parameter deals with the amount of material deployed per unit time. The greater the print speed, the lesser the time to print is required. However, very high print speed may result in poor surface quality of the printed part. Print speed can be adjusted from 10 mm/s to 100 mm/s.
Print temperature	This parameter deals with the nozzle temperature from where the material is extruded. The melting temperature for PLA is 185 Celsius, and the optimum temperature defined by the material manufacturer is 195 to 240 Celsius. Print temperature affects the viscosity of the material.
Fan speed	This parameter deals with the cooling part of the system. The nozzle head and the printed part get heated up and there is a need of fans to exhaust the heat. The fan speed can be adjusted from 0 to 100%.
Layer height	This parameter deals with the height of each deployed layer. It varies from 0.04 to 0.32 mm. By changing the value from low to high, quality parts in terms of surface smoothness are printed. The layer height also affects the time of printing. A smaller layer height will result in more time required to print the part.

parameter is briefed below in Table 2 to give the reader a general overview of the effective parameters.

Specimen

The specimen selected to test the parametric influence on inter-layer sticking were all of the same design. The Figure 1 shows the dimensions of the specimen used in the experiments. Moreover, to ensure better accuracy all the specimens were printed on the same printer using the same material batch, and at the same positional augment. The positional coordinates were the center of the build plate. All the specimens were printed with 100% scaling in all 3 axes.

Experimental Details

All the materials were 3D printed with the dimensions explained above. In order to determine the effects of each different process parameter on the

Figure 1. Technical drawing of the specimen

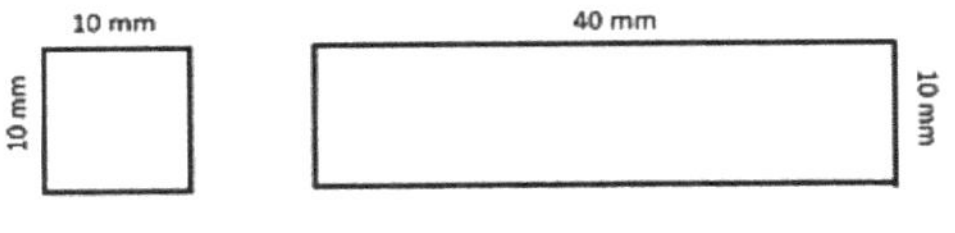

strength of the printed materials were tested. Standard 3 Point Bending Tests (3-PBT) were conducted using Shimadzu 3 point bending experimental setup. In 3PBT, the specimen is placed flat symmetrically on 2 edge-rests and the third edge applies the force at the center of the specimen. The tests were displacement controlled and the rate of change in the force per distance was set to 2N/mm for each specimen. After the tests, the force at the fracture was recorded for each specimen.

All specimens were printed with the same inner mesh pattern. The mesh pattern was coplanar and continuum. Therefore, it enabled the inter-layer fracture in order to obtain the strength of the part due to the layer sticking only. Each fractured specimen was observed and analyzed afterwards to ensure that the fracture occurred at the inter layer and at the middle of the part. It is essential to ensure that inter layer fracture takes place because if the fracture occurs at the intra layer then the material strength is also an active parameter in showing resistance to the failure. The Figure 2 shows the 3 Point Bending Test machine used in the experiment.

Results

This section of the chapter focuses, in particular, on the effect of each effective parameter on the strength of the printed part for inter layer sticking. In the following Figures from 6 to 9 and the equations from 1 to 4 formulate the current graphs.

Figure 2. 3 Point bending test a) Machine, b) Specimen during testing

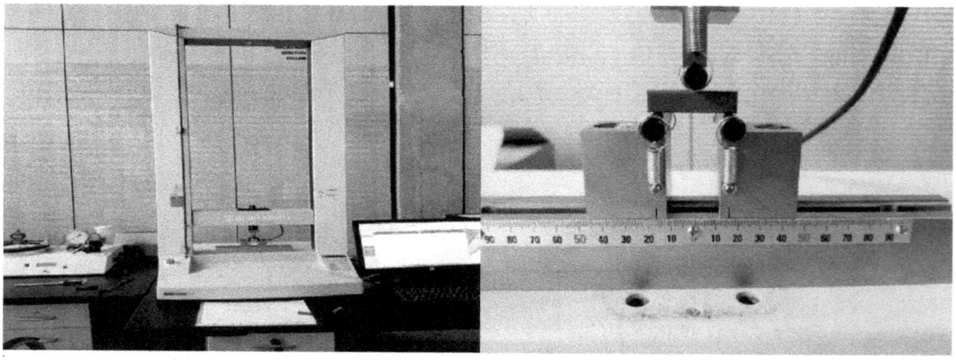

(a) (b)

Print Speed

As shown in Figure 3, the print speed is tested at 4 different values: 10, 35, 60 and 100 mm/s. A general trend can be discerned from the average values. When the print speed decreases, the strength of the part also decreases. However, in the middle values the specimen exhibited a higher resistance to fracture. The Figure 3 shows the correlation of the force at fracture vs print speed. A third order polynomial is also formulated as shown in the Equation 1 below:

$$F(P_s) = 0{,}0049\, P_s^{\,3} - 1{,}1247\, P_s^{\,2} + 74{,}229\, P_s - 265{,}77 \quad (1)$$

where P_s is the print speed and $F(P_s)$ is the force at fracture due to print speed.

When the print speed is slower than 40mm/sec, the previous layer dries when the new layer is being poured and this causes a loss in the connecting

Figure 3. Force vs. print speed

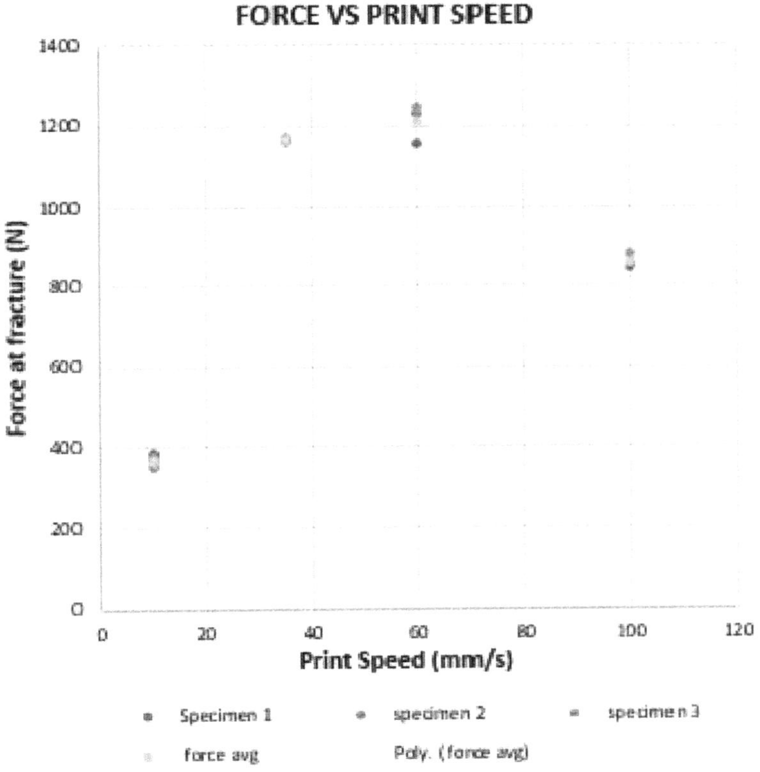

bonds. When the print speed is higher than 60mm/sec, the previous layer does not have enough time to solidify causing loss in the layer connection.

Print Temperature

As shown in Figure 4, the print temperature is tested at 4 different values: 195, 210, 225 and 240ºCelsius. Increase in the print temperature at first increases the strength and then decreases it. However, in the middle values the specimen exhibited an increasing resistance to fracture. A third order polynomial is also formulated as shown in Equation 2:

$$F(P_t) = -0{,}0437\ P_t^3 + 28{,}181\ P_t^2 - 6025{,}2\ P_t + 428528 \quad (2)$$

where P_t is the print speed and $F(P_t)$ is the force at fracture due to print temperature.

At lower print temperature the material dries too fast for the following layer to stick on it. Whereas at the higher print temperatures (greater than

Figure 4. Force vs. print temperature

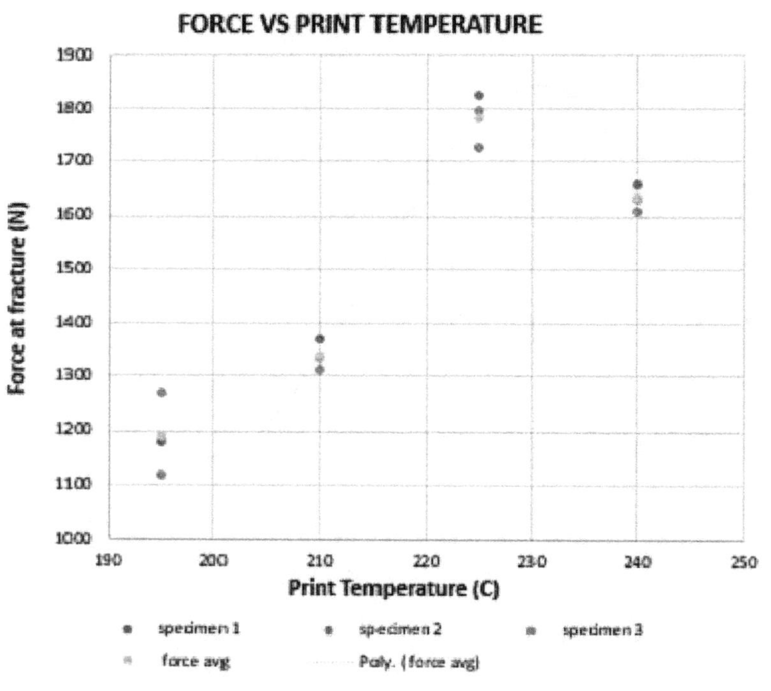

230° Celsius) the material does not have time to dry before the following layer is extruded.

Fan Speed

The 4 values that were chosen for fan speed were: 0, 25, 50 and 100%. Since the deployed layer could not dry before the next layer is printed over it, at 0% fan speed the part cannot be manufactured. Figure 5 shows test values starting from 25% and ending with 100%. Figure 5 and Equation 3 below represent the general trending behavior of fan speed on fracture force:

$$F(f_s) = 0{,}067\, f_s^2 - 11{,}306\, f_s + 1800{,}5 \qquad (3)$$

where f_s is the fan speed and $F(f_s)$ is the force at the time of fracture due to fan speed. Increasing the fan speed decreases the fracture force proportionally.

Layer Height

The layer height was tested at values of 0.05, 0.1, 0.2, and 0.3 mm. The general trend, as shown in Figure 6, is exponential decrease. However, in the middle range of 0.2 to 0.3 the values undergo a saturation. This means that the decrease in layer height results in a n increase in strength of the part. The relation in between the strength of the part and the layer height is given in Equation 4:

$$F(l_h) = 196169\, l_h^3 - 70216\, l_h^2 - 146{,}4\, l_h + 1728{,}6 \qquad (4)$$

where l_h is the layer height speed and $F(l_h)$ is the force at fracture due to layer height. It is observed that it is better to have a smaller layer height for higher fracture resistance.

By using the Equations from 1 to 4, which were found based on experimental findings, a user can find the corresponding force at fracture for any given values of process parameters. These parameters are namely; print speed, print temperature, fan speed and layer height. Therefore, a user can determine the force at fracture before printing the part and can prevent unexpected failures and improve the integrity of the manufactured parts.

Figure 5. Force vs. fan speed

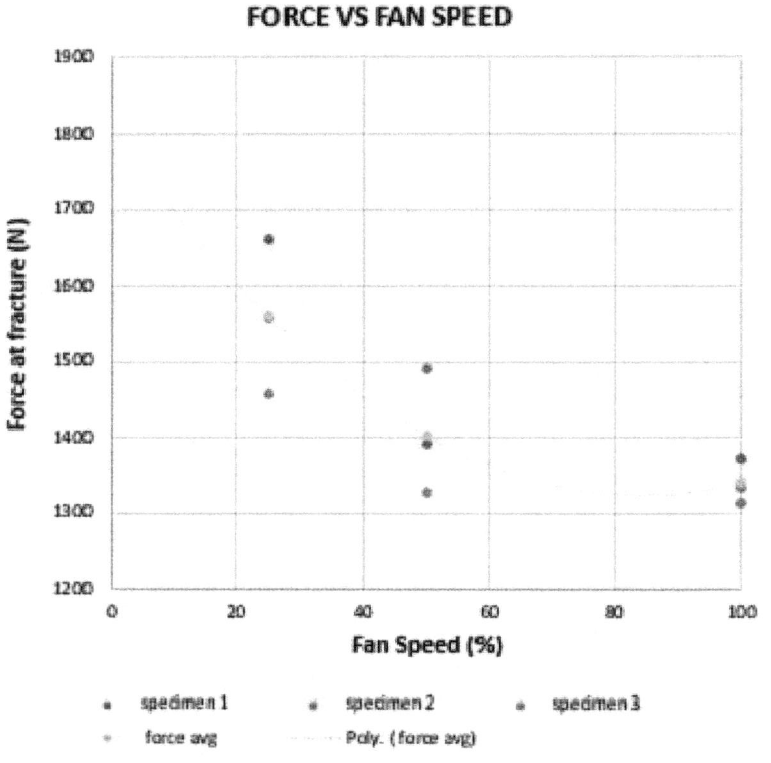

Dimensional Accuracy

The dimensional tolerance is defined as the degree of disagreement in terms of dimensions between the fed design and the manufactured product. In FDM 3D printing, the tolerance is dependent on the printer and the nozzle size of the printer, and the geometry and size of the part. The dimensional accuracy in this research was distributed over two sections.

1. The conventional tolerance error range of the printer.
2. The tolerance error caused by the effective parameters.

The smaller printers usually have less tolerance error compared to bigger printers. In the experiment, the GO 2 by Ultimaker was used which is the smallest printer in the Ultimaker 3D printer family. There was no conventional

Figure 6. Force vs. layer height

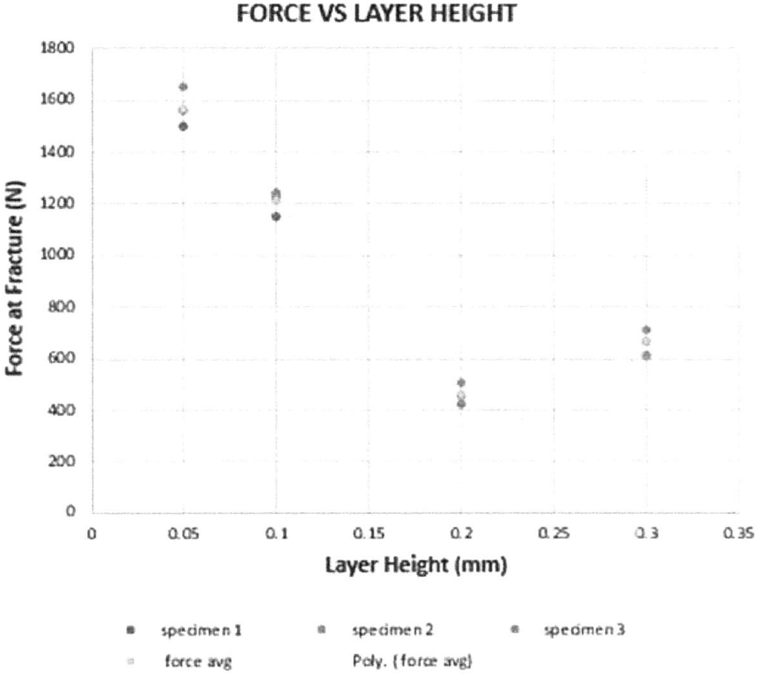

tolerance error found in printed specimen. This further implies that there was no tolerance error found caused by effective parameters. The size of the specimen was small enough to experience zero tolerance error.

CONCLUSION AND FUTURE WORK

3D printing is getting more common with the advancements of printing machines and materials. When the mechanical engineering design process is considered, the designer should know the properties of the material to be used at the time of designing the part. For conventional engineering materials such as aluminum or steel, properties such as yield strength, density or thermal resistance is very well known. However, there is not enough information in the literature about the physical properties of 3D printed parts. For the material, such as a given filament, properties are known but the properties of the manufactured part show different characteristics depending on the selection of the printing parameters. In this study layer sticking is considered as the strength of the part and the relations in between the 4 different printing parameters to strength are experimentally found. From this work the following conclusions can be drawn:

1. Print speed of 40 mm/sec to 60 mm/sec gives the best layer sticking result.
2. Print temperature of 230° Celsius causes better layer sticking.
3. Layer sticking is inversely proportional to the fan speed.
4. The smaller the layer height, the higher the layer sticking.

As future work, we believe that there is a need for a study to understand the effects of 3D printing parameters on the final product to form graphics like Ashby Charts. These charts will help the designers to choose 3D printing parameters more consciously.

REFERENCES

Behzadnasab, M., & Yousefi, A. (2016). Effects of 3D printer nozzle head temperature on the physical and mechanical properties of PLA based product. *12th International Seminar on Polymer Science and Technology*.

Boschetto, A. (2017). Surface Characterization in Fused Deposition Modeling. *3D Printing: Breakthroughs in Research and Practice*, 2-47. doi:10.4018/978-1-5225-1677-4.ch002

Chacon, J. M., Caminero, M. A., Garcia-Plaza, E., & Nunez, P. J. (2017). Additive manufacturing of PLA structures using fused deposition modelling: Effect of process parameters on mechanical properties and their optimal selection. *Materials & Design*, *124*, 143–157. doi:10.1016/j.matdes.2017.03.065

Christiyan, K. J., Chandrasekhar, U., & Venkateswarlu, K. (2016). A study on the influence of process parameters on the mechanical properties of 3D printed ABS composite. *Materials Science and Engineering*, *114*.

Hanzl, P., Zetek, M., Baksa, T., & Kroupa, T. (2015). The influence of processing parameters on the mechanical properties of SLM parts. *Procedia Engineering*, *100*, 1405–1413. doi:10.1016/j.proeng.2015.01.510

Jong, J. P., & Bruijn, E. D. (2013). *Innovation Lessons From 3-D Printing*. Retrieved from https://sloanreview.mit.edu/article/innovation-lessons-from-3-d-printing/

Kocovic, P. (2017). *From Modeling to 3D Printing, IGI Global. 3D Printing and Its Impact on the Production of Fully Functional Components: Emerging Research and Opportunities*. doi:10.4018/978-1-5225-2289-8.ch002

Lee, C. Y., & Liu, C. Y. (2019). The influence of forced-air cooling on a 3D printed PLA part manufactured by fused filament fabrication. *Additive Manufacturing*, *25*, 196–203. doi:10.1016/j.addma.2018.11.012

Mahamood, R.M. (2017). Laser Metal Deposition Process. *3D Printing: Breakthroughs in Research and Practice*, 172-182. doi:10.4018/978-1-5225-1677-4.ch009

Mahamood, R.M., & Akinlabi, E.T. (2017). Laser Additive Manufacturing. *3D Printing: Breakthroughs in Research and Practice*, 154-171. doi:10.4018/978-1-5225-1677-4.ch008

Miyanaji, H., Zhang, S., Lassell, A., Zandinejad, A. A., & Yang, L. (2016). Optimal process parameters for 3D printing of porcelain structures. *Procedia Manufacturing*, *5*, 870–887. doi:10.1016/j.promfg.2016.08.074

Pritish, S., Arnab, S., & Teg, C. (2016). The influence of layer thickness on mechanical properties of the 3D printed ABS polymer by fused deposition modeling. *Key Engineering Materials*, *706*, 63–67. doi:10.4028/www.scientific.net/KEM.706.63

Rankouhi, B., Javadpour, S., Delfanian, F., & Letcher, T. (2016). Failure analysis and mechanical characterization of 3D printed ABS with respect to layer thickness and orientation. *Journal of Failure Analysis and Prevention*, *16*(: 3), 467–481. doi:10.100711668-016-0113-2

Suresh, G., & Narayana, K. L. (2017). A Review on Fabricating Procedures in Rapid Prototyping. *3D Printing: Breakthroughs in Research and Practice*, 1-21. doi:10.4018/978-1-5225-1677-4.ch001

Vaezi, M., & Chua, C. K. (2011). Effects of layer thickness and binder saturation level parameters on 3D printing process. *International Journal of Advanced Manufacturing Technology*, *53*(1-4), 275–284. doi:10.100700170-010-2821-1

Chapter 10
Parameters Optimization of FDM for the Quality of Prototypes Using an Integrated MCDM Approach

Jagadish
https://orcid.org/0000-0001-5699-5674
National Institute of Technology Raipur, India

Sumit Bhowmik
https://orcid.org/0000-0002-7787-756X
National Institute of Technology Silchar, India

ABSTRACT

Fused deposition modeling (FDM) is one of the emerging rapid prototyping (RP) processes in additive manufacturing. FDM fabricates the quality prototype directly from the CAD data and is dependent on the various process parameters, hence optimization is essential. In the present chapter, process parameters of FDM process are analyzed using an integrated MCDM approach. The integrated MCDM approach consists of modified fuzzy with ANP methods. Experimentation is performed considering three process parameters, namely layer height, shell thickness, and fill density, and corresponding response parameters, namely ultimate tensile strength, dimensional accuracy, and manufacturing time are determined. Thereafter, optimization of FDM process parameters is done using proposed method. The result shows that exp.no-4 yields the optimal process parameters for FDM and provides optimal parameters as layer height of 0.08 mm, shell thickness of 2.0 mm and fill density of 100%. Also, optimal setting provides higher ultimate TS, good DA, and lesser MT as well as improving the performance and efficiency of FDM.

DOI: 10.4018/978-1-5225-9167-2.ch010

INTRODUCTION

Rapid prototyping (RP) or additive manufacturing (AM) is one of the important techniques to build a prototype for components using while product development. These components can be used in assemblies, product testing and tooling for the short or medium run production. There are various additive manufacturing processes are available which include selective laser sintering (SLS), stereo lithography (SLA), ink jet modeling (IJM), direct metal deposition (DMD), fused deposition modeling (FDM), laminated object manufacturing (LOM), solid ground curing (SGC) and 3D plotting etc. Among the various additive manufacturing processes mentioned above, fused deposition modeling (FDM) is most used additive manufacturing process for building prototypes because of the less time consuming and its ease of operation. Since there are wide applications are involved by using the FDM process, the quality of the prototypes and cost of the product development are become important factors. In many engineering applications, qualities of the prototype like surface finish, strength, dimensional accuracy are important factors. Since the properties of the fabricated products changing by varying the process parameters, the optimal combination of the process parameters is required for better quality products in both technological and economical view. So, the optimization of the FDM process parameters is required for quality product fabrication in less time. The variation of the quality and other functional properties of the fabricated products according to the usability can be done by using optimization.

The evolutionary approach, bacterial foraging technique, was used to predict the optimal parameter settings and also studied five important process parameters such as layer thickness, orientation, raster angle, raster width and air gap have been considered to study their effects on three responses viz., tensile, flexural and impact strength of the test specimen. The bacterial foraging technique is used to suggest theoretical combination of parameter settings to achieve good strength simultaneously for all responses (Panda et al., 2009). Another researcher (Mohamed et al., 2016) described the effects on build time, feedstock material consumption and dynamic flexural modulus using Q-response methodology were studied influence of critical FDM parameters are layer thickness, air gap, raster angle, build orientation, road width, and number of contours are studied. The results show that Q-optimal design is a very promising method in FDM process parameter optimization and also confirms the adequacy of the developed models.

An experimental study carried out to investigate the independent effect of each processing parameter on the mechanical properties and dimensional accuracy repeatability of FDM parts. This research approach utilizes a tensile test per ASTM D638 standards to obtain the mechanical properties of each fabricated sample. In addition, the research work provides a Finite Element Analysis (FEA) model for AM parts (Alafaghani et al., 2017). Many researchers in the past (Kumar et al. 2016; Raju et al., 2014; Lee et al., 2005; Nidagundi et al., 2015; and Basavaraj et al., 2017) have deliberated the experiments using Taguchi method and studied the effectiveness of response parameters using analysis of variance (ANOVA). The study carried out (Sood et al., 2010) five significant process constraints such as layer thickness, orientation, raster angle, raster width and air gap are measured and their influence on the reactions such as tensile, flexural and impact strength of test undergone. The validity of the models are tested using analysis of variance (ANOVA) and the concept of desirability function is used for maximizing all responses simultaneously. The experimental work on the cause of the main FDM process variable parameters namely, layer thickness (A), air gap (B), raster width (C), contour width (D), and raster orientation (E). The novel ABS-M30i biomedical material was used in this research work to build parts. Experiments were conducted using Taguchi's design of experiments with two levels for each factor. The results are analyzed statistically to determine the significant factors and their interactions (Kumar et al. 2014). Taguchi's L9 orthogonal array with one replication has been used as an experimental design and then analysis of variance (ANOVA) has been applied to determine the significant parameters and their contributions on response variables. The Taguchi method has been combined with grey relational analysis for multi-characteristic optimization and also, the results are validated with the help of confirmation experiments (Pradeep et al., 2018). Additive manufacturing processes are employed to create physical models from three-dimensional (3D) computer-aided design (CAD) math data. A model that generates an optimal solution (minimum material, minimum build time, etc.) needs to be developed by using the Genetic Algorithm approach (Villalpando et al. 2014). The Genetic algorithm based strategy is used to obtain optimum orientation of the parts for additive manufacturing process (Phatak & Pandee, 2012).

A few researchers have utilized grey Taguchi method to obtain optimize the process parameters for good dimensional accuracy (Anoop et al., 2009). Grey relational analysis and fuzzy technique of order preference by a similarity to ideal solution (fuzzy TOPSIS) methods are used for selection of rapid prototyping systems including both benefit and non-benefit criteria. It has been concluded that selective laser sintering (SLS 2500) is the most appropriate RP system for better dimensional accuracy and surface quality whereas 3-D printing (Z 402) is an appropriate RP system for better build time (Siba & Narayan, 2013).The effect of five process parameters such as layer thickness, part build orientation, raster angle, air gap, and raster width along with their interactions has been studied using Taguchi's L27 orthogonal array. It has been observed that optimal factor settings for each performance characteristic such as percentage change in length, width, thickness, and diameter are different. The results show overall dimensional accuracy is predicted using artificial neural network (Sood et al., 2014). Ananya et al., 2018, discusses micromachining on Electric Discharge Machining, its working principle, and problems associated with it. Solution to those problems is suggested with the addition of powder in dielectric fluid. The optimization of Material Removal Rate (MRR) is done with the help of ANN toolbox in MATLAB. An inference engine is developed to perform the inference operations on the rules for fuzzy prediction model based on Mamdani method. The predicted results are in good agreement with the values from the experimental data with average percentage error of less than 4.5 (Sahu et al., 2013). The study focused (Vinodh et al., 2011) on fuzzy analytic network process (fuzzy ANP) approach has been used for the supplier selection process. The results of the validation study indicated that the application of fuzzy ANP is practically feasible and adaptable in the contemporary industrial scenario. The researcher (Jagadish & Ray, 2014) proposed a novel method in multi-objective optimization on the basis of simple ratio analysis (MOOSRA). A case study of cutting fluid selection for gear hobbing process was presented to validate the proposed model. The obtained result using MOOSRA has been compared with Analytical Hierarchical Process (AHP) and Decision Making Framework (DMF). The result shows that Syntilo 9930c is optimal in comparison with other. A few researchers illustrated an overview of applications of some MCDM methods for optimization followed by detailed fundamental aspects of optimization issues in green manufacturing. The work proposed an integrated method consisting of AHP coupled with MOORA and validated through an experimental case study carried out (Bhowmik & Jagadish, 2017) and AHP (Borah & Jagadish, 2018) have been used for the optimization.

The above literature review suggests that the optimal combination of process parameters results the better quality and functional properties of the products. Also, as we calculate the different combination of the criteria / response parameters like strength, surface finish, process time, accuracy of the products and process energy the problem become more complex. The above optimization methods have not considered inter-relationship of the criteria while fabrication using AM techniques. The complexity of the combination of criteria / response parameters occurs because of the degree of requirement of certain criteria / response parameter while fabrication of the products. For example, there is a need of good dimensional accuracy while fabricating the prototype for precision measuring instrument and poor dimensional accuracy is sufficient for the prototypes use for daily needs etc. Likewise, as we consider each criteria / response parameter while fabricating the products in FDM there is a necessity of flexibility in decision making. So multi-criteria decision making is required for optimizing the process parameters for better outcome of the results. In the present study a hybrid multi-criteria decision-making technique is utilized to overcome the above limitations referred from above literature.

The chapter sections are organized as follows: Proposed Methodology, Experimental Details, Testing and Analysis of the Fabricated Samples, Modeling of the FDM Process Using Modified Fuzzy Analytical Network Process, Results and Discussion, and Conclusion.

PROPOSED METHODOLOGY

In this section, modified Fuzzy-ANP (M-FANP) for optimizing the process parameters of FDM process for better quality product fabrication is discussed. Here, modified Fuzzy theory is used to define the flexibility of criteria weights with the help of linguistic variables and to handle vagueness information while ANP is used optimization of AM process considering the inter-dependency relationship among the criteria / response parameters. The detailed steps of the proposed method are explained as follows:

Step 1: Evaluation of Pair-Wise Comparison Matrix

Optimization of AM process is starts with the formulation of pairwise comparison matrix. It includes no. of criteria's as process parameters and corresponding alternatives as response parameters. The pairwise comparison is done between the criteria and criteria, criteria and alternatives, and alternatives and alternatives using modified Fuzzy theory. The pairwise comparison between the *n* no. of criteria is determined using following equation.

$$W = \begin{bmatrix} 1 & a_{12} & \cdots & a_{1n} \\ a_{21} & 1 & \cdots & a_{2n} \\ \vdots & \vdots & \ddots & \vdots \\ a_{1n} & a_{n2} & \cdots & 1 \end{bmatrix} (\text{n=1, 2, 3....x}) \tag{1}$$

where, a_{12} represents relative significance of criteria 1 on criteria 2; a_{12} denotes the relative significance of criteria 2 on criteria 1; n is the number of criteria.

Step 2: Determination of Relative Fuzzy Weights

In this step, the relative fuzzy weights for each of the criteria are done using normalization process. Geometric mean method is employed for the normalization via geometric aggregation of the criteria's values [17] using Eq. (2). Then, aggregated fuzzy values for each of the criteria are converted in to best non-fuzzy aggregated values using Eq. (3). In the last step, relative weights of the criteria AM process is done using Eq. (4).

$$GA = (l_{ij} = \prod_{j=1}^{k} l_{ij}, m_{ij} = \prod_{j=1}^{k} m_{ij}, u_{ij} = \prod_{j=1}^{k} u_{ij} \tag{2}$$

$$BNP = \frac{\left[\left(c - a\right) + b - a\right]}{3} + a \tag{3}$$

$$w_i = \frac{BNP_i}{\sum BNP_i} (\text{For } i=1, 2, 3....n) \tag{4}$$

where, *l* is smallest likely value, *m* is most likely value and *u* is largest value of the fuzzy event. a, b, c represents smallest, most likely and largest possible fuzzy weights respectively. The BNP indicates best non-fuzzy aggregated value; w_i is the precise fuzzy weights for each of the criteria.

Step 3: Consistency Index Check

The calculation of the consistency index is required to know the pairwise comparison matrix whether accepted or not. Saaty [18] proposed the following Eq. (5) and Eq. (6) to measure the consistency of the reciprocal matrix called consistency index (CI) and constancy ratio (CR) respectively.

$$CI = \frac{\lambda_{\max} - n}{n - 1} \tag{5}$$

$$CR = \frac{CI}{RI} \tag{6}$$

where, CI represents the consistency index, n is order of the matrix and λ_{max} represents maximum Eigen value in the pairwise comparison matrix. The random index varying by the order of the matrix. As mentioned in the literature (Barzilai, 1997), the pairwise comparison matrix is accepted when CR < 0.1.

Step 4: Representation of the Comparison Matrix of Alternatives With the Criteria / Response Parameters

This step demonstrates the comparison of the alternatives with the criteria / response parameters by taking experimental data into consideration. The first approach is experimental data represented according to the respective criteria denoted in the following Eq. (7). After that, the normalization of the experimental data is done to make all elements in the comparison matrix dimensionless by using in Eq. (8). Second approach, the fuzzy decision matrix is demonstrated by considering the decision makers opinion with the help of linguistic variables as shown in Eq. (9). Further, the fuzzy weights are converted to non-fuzzy weights using Eq. (10). The defuzzified values i.e. crisp weights calculated by Eq. (10) used to compare the relative importance among the criteria / response parameter. Finally, the weighted normalized comparison matrix is calculated by using Eq. (11). The weighted normalized comparison matrix is further used in the super matrix for comparing the alternatives with respect to each criteria / response parameter.

$$M_{ij} = \begin{bmatrix} & C_1 & C_2 & \cdots & C_j \\ A_1 & x_{11} & x_{12} & \cdots & x_{1n} \\ A_2 & x_{21} & x_{22} & \cdots & x_{2n} \\ \vdots & \vdots & \cdots & \cdots & \vdots \\ A_i & x_{m1} & x_{m2} & \cdots & x_{mn} \end{bmatrix} (\text{For } i = 1,2,3,4..m \text{ , } j = 1,2,3,4...n \text{ }) \quad (7)$$

where M_{ij} is comparison matrix with respect to each ith alternative and j[th] criteria / response parameter, $A_1, A_2, A_3, \ldots A_i$ are representing alternatives, $C_1, C_2, C_3 \ldots C_j$ represents criteria / response parameters and $x_{11}, x_{12} \ldots\ldots\ldots x_{mn}$ represents performance values of the criteria / response parameters corresponding to each process parameter setting of the FDM process.

$$N_{ij} = \frac{C_i}{C_{i\max}} \text{ (For } i=1, 2, 3....n) \quad (8)$$

where N_{ij} represents the normalized decision matrix, C_i denotes the column of the comparison matrix of the respective criteria / response parameter with respect to corresponding experimental settings, C_{max} represents the maximum performance value of the corresponding column of the comparison matrix.

$$f = \begin{bmatrix} x_{11} & x_{12} & \cdots & x_{1n} \\ x_{21} & x_{22} & \cdots & x_{2n} \\ \vdots & \cdots & \cdots & \vdots \\ x_{m1} & x_{m2} & \cdots & x_{mn} \end{bmatrix} (\, i = 1,2,3,4..m \text{ , } j = 1,2,3,4...n \,) \quad (9)$$

where, x_{ij} is the criteria / response parameter weight by the decision maker expressed in triangular fuzzy number.

$$DF_j = \frac{\left[\left(Ux_{ij} - Lx_{ij}\right) + \left(Mx_{ij} - Lx_{ij}\right)\right]}{3} + Lx_{ij} \quad (10)$$

where *U, M, L* represents higher, medium and lower limitations of the fuzzy weights of the response parameters in the aggregated fuzzy decision matrix.

$$W_j = M_{ij} \times DF_j \tag{11}$$

where, W_j indicates weighted normalized decision matrix, G_{ij} is normalized decision matrix and w_j represents priority weight of the response parameter.

Step 5: Determination of the Limited Super Matrix

The demonstration of the limited super matrix is calculated by the following representation of un-weighted super matrix and weighted super matrix. The un-weighted super matrix represented in Eq. (12). Where G, C, A in the following Eq. (12) denotes goal, criteria and alternatives respectively. Then the weighted super matrix is calculated by adjusting the un-weighted super matrix to column stochastic so that the sum of the elements in each column is equal to one. Finally, the limited super matrix is obtained by raising the weighted super matrix to power k, where k is an arbitrary large number until the row elements converge to the same value for each column for each column of the matrix.

$$S = \begin{bmatrix} & G & C & A \\ G & 0 & 0 & 0 \\ C & W_{21} & W_{22} & 0 \\ A & 0 & W_{32} & I \end{bmatrix} \tag{12}$$

where, W_{21} is the pairwise comparison matrix of the criteria / response parameter with respect to goal, W_{22} is the pairwise comparison matrix comprised of criteria / response parameter while considering the interdependency among, W_{32} is the comparison matrix of the alternatives with respect to the criteria represented in step 4 and I denotes identity matrix.

$$S = \lim_{k \to \infty} S^k \tag{13}$$

Step 6: Ranking of Alternatives

The ranking of the alternatives is obtained after converging the values of the limited super matrix in each column is same. Then the values of alternative values in each column with respect to each criteria / response parameter are ranked in ascending order.

EXPERIMENTAL DETAILS

Material and Equipment

The quality or other aesthetic properties of the fabricated parts in the FDM process is mainly depend upon the type of material used throughout the process. In the current study, poly lactic acid (PLA) is used as a material for fabrication of the parts. The Poly lactic acid (PLA) materials is copolymer of poly L-lactic acid (PLLA) and poly D, L-lactic acid which are produced from L-lactides and D, L-lactides respectively. The chemical structure of the PLA, atom representation as spheres with conventional color coding, and filament of PLA are represented as shown in Figure 1 (a), (b), and (c) respectively.

The PLA is second most used bio plastic in the world which is usually derived from renewable resources like corn starch, tapioca roots, starch and sugarcane etc. Also, it is most used material in Additive Manufacturing

Figure 1. (a) Chemical structure of PLA material (b) Representation of atoms as spheres of PLA material (c) Filament uses in FDM process

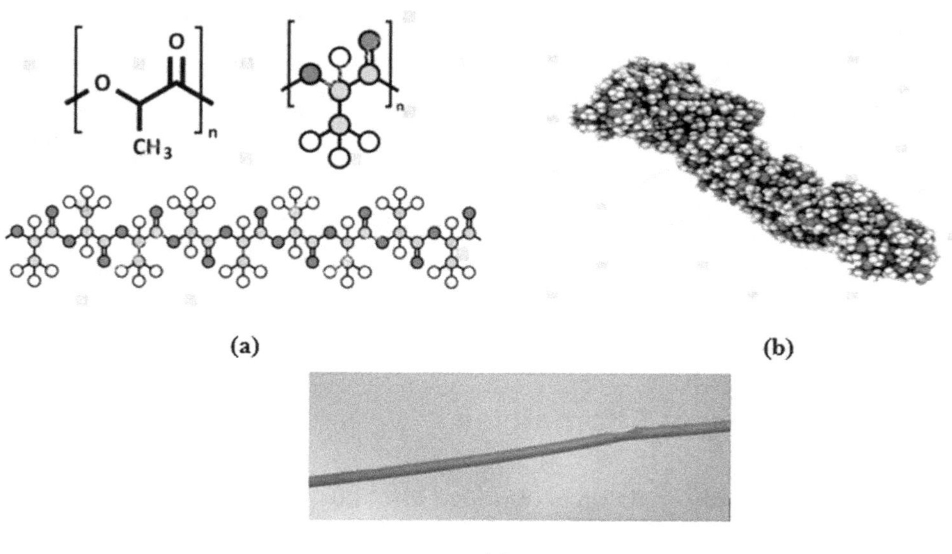

processes due to the higher tensile strength of 57 MPa and heat deflection temperature of 55^{0}C. In the FDM process, PLA material usually wounded on the spool in the form of filament of diameter 1.75mm for fabricating of the parts layer by layer by heating and extruding of the filament.

The present study is used the model R x P 2200 I cube for fabrication of the products by using the technology called fused filament fabrication usually known as fused deposition modeling (FDM). The equipment has a build volume of 305x305x305 mm and having nozzle of diameter of 0.4 mm is installed at head. Also, it has open filament system i.e. different type of materials like ABS, Nylon etc. can be used according to the use for fabrication of the products in the FDM process. The CURA software is used for converting the 3D models into 3D print files which are consists of G-codes. Also, CURA software is used for varying the process parameters and other manual settings used in the FDM process. The experimental setup of the FDM process of Model RxP 2200 I and the close view of FDM head while fabricating the products is shown in Figure 2 (a), (b) respectively.

Figure 2. (a) Experimental setup of the FDM of model RxP2200I (b) close view of the FDM head

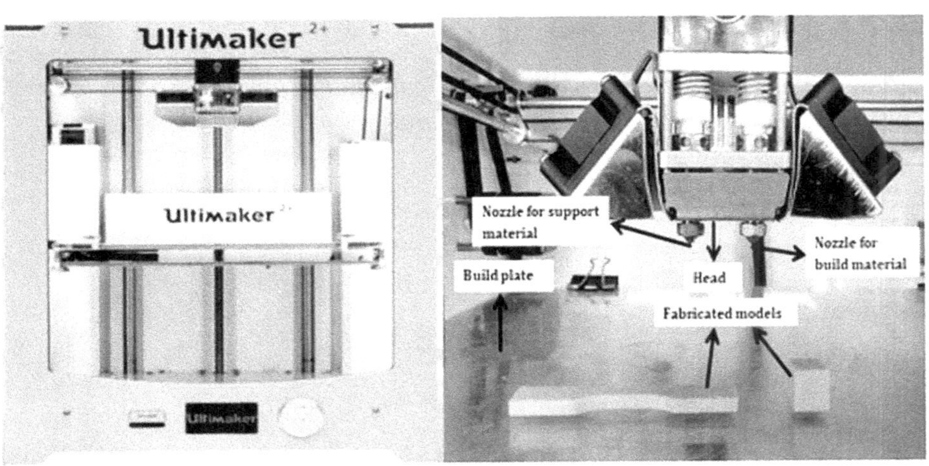

(a) (b)

TESTING AND ANALYSIS OF THE FABRICATED SAMPLES

The fabrication of the experimental samples is carried out in three steps; firstly, CAD models are designed for calculating the ultimate tensile strength according to ASTM D-638 and the cube samples of volume 1x1x1cm for calculating the dimensional accuracy is done. Second, these CAD models are installed in CURA software for converting 3D CAD data into 3D print files comprised of G-codes. Finally, the fabrication of the experimental samples is done by designing the experiments with Taguchi's L-16 orthogonal array with the process parameters considered in the study such as layer height, shell thickness and fill density as shown in Table 1. The fabricated tensile specimen samples and cube samples are shown in Figure 3 (a), (b) respectively.

After fabrication of the samples, the experimentation is carried out for tensile specimen samples for calculating the ultimate tensile strength with the help of universal testing machine of the capacity 20 tons. Also, the dimensional accuracy for each sample is calculated by measuring the specimen dimensions in the directions of X, Y and Z axes. Another criterion manufacturing time is calculated for each of the experimental setting while fabricating the tensile specimen and cube specimen together. The experimental data obtained in the present study is shown in Table 1.

Figure 3. (a) Tensile specimen samples to calculate ultimate tensile strength (b) cube samples to calculate dimensional accuracy

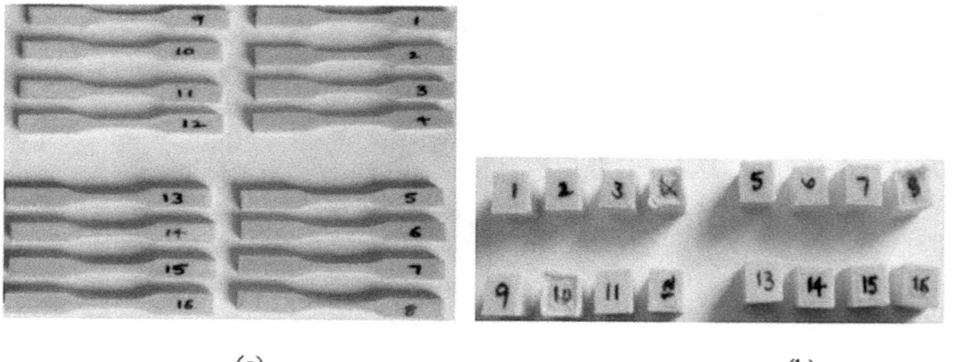

(a) (b)

Table 1. Experimental results of FDM process

Exp. No	LH (mm)	ST (mm)	Fill density (%)	Ultimate Tensile Strength (UTS) in (MPa)	Dimensional Accuracy (DA) in (mm^3)	Manufacturing Time (MT) in (min)
1	0.08	0.8	25	52.75	1095.29	46
2	0.08	1.2	50	57.10	1050.18	58
3	0.08	1.6	75	50.42	1114.74	67
4	0.08	2.0	100	56.78	1150.50	68
5	0.12	0.8	50	39.13	1050.63	37
6	0.12	1.2	25	42.09	1037.06	33
7	0.12	1.6	100	56.72	1161.26	50
8	0.12	2.0	75	55.50	1140.16	46
9	0.16	0.8	75	38.46	1060.65	33
10	0.16	1.2	100	49.98	1107.57	38
11	0.16	1.6	25	44.60	1019.06	27
12	0.16	2.0	50	46.35	1040.43	32
13	0.20	0.8	100	52.77	1192.26	32
14	0.20	1.2	75	41.17	1028.21	28
15	0.20	1.6	50	50.39	1019.04	26
16	0.20	2.0	25	49.85	1017.03	24

MODELING OF THE FDM PROCESS USING MODIFIED FUZZY ANALYTICAL NETWORK PROCESS

In the present section, a hybrid multi-criteria decision-making method M-Fuzzy analytical network process is used to validate the experimental data for getting the optimal setting among the experimental settings in the FDM process. Due to advantage of F-ANP for the combined effect of interdependency of the criteria consideration and flexibility of the decision making, the selection of the experimental setting become more optimal. In the present study, three process parameters considered namely layer height, shell thickness and fill density while ultimate tensile strength, dimensional accuracy and manufacturing time are considered as criteria. The modeling of the FDM process using Fuzzy-ANP starts with the calculation of the priority weights of criteria while pairwise comparison with respect to goal. After pairwise comparison of the criteria the normalization of the geometric mean method (NGM) is applied to calculate the priority weights. Then consistency ratio (CR) is calculated for checking the consistency of pairwise comparison matrix. The pairwise comparison matrix of the criteria matrix with respect to goal is shown in Table 2. The priority weights 0.366, 0.366 and 0.267 in Table 2 indicates that ultimate tensile strength and dimensional accuracy have the equal weightage while manufacturing time has lower weightage with respect to goal of quality products fabrication.

Similarly, the calculation of the priority weights of pairwise comparison matrices for inner dependence of criteria with respect to each criterion is obtained by NGM method followed by consistency ratio checking. The inner dependence of the criteria with respect to ultimate tensile strength, dimensional accuracy and manufacturing time with their calculated priority weights are shown in Table 3, Table 4, Table 5 respectively. The priority weights 0.7343 and 0.2656 in Table 3 and Table 4 indicates that dimensional accuracy, ultimate tensile strength has higher priority than manufacturing time with respect to getting good strength and better dimensional accuracy respectively. Also, priority weights in Table 5 indicate that ultimate tensile strength and dimensional accuracy has equal weightage for getting optimal manufacturing time.

After pairwise comparison of the criteria, the pairwise comparison of alternatives with the criteria is obtained by consideration of the experimental data obtained in the FDM process and fuzzy weights obtained by decision makers. Then the weighted pairwise comparison matrix is calculated by multiplying weights of the criteria with the respective experimental data obtained in the FDM process. The weighted normalized pairwise comparison matrix is shown in Table 6.

Finally, the limited super matrix for ranking the alternatives is obtained by the following un-weighted and weighted super matrix calculation. The

Table 2. Pairwise comparison of criteria / response parameter matrix with respect to goal

GOAL	Ultimate tensile strength (UTS)	Dimensional accuracy (DA)	Manufacturing time (MT)	Priority weights
Ultimate tensile strength (UTS)	(1,1,1)	(0.3,0.5,0.7)	(0.1.0.3,0.5)	0.366
Dimensional accuracy (DA)	(0.3,0.5,0.7)	(1,1,1)	(0.1.0.3,0.5)	0.366
Manufacturing time (MT)	(0.1.0.3,0.5)	(0.1.0.3,0.5)	(1,1,1)	0.267

Table 3. Inner dependence of the criteria matrix with respect to ultimate tensile strength (UTS)

Ultimate tensile strength (UTS)	Dimensional accuracy (DA)	Manufacturing time (MT)	Priority weights
Dimensional accuracy (DA)	(1,1,1)	(0.5, 0.7,0.9)	0.7343
Manufacturing time (MT)	(0.1.0.3,0.5)	(1,1,1)	0.2656

Table 4. Inner dependence of the criteria matrix with respect to dimensional accuracy (DA)

Dimensional accuracy (DA)	Ultimate tensile strength (UTS)	Manufacturing time (MT)	Priority weights
Ultimate tensile strength (UTS)	(1,1,1)	(0.5, 0.7, 0.9)	0.7343
Manufacturing time (MT)	(0.1, 0.3,0.5)	(1,1,1)	0.2656

Table 5. Inner dependence of the criteria matrix with respect to Manufacturing time (MT)

Manufacturing time (MT)	Ultimate tensile strength (UTS)	Dimensional accuracy (DA)	Priority weights
Ultimate tensile strength (UTS)	(1,1,1)	(0.3, 0.5,0.7)	0.5
Dimensional accuracy (DA)	(0.3, 0.5,0.7)	(1,1,1)	0.5

un-weighted super matrix is formed with the pairwise comparison matrices obtained using step.1-4 in the research methodology. Then weighted super matrix is obtained by adjusting the un-weighted super matrix to column stochastic i.e. the sum of the elements in each column equal to one. Further the limited super matrix is calculated by raising the weighted super matrix to power k (any arbitrary number). The obtained limited super matrix is shown in Table 7 for ranking the alternatives in the FDM process.

Thereafter, the ranking of the alternatives is done to know the relative significance of each experimental setting used in the FDM process in the present study. Any column in the limited super matrix can be taken in ascending order of the values for raking of the alternatives. The experimental setting and the ranking of alternatives is shown in Table 8.

Table 6. The weighted normalized pairwise comparison matrix

Exp. No	Ultimate tensile Strength (UTS)	Dimensional accuracy (DA)	Manufacturing Time (MT)
1	0.7852	0.7426	0.1015
2	0.8500	0.7120	0.1279
3	0.7506	0.7558	0.1478
4	0.8452	0.7800	0.1500
5	0.0816	0.7123	0.0816
6	0.0728	0.7031	0.0728
7	0.1103	0.7873	0.1103
8	0.1015	0.7730	0.1015
9	0.0728	0.7191	0.0728
10	0.0838	0.7509	0.0838
11	0.0596	0.6909	0.0596
12	0.0706	0.7054	0.0706
13	0.0706	0.8083	0.0706
14	0.0618	0.6971	0.0618
15	0.0574	0.6909	0.0574
16	0.0529	0.6895	0.0529

RESULTS AND DISCUSSION

The hybrid multi criteria decision making method M-Fuzzy ANP is used to determine the optimal result for fabricating the quality of prototypes in the FDM process. The optimal experimental settings are done by several runs (trails) to find out ranking of the alternatives. In this chapter the experimental setting no.4 of alternative is ranked as first, i.e. the corresponding process parameters of this experimental setting shows the better quality of fabricated products compared to any other experimental settings used in the present experimentation and also experimental setting no.9 is the worst quality of fabricated products compared to any other experimental settings shown in Table 8. The experimental sample parameters such as layer height 0.08mm, shell thickness 2.0mm and fill density is 100 percentages gives the optimal value of the criteria/response parameters namely ultimate tensile strength (UTS) is 56.78MPa, dimensional accuracy (DA) is 1150.50mm^3 and manufacturing time

Table 7. Limited super matrix

	Goal	(UTS)	(DA)	(MT)	A1	A2	A3	A4	A5	A6	A7	A8	A9	A10	A11	A12	A13	A14	A15	A16
Goal	0	0	0	0	0	0	0	0	0	0	0	0	0	0	0	0	0	0	0	0
(UTS)	4.1949	6.4293	6.4293	5.7169	0	0	0	0	0	0	0	0	0	0	0	0	0	0	0	0
(DA)	4.1949	6.4293	6.4293	5.7169	0	0	0	0	0	0	0	0	0	0	0	0	0	0	0	0
(MT)	2.2968	3.5201	3.5201	3.1300	0	0	0	0	0	0	0	0	0	0	0	0	0	0	0	0
A1	5.3170	8.1490	8.1490	7.2460	0	0	0	0	0	0	0	0	0	0	0	0	0	0	0	0
A2	5.4804	8.3994	8.3994	7.4687	0	0	0	0	0	0	0	0	0	0	0	0	0	0	0	0
A3	5.3299	8.1689	8.1689	7.2636	0	0	0	0	0	0	0		0	0	0	0	0	0	0	0
A4	5.7333	8.7871	8.7871	7.8134	0	0	0	0	0	0	0	0	0	0	0	0	0	0	0	0
A5	4.4980	6.8938	6.8938	6.1298	0	0	0	0	0	0	0	0	0	0	0	0	0	0	0	0
A6	4.5988	7.0483	7.0483	6.2673	0	0	0	0	0	0	0	0	0	0	0	0	0	0	0	0
A7	5.6818	8.7082	8.7082	7.7432	0	0	0	0	0	0	0	0	0	0	0	0	0	0	0	0
A8	5.5566	8.5162	8.5162	7.5725	0	0	0	0	0	0	0	0	0	0	0	0	0	0	0	0
A9	4.4711	6.8525	6.8525	6.0932	0	0	0	0	0	0	0	0	0	0	0	0	0	0	0	0
A10	5.1740	7.9299	7.9299	7.0511	0	0	0	0	0	0	0	0	0	0	0	0	0	0	0	0
A11	4.6590	7.1405	7.1405	6.3492	0	0	0	0	0	0	0	0	0	0	0	0	0	0	0	0
A12	4.8154	7.3802	7.3802	6.5624	0	0	0	0	0	0	0	0	0	0	0	0	0	0	0	0
A13	5.4820	8.4019	8.4019	7.4708	0	0	0	0	0	0	0	0	0	0	0	0	0	0	0	0
A14	4.5124	6.9159	6.9159	6.1495	0	0	0	0	0	0	0	0	0	0	0	0	0	0	0	0
A15	4.9443	7.5778	7.5778	6.7381	0	0	0	0	0	0	0	0	0	0	0	0	0	0	0	0
A16	4.9046	7.5170	7.5170	6.6840	0	0	0	0	0	0	0	0	0	0	0	0	0	0	0	0

Table 8. The ranking of the alternatives

Exp. No	1	2	3	4	5	6	7	8	9	10	11	12	13	14	15
Rank	7	5	6	1	15	13	2	3	16	8	12	11	4	14	9

(MT) required is 68 minutes for experimental no.4 and worst experimental sample parameters such as layer height 0.16mm, shell thickness 0.8mm and fill density is 75 percentages gives value of the criteria/response parameters namely ultimate tensile strength (UTS) is 38.46MPa, dimensional accuracy (DA) is 1060.65mm^3 and manufacturing time (MT) required is 33minutes for experimental no.9 as shown in Table 1. Based on the results obtained in the present chapter, the higher value of ultimate tensile strength, optimal dimensional accuracy values and manufacturing time also higher are obtained for experimental no.4. These experimental values describe the fabrication of the components using the obtained experimental setting gives better quality products in the technical and economical point of view.

CONCLUSION

The present chapter described the modeling and optimization of the process parameters in FDM process using the proposed hybrid multi-criteria decision-making method i.e. M-Fuzzy ANP. The result shows that experimental no 4 gives the optimal setting among 16 experimental runs. The following conclusions are explained as follows:

- The selection of layer thickness is more important while fabricating products in FDM. Lower layer thickness gives the better surface finish, better dimensional accuracy and gives good strength.
- The optimal experimental sample parameters such as layer height 0.08mm, shell thickness 2.0mm and fill density is 100 percentages gives the optimal value of the criteria/response parameters namely ultimate tensile strength (UTS) is 56.78MPa, dimensional accuracy (DA) is 1150.50mm^3 and manufacturing time (MT) required is 68 minutes for experimental no.4.
- The 100 percent fill density is required to get the higher strengthen fabricated parts.

- The combination of layer height 0.08mm, shell thickness 2.0mm and fill density 100 percent gives quality products than any other experimental setting used in the present study.
- The worst experimental sample parameters such as layer height 0.16mm, shell thickness 0.8mm and fill density is 75 percentages gives value of the criteria/response parameters namely ultimate tensile strength (UTS) is 38.46MPa, dimensional accuracy (DA) is 1060.65mm^3 and manufacturing time (MT) required is 33minutes for experimental no.9.

The present study concluded that, to fabricating the quality prototypes, the proposed hybrid method M-Fuzzy ANP gives the viable and optimal solution than existing fuzzy ANP methods. Also, due to the use of M-Fuzzy ANP for modeling of the FDM process can achieve good (higher) strength and accuracy characteristics while fabrication. Hence this proposed hybrid method can be used for any other manufacturing systems for obtaining the optimal experimental settings considering technical and economical point of view.

REFERENCES

Alafaghani, A., Qattawi, A., Alrawi, B., & Guzman, A. (2017). Experimental Optimization of Fused Deposition Modelling Processing Parameters: A Design for Manufacturing Approach, 45th SME NAMRC conference. *Procedia Manufacturing*, *10*, 791 – 803.

Anoop, K. S., Ohdar, R. K., & Siba, S. M. (2009). Improving dimensional accuracy of fused deposition modelling processed part using grey Taguchi method. *4th International Conference on Materials Processing and Characterization Proceedings*, *2*, 4243 – 4252.

Barzilai. (1997). Deriving weights from pairwise comparison matrices. *Journal of Operational Research Society*, *48*, 1226–1232.

Basavaraj, C.K., & Vishwas, M. (2017). Studies on Effect of Fused Deposition Modelling Process Parameters on Ultimate Tensile Strength and Dimensional Accuracy of Nylon. *Journal of Procedia Manufacturing*, *10*, 791-801.

Bhowmik, S., & Jagadish. (2017). Multi-criteria decision making for optimization of product development under green manufacturing environment. In *Design and Optimization of Mechanical Engineering Products*. Hershey, PA: IGI Global.

Borah, A., & Jagadish. (2018) A Multi Criteria Decision Making Approach for Rapid Prototyping Process Selection. *Proceedings of IRF International Conference*, 1-5.

Jagadish & Ray, A. (2014). Green cutting fluid selection using MOOSRA method. *International Journal of Research in Engineering and Technology*, *3*(3), 559–563.

Karsh, P. K., & Singh, H. (2018). Multi-Characteristic Optimization in Wire Electrical Discharge Machining of Inconel-625 by Using Taguchi-Grey Relational Analysis (GRA) Approach: Optimization of an Existing Component/ Product for Better Quality at a Lower Cost. In Design and Optimization of Mechanical Engineering Products (pp. 283–303). Hershey, PA: IGI Global. doi:10.4018/978-1-5225-3401-3.ch014

Kumar, N., Kumar, H., & Khurmi, J.S. (2016). Experimental Investigation of process parameters for rapid prototyping technique (Selective Laser Sintering) to enhance the part quality of prototype by Taguchi method. *3rd ICIAME Procedia Technology*, *23*, 352 – 360.

Kumar, S., Kannan, V. N., & Sankaranarayanan, G. (2014). Parameter Optimization of ABS-M30i Parts Produced by Fused Deposition Modeling for Minimum Surface Roughness. *International Journal of Current Engineering and Technology, 3*, 93–97.

Lee, B. H., Abdullah, J., & Khan, Z. A. (2005). Optimization of rapid prototyping parameters for production of flexible ABS object||. *Journal of Materials Processing Technology, 169*(1), 54–61. doi:10.1016/j.jmatprotec.2005.02.259

Mohamed, O.A., Masood, S.H., & Bhowmik, J.L. (2016). Mathematical modeling and FDM process parameters optimization using response surface methodology based on Q-optimal design. *Journal of Applied Mathematical Modelling, 40*, 10052–10073.

Nidagundi, V. B., Keshavamurthy, R., & Prakash, C. P. S. (2015). Studies on Parametric Optimization for Fused Deposition Modelling Process. *4th International Conference on Materials Processing and Characterization Proceedings, 2*, 1691 – 1699. 10.1016/j.matpr.2015.07.097

Panda, S.K., Padhee, S., Anoop K.S., & Mahapatra, S.S. (2009). Optimization of fused deposition modeling (FDM) process parameters using Bacterial Foraging Technique. *Journal of Intelligent Information Management, 1*, 89.

Phatak, A. M., & Pandee, S. S. (2012). Optimum part orientation in Rapid Prototyping using generic algorithm. *Journal of Manufacturing Systems, 31*(4), 395–402. doi:10.1016/j.jmsy.2012.07.001

Raju, B.S., Shekar, U.C., Venkateswarlu, K., & Drakashayani, D.N. (2014). Establishment of Process model for rapid prototyping technique (Stereolithography) to enhance the part quality by Taguchi method. *2nd ICIAME 2014 Procedia Technology, 14*, 380 – 389.

Sahu, R., Mahapatra, S., & Sood, A. (2013). A Study on Dimensional Accuracy of Fused Deposition Modeling (FDM) Processed Parts using Fuzzy Logic. *Journal for Manufacturing Science & Production, 13*(3), 183–197. doi:10.1515/jmsp-2013-0010

Siba, S. M., & Narayan, P. B. (2013). Benchmarking of rapid prototyping systems using grey relational analysis. *International Journal of Services and Operations Management, 16*(4), 460–477. doi:10.1504/IJSOM.2013.057509

Sood, A.K., Ohdar, R.K., & Mahapatra, S.S. (2010). Parametric appraisal of mechanical property of fused deposition modelling processed parts. *Journal of Materials and Design*, 287-295.

Sood, A. K., Ohdar, R. K., & Mahapatra, S. S. (2014). Parametric appraisal of fused deposition modelling process using the grey Taguchi method. *Proceedings of the Institution of Mechanical Engineers. Part B, Journal of Engineering Manufacture*, *224*(1), 135–145. doi:10.1243/09544054JEM1565

Upadhyay, A., Prakash, V., & Sharma, V. (2018). Optimizing Material Removal Rate Using Artificial Neural Network for Micro-EDM. In Design and optimization of Mechanical Engineering Products (pp. 209–233). India: IGI global.

Villalpando, L., Eiliat, H., & Urbanic, R. J. (2014). An optimization approach for components built by fused deposition modeling with parametric internal structures, Product Services Systems and Value Creation. *Proceedings of the 6th CIRP Conference on Industrial Product-Service Systems*, 800-805.

Vinodh, S., Anesh, R. R., & Gautham, S. G. (2011). Application of fuzzy analytic network process for supplier selection in a manufacturing organization. *Expert Systems with Applications*, *38*(1), 272–280. doi:10.1016/j.eswa.2010.06.057

Chapter 11
Determination of Optimum Process Parameter Values in Additive Manufacturing for Impact Resistance

Fasih Munir Malik
Abdullah Gül University, Turkey

Syed Faiz Ali
Abdullah Gül University, Turkey

Burak Bal
Abdullah Gül University, Turkey

Emin Faruk Kececi
Abdullah Gül University, Turkey

ABSTRACT

3D printing as a manufacturing method is gaining more popularity since 3D printing machines are becoming easily accessible. Especially in a prototyping process of a machine, they can be used, and complex parts with high quality surface finish can be manufactured in a timely manner. However, there is a need to study the effects of different manufacturing parameters on the materials properties of the finished parts. Specifically, this chapter explains the effects of six different process parameters on the impact resistance. In particular, print temperature, print speed, infill ratio, infill pattern, layer height, and print orientation parameters were studied, and their effects on impact resistance were measured experimentally. Moreover, the optimum values of the process parameters for impact resistance were found. This

DOI: 10.4018/978-1-5225-9167-2.ch011

chapter provides an important guideline for 3D manufacturing in terms of impact resistance of the printed parts. Furthermore, by using this methodology the effects of different 3D printing process parameters on the other material, properties can be determined.

INTRODUCTION

Additive manufacturing (AM), also known as 3D printing, is receiving great attention due to the fact that it allows easy and time efficient realization of designs from prototypes to functional products with complex geometries (Zotti et al., 2018), (León-Cabezas et al., 2017), (Castro et al., 2015). The process of AM starts with computer aided design and continues with printing successive layers on top of each other to obtain the desired final shape. 3D printing technology started to be used in mid-80's with a process known as Stereolithograpgy and was followed by powder bed fusion, fused deposition modelling, inkjet printing and contour crafting, chronologically (Spoerk et al., 2018). The main printed parts using Stereolithograpgy 3D printing technology are patterns, mold and models by photopolymerization. Powder bed fusion (or selective laser melting) allows the manufacturing of 3D parts by selectively melting and layer by layer fusing metallic powder materials (Arisoy et al., 2019). For the process of fused deposition modelling, 3D computer models can be printed without requiring a mold (Daver et al., 2018). In contour crafting, the parts are printed by integrating material delivery and installation in to one system (Zareiyan & Khoshnevis, 2017). With the aid of this rapidly developed technology, different material groups from polymers to metals with very complex geometries are now able to be printed. However, in order to use these 3-D printed materials in design, the mechanical characterization of them is of utmost importance.

The mechanical characterization of 3-D printed parts has been carried out by using several experimental methods. In order to determine the material properties, tensile tests, compression tests and indentation tests have been conducted at different loading conditions (Spoerk et al., 2018; Ahn et al., 2002; Rankouhi et al., 2016). Particularly in these studies, strength, ductility, modulus and hardness values of 3-D printed materials have been measured. In addition, as a conclusion of these studies, it has been proven that the mechanical properties of 3-D printed materials depend on both the unprinted material and 3-D process parameters. For instance, optimum tensile properties are obtained when the filaments are parallel to the loading direction (Dizon et al.,

2018). However, in spite of many works on the mechanical properties of 3-D printed materials under tensile and compressive loads, the number of works on the impact response of 3-D printed materials are very limited. In these limited studies, effects of building orientation, layer thickness, fiber volume content, lattice topology and build direction on the impact performance of 3-D printed materials have been investigated (Caminero et al., 2018; Bourell et al., 2017; Ngo et al., 2018; Lou et al., 2018). However, there is a lack of research about how 3-D printed materials are subject to impact loadings during operation. Therefore, the impact response of these materials should also be investigated in order to utilize them in safe design.

As opposed to tensile and compressive loadings, an impact test demonstrates the material's behavior at an instant high strain rate shock loading. In addition, it was proven that, it enhances only one deformation mechanism over complex microstructural interactions (Bal et al., 2015; Toker et al., 2014). Moreover, with the aid of an impact test, the ductile to brittle transition temperature of the materials can be determined. Specifically, a ductile material at room temperature or high temperatures may exhibit brittle behavior at low temperatures (Das et al., 2011). That is why temperature is very crucial to consider in design if the material will be subject to different temperatures. In addition, even though impact performance of 3-D printed materials is very important in the manufacturing of high quality and reliable end-use 3-D printed materials in terms of impact performance, the detailed studies on this problem is limited in the literature and to the best of the authors' knowledge, there is no study, which shows the detailed effects of 3-D process parameters on the impact response of 3-D printed materials. Moreover, even though the effects of different factors, such as yield strength (Figure 1) on impact energy of several metals are well known, these kind of studies for 3-D printed materials are very limited.

This chapter focuses on the optimization of 3-D process parameters that result in optimum impact energy of 3-D printed materials. For this purpose, an experimental approach is adopted, and obtained results are theoretically analyzed to find the effect of 3-D process parameters on the impact response of 3-D printed materials. The 3D printing mechanism adopted is Fused Deposition Modeling (FDM). The material used in experiments is PLA. The CURA slicing software is used and the specimen are printed by an Ultimaker GO2 3D printer. The following part of the chapter explains the experiments and the test results showing the effect of different parameters on impact response.

Figure 1. The relationship between yield strength and impact energy of several materials
Source: (Inoue et al., 2012)

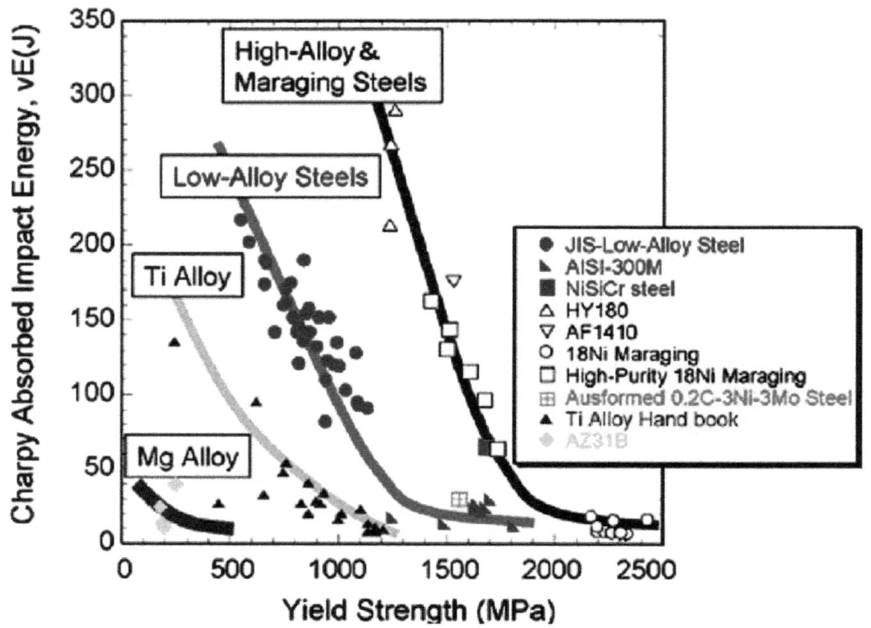

METHODOLOGY

A standard Charpy impact test machine with 300 Joules impact energy capacity (Figure 2) was utilized for impact testing of the specimens. Since specimen were 3D printed from PLA filament of 2.85 mm therefore, the impact resistance observed was negligible when the impact tests were conducted according to standards. In order to increase the sensitivity of the results, the mass of the hammer was reduced by removing the metal side plates. The starting position of the hammer was also lowered. The tests were carried out with the same conditions to ensure that same impact energy was given to all of the 3D printed specimens.

The standard dimensions for the charpy impact test specimen are $55 \times 10 \times 10$ mm and there is a notch at the center of the specimen. In this research, the dimensions of the specimens were increased for the test. In order to enhance the impact resistance strength of the specimens, specimens were

Figure 2. Charpy impact test machine

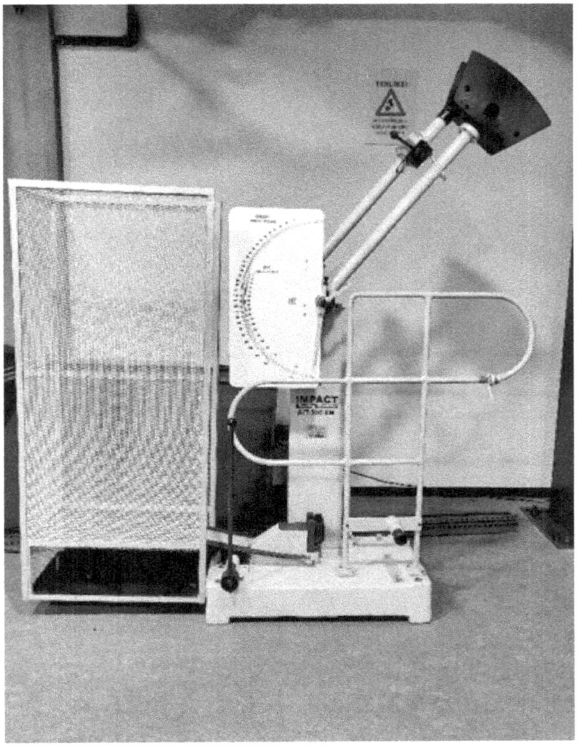

printed with the dimensions 55 × 20 × 10 mm without a notch. All the specimens were printed in same print orientation as shown in Figure 3.a and in order to determine the effect of print orientation on the impact resistance of 3D printed specimens, the orientation was changed as shown in Figure 3b and Figure 3c. Figure 4 shows the specimen dimensions.

The process parameters that were changed in order to determine the effects on the impact resistance of the 3D printed part were print temperature, print speed, infill ratio, infill pattern, layer height and print orientation. The process parameters were tested with different values. The maximum, average and minimum values were changed while the rest of the parameters were constant. Each test at a given value for the parameters were carried out 3 times to ensure the consistency of the results and the average values were plotted.

Figure 3. The print orientations of impact specimens. Figure 3a) Initial print orientation, Figure 3b) Second print orientation, Figure 3c) Third figure orientation

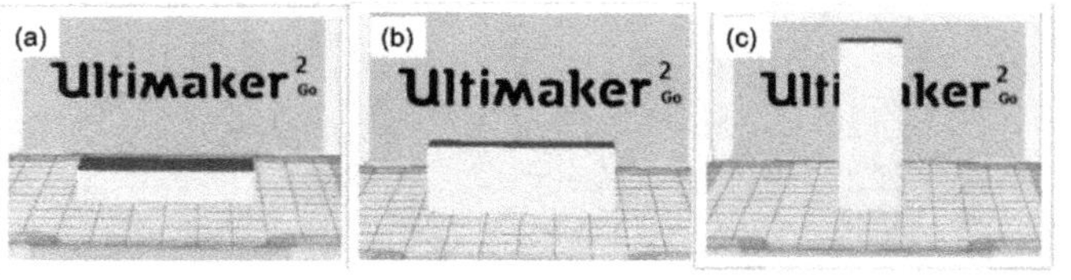

Figure 4. Technical drawing of the impact test specimens (units are in mm)

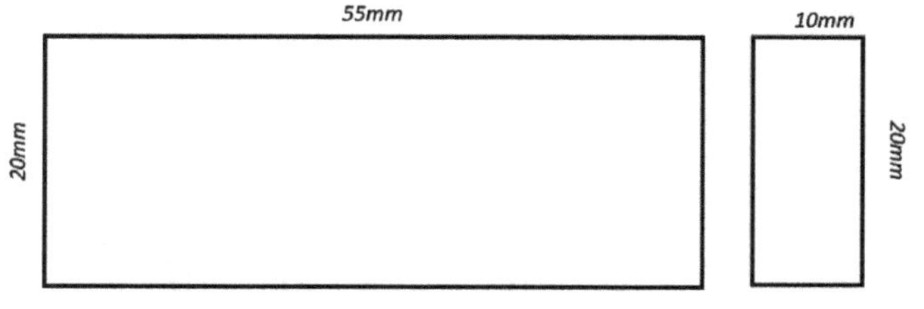

The energies required to fracture the samples were recorded and the graphs were plotted to exhibit the trend of the process parameters' effects on the impact resistance of the part. The graphs were plotted in Igor advanced plotting program. The specimen parts were designed in Solidworks and Cura software was used to compile the G-Code. To manufacture the specimens an Ultimaker Go2 3D printers were used. The material used to 3D print the specimen parts was PLA with 2.85mm diameter.

RESULTS

By using the impact test machine, different specimens are tested and the energy causing fracture is shown for each parameter as a figure in the following section. While testing one of the process parameters, the other parameters were set constant in order to measure only the effect of the interested parameter. Print temperature, print speed, infill ratio, infill pattern, layer height and print orientation were the measured parameters.

PRINT TEMPERATURE

The print temperature parameter was tested on 4 different values; 185, 210, 225 and 240 degree Celsius. The testing range, from 185 to 240, was selected as advised by the material manufacturer as an optimum temperature range. As shown in Figure 5, the temperature agrees to almost a positive linear relation with the impact resistance. The highest impact resistance was observed at 240°C. Therefore, it is observed that, within the optimum temperature range, increasing the printing temperature increases the impact resistance almost linearly.

PRINT SPEED

The print speed is the speed of the nozzle head during the printing process and it was tested at 3 different values; 20, 60 and 100 mm/s. As shown in Figure 6, the highest resistance to impact was achieved at 20 mm/s print speed, where at the 100mm/s the impact resistance was the lowest. Figure 6 shows the decrease in impact resistance with the increase in print speed. However, the gradient trajectory suggests that the highest impact resistance will be accomplished at the print speed between 30 and 40 mm/s. The higher print speed allows less time to manufacture but on the other hand, less time for cooling of the layer before the following layer is extruded.

Figure 5. The effect of print temperature on impact resistance

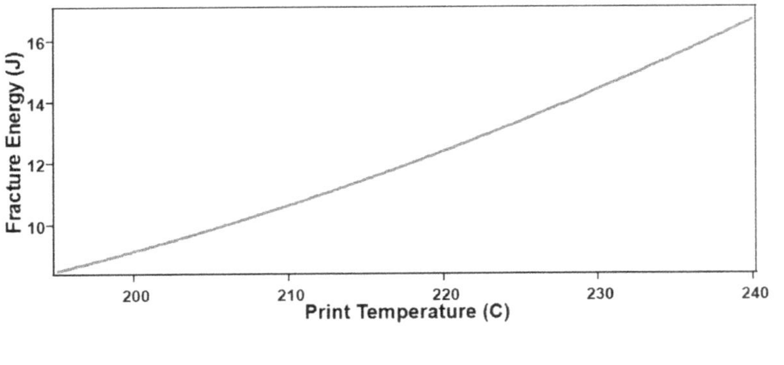

Figure 6. Print speed effect on the impact resistance

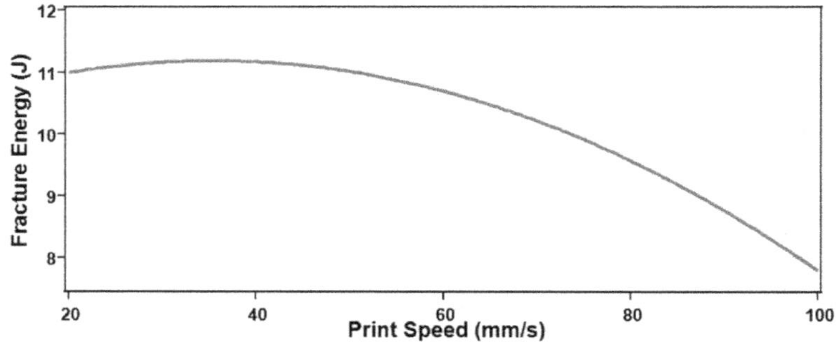

INFILL RATIO

The infill ratio defines what percentage of the inner section of the part will be filled. Increasing the infill ratio makes a part denser; however, this causes more time for the part to be manufactured. The infill ratio was tested on 4 different values: 20, 45, 70 and 100%. As shown in Figure 7, there is an increasing in impact resistance as the infill ratio increases. However, the highest energy value was not obtained at 100% infill ratio. The figure suggests that the highest impact resistance because of infill ratio can be achieved at 90% infill ratio. It is estimated that while it is better for the impact resistance to have a denser part, having some gaps in the part allow the test specimen to absorb some of the impact energy with the elastic shape change.

INFILL PATTERN

Infill pattern parameter defines the inner part structure and if the pattern allows a cage structure, the impact resistance of the part increases. The slicing software allows 6 different patterns which are grid, triangle, cubic, cross, lines and zig zag. There was a drastic change in the impact resistance depending on the infill pattern, Figure 8. The highest impact resistance was measured with grid pattern while the lowest of that is seen with the cross pattern.

Figure 7. The effect of infill ratio on impact resistance

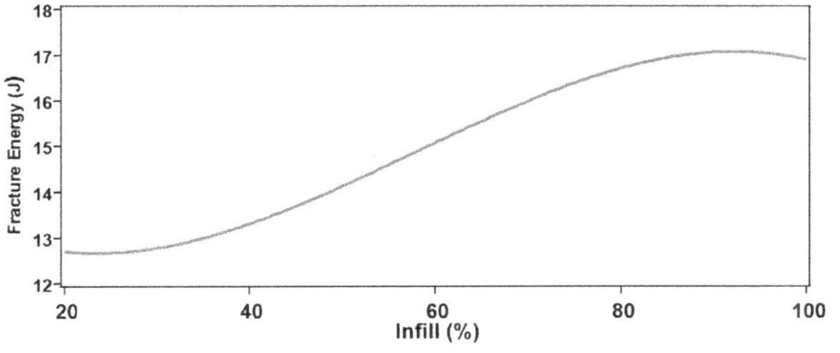

Figure 8. Impact resistance values for 6 different infill patterns

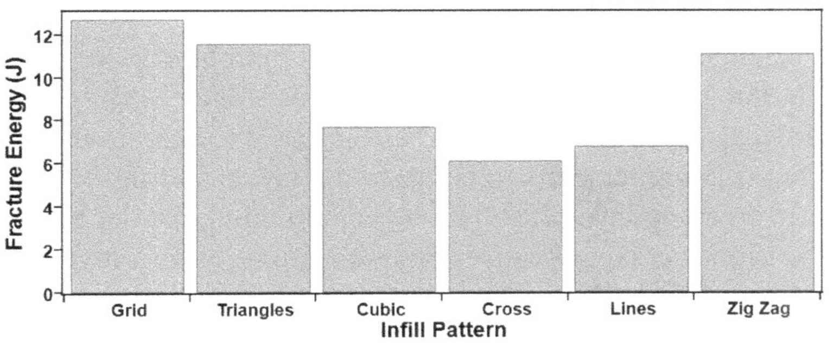

LAYER HEIGHT

The layer height is a process parameter which primarily affects the print-time and surface quality of the part. Increasing the layer height results in less printing time but it also results in a less smooth surface quality, and vice versa. However, in terms of impact resistance the layer height does not have a linear effect. As shown in Figure 9, the lowest impact resistance was observed by the part with 0.2 mm layer height and the most impact resistance was observed by the parts which had the layer height of 0.1 mm and 0.3 mm. The layer-height was tested in 3 different values: 0.1, 0.2 and 0.3 mm. It is concluded that the optimum value of layer height is 0.3(mm). Although the 0.1 mm layer height also rendered the same impact resistance,however, more time is required to print the part with the 0.1 mm layer height and this excludes it from being an optimum value.

Figure 9. The effect of layer height on impact resistance

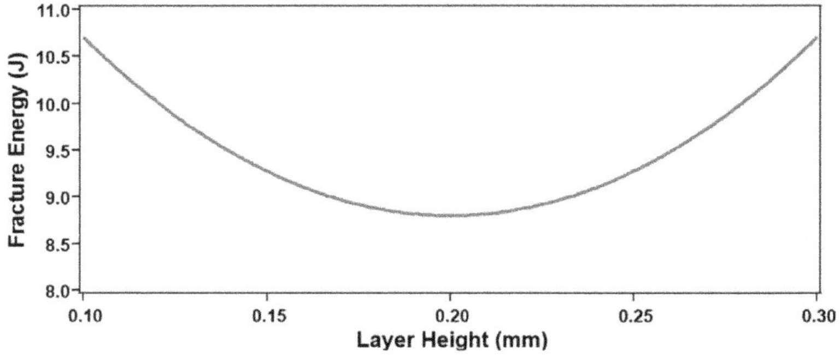

PRINT ORIENTATION

The print orientation was tested for 3 different options shown in Figure 10. The orientation 2 showed the highest impact resistance, while orientation 3 showed the least resistance. Depending on the print orientation and the impact direction the fracture happens in between the layers and in the material on that specific breaking layer. It should be kept in mind that depending on the impact direction the same print orientation might have a higher or lower impact resistance. This shows that when the part is being manufactured the direction of impact force should be known in order to choose the suitable print orientation.

Figure 10. The change in impact resistance caused by the print orientation

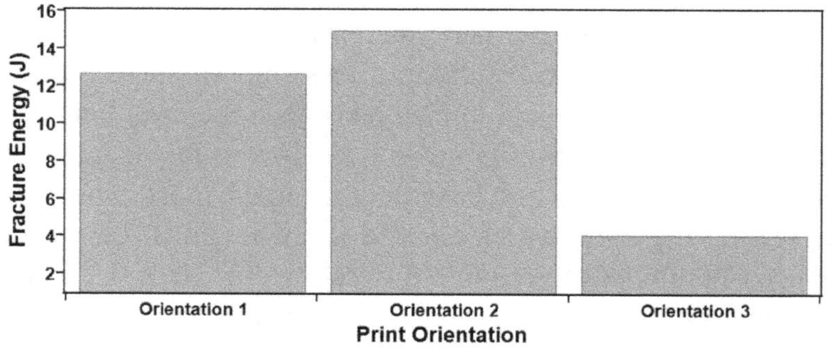

CONCLUSION AND FUTURE WORK

In this chapter the effects of the printing parameters namely, print temperature, print speed, infill ratio, infill pattern, layer height and print orientation on the impact resistance of 3D printed parts has been investigated. As a result of this study; the optimum values of the parameters were found experimentally. From this work the following conclusions can be drawn:

1. Increasing the print temperature results in the increase of impact resistance.
2. Impact resistance is inversely proportional to the print speed.
3. There are 3 main stages for the effect of infill ratio. Stage 1 is the initial stage where the effect of infill ratio on the impact resistance is negligible. In Stage 2 after a critical infill ratio an abrupt change, which is directly proportional with the infill ratio, on the impact resistance is measured. Stage 3 is the saturation stage and in this stage the impact resistance reaches the steady state and does not change with the infill ratio.
4. Infill pattern can change the impact resistance up to 100% (Cross infill pattern has 6 Joules whereas Grid infill pattern has 12 Joules).
5. The effect of layer height on the impact resistance is a parabolic function where at 0.2 mm layer height results in the least impact resistance.
6. The print orientation depends on the impact force direction.

This study opens a new venue for the manufacturing of 3D printed parts in terms of their impact resistance. There are also other parameters; such as nozzle size and material feed speed, that should be studied. As future work, the effects of these parameters will be investigated. In addition, other than the impact resistance, the effect of aforementioned parameters on different material properties needs to be investigated in order to manufacture functional 3D printed parts.

REFERENCES

Ahn, S., Montero, M., Odell, D., Roundy, S., & Wright, P. (2002). Anisotropic material properties of fused deposition modeling ABS. *Rapid Prototyping Journal*, *8*(4), 248–257. doi:10.1108/13552540210441166

Arısoy, Y., Criales, L., & Özel, T. (2019). Modeling and simulation of thermal field and solidification in laser powder bed fusion of nickel alloy IN625. *Optics & Laser Technology*, *109*, 278–292. doi:10.1016/j.optlastec.2018.08.016

Bal, B., Gumus, B., Gerstein, G., Canadinc, D., & Maier, H. (2015). On the micro-deformation mechanisms active in high-manganese austenitic steels under impact loading. *Materials Science and Engineering A*, *632*, 29–34. doi:10.1016/j.msea.2015.02.054

Bourell, D., Kruth, J., Leu, M., Levy, G., Rosen, D., Beese, A., & Clare, A. (2017). Materials for additive manufacturing. *CIRP Annals*, *66*(2), 659–681. doi:10.1016/j.cirp.2017.05.009

Caminero, M., Chacón, J., García-Moreno, I., & Rodríguez, G. (2018). Impact damage resistance of 3D printed continuous fibre reinforced thermoplastic composites using fused deposition modelling. *Composites. Part B, Engineering*, *148*, 93–103. doi:10.1016/j.compositesb.2018.04.054

Castro, G., Rodríguez, J., Montealegre, M., Arias, J., Yañez, A., Panedas, S., & Rey, L. (2015). Laser Additive Manufacturing of High Added Value Pieces. *Procedia Engineering*, *132*, 102–109. doi:10.1016/j.proeng.2015.12.485

Das, P., Jayaganthan, R., & Singh, I. (2011). Tensile and impact-toughness behaviour of cryorolled Al 7075 alloy. *Materials & Design*, *32*(3), 1298–1305. doi:10.1016/j.matdes.2010.09.026

Daver, F., Lee, K., Brandt, M., & Shanks, R. (2018). Cork–PLA composite filaments for fused deposition modelling. *Composites Science and Technology*, *168*, 230–237. doi:10.1016/j.compscitech.2018.10.008

Dizon, J. R., Espera, A. H. Jr, Chen, Q., & Advincula, R. C. (2018). Mechanical characterization of 3D-printed polymers. *Additive Manufacturing*, *20*, 44–67. doi:10.1016/j.addma.2017.12.002

Inoue, T., Kimura, Y., & Ochiai, S. (2012). Shape effect of ultrafine-grained structure on static fracture toughness in low-alloy steel. *Science and Technology of Advanced Materials*, *13*(3), 035005. doi:10.1088/1468-6996/13/3/035005 PMID:27877493

León-Cabezas, M., Martínez-García, A., & Varela-Gandía, F. (2017). Innovative advances in additive manufactured moulds for short plastic injection series. *Procedia Manufacturing*, *13*, 732–737. doi:10.1016/j.promfg.2017.09.124

Lou, X., Andresen, P. L., & Rebak, R. B. (2018). Oxide inclusions in laser additive manufactured stainless steel and their effects on impact toughness and stress corrosion cracking behavior. *Journal of Nuclear Materials*, *499*, 182–190. doi:10.1016/j.jnucmat.2017.11.036

Ngo, T., Kashani, A., Imbalzano, G., Nguyen, K., & Hui, D. (2018). Additive manufacturing (3D printing): A review of materials, methods, applications and challenges. *Composites. Part B, Engineering*, *143*, 172–196. doi:10.1016/j.compositesb.2018.02.012

Rankouhi, B., Javadpour, S., Delfanian, F., & Letcher, T. (2016). Failure Analysis and Mechanical Characterization of 3D Printed ABS With Respect to Layer Thickness and Orientation. *Journal of Failure Analysis and Prevention*, *16*(3), 467–481. doi:10.100711668-016-0113-2

Spoerk, M., Savandaiah, C., Arbeiter, F., Traxler, G., Cardon, L., Holzer, C., & Sapkota, J. (2018). Anisotropic properties of oriented short carbon fibre filled polypropylene parts fabricated by extrusion-based additive manufacturing. *Composites. Part A, Applied Science and Manufacturing*, *113*, 95–104. doi:10.1016/j.compositesa.2018.06.018

Toker, S., Canadinc, D., Taube, A., Gerstein, G., & Maier, H. (2014). On the role of slip–twin interactions on the impact behavior of high-manganese austenitic steels. *Materials Science and Engineering A*, *593*, 120–126. doi:10.1016/j.msea.2013.11.033

Zareiyan, B., & Khoshnevis, B. (2017). Interlayer adhesion and strength of structures in Contour Crafting - Effects of aggregate size, extrusion rate, and layer thickness. *Automation in Construction*, *81*, 112–121. doi:10.1016/j.autcon.2017.06.013

Zotti, A., Zuppolini, S., Tábi, T., Grasso, M., Ren, G., Borriello, A., & Zarrelli, M. (2018). Effects of 1D and 2D nanofillers in basalt/poly(lactic acid) composites for additive manufacturing. *Composites. Part B, Engineering*, *153*, 364–375. doi:10.1016/j.compositesb.2018.08.128

Chapter 12
Multi-Criterion Decision Method for Roughness Optimization of Fused Deposition Modelled Parts

Azhar Equbal
RTC Institute of Technology, India

Md. Asif Equbal
Cambridge Institute of Technology, India

Md. Israr Equbal
J. B. Institute of Engineering and Technology, India

Anoop Kumar Sood
National Institute of Foundry and Forge Technology, India

ABSTRACT

Fused deposition modelling is an extrusion-based automated fabrication process for making 3D physical objects from part digital information. The process offers distinct advantages, but the quality of part lacks in surface finish when compared with other liquid or powder based additive manufacturing processes. Considering the important factors affecting the part quality, the chapter attempted to optimize the raster angle, air gap, and raster width to minimize overall part roughness. Experiments are designed using face-centered central composite design and analysis of variance provides the effects of processing parameters on roughness of part. Suitability of developed model is tested using Anderson-darling normality test. Desirability method propose that roughness of different part faces are affected differently with chosen parameters, and thus, hybrid approach of WPCA based TOPSIS is used to break the correlation between part faces and reduce the overall part roughness. Optimizing shows that lower raster angle, lower air gap, and larger raster width minimizes overall part roughness.

DOI: 10.4018/978-1-5225-9167-2.ch012

INTRODUCTION

Additive manufacturing (AM) is a layer based automated fabrication process for making scaled three dimensional physical objects directly from 3D-CAD (computer aided design) data (Equbal et al., 2015). The part is fabricated by depositing the part material in a layerwise deposition principle following bottom up approach even in an office friendly environment (Mohamed et al., 2016). AM processes have the ability to produce any complex geometry of part in a less time span without any specific tooling. As per ASTM F2792-12a, AM processes are categorized into seven different categories (ASTM Designation, 2012). Fused deposition modelling (FDM) is an extrusion-based AM process that construct the 3D object directly from its part digital information. The part material is used in the form of strand or filament which is heated in a liquifier to semi-molten state before extruding it through a nozzle onto a table or a plateform provided in a bulid chamber. While depositing the material on plateform, the nozzle moves in three different axes thus creating a cross section of three-dimensional object (Sood et al., 2009). The material on deposition gets cools, hardens and bonds to the layer beneath it. Based on the layer thickness used the process is repeated up to the last layer. The different parts materials used for fabrication are ABS (acrylonitrile butadiene styrene), PC (polycarbonate), ABSi (high impact grade of ABS) and PC-ABS. The process also uses water works soluble support for ABS, ABSi and PC-ABS and breakaway support for PC and PC-ISO (BASS™). Support material use can be easily breakaway by hand. It can build part in three different layer thicknesses that are 0.127mm, 0.178mm and 0.254mm. Process chain of part fabrication is as follows:

Step 1: 3D CAD model of part is made using any suitable CAD software.
Step 2: CAD model is converted into STL file format.
Step 3: Different processing parameters for part building and location of support structure is decided. The data generated is then changed to SML (Stratasys machine language) format and sent to hardware of FDM for part fabrication.
Step 4: Part is fabricated by layerwise deposition of materials in the build chamber.
Step 5: Once the part build is finished, it is taken out from the chamber and support material is removed by hand or some time by vibrations generated in ultrasonic bath.

FDM parts have good thermal and chemical resistant properties and excellent strength-to-weight ratios. It has very wide application in automobiles, medical and pattern for investment casting. FDM process offers distinct advantages like fabrication of complex part geometries, reduced product development time and cost, no process plan is required and absence of any specific tooling. Although the process offers numerous advantages, however when compared with other AM processes the surface finish of FDM parts are very poor. Visible rough patterns are inbuilt fabrication constraints of FDM parts which limits their industrial scopes. Surface roughness of FDM part is mainly affected by raster pattern deposition and staircase effect at the vertical plane. In a FDM system, the layer thickness could not be very small because the diameter of extruded filament material has hundreds of micrometers. Hence, the staircase at the vertical surface is quite high (Kim et al., 2018). Surface finish is an important parameter defining the FDM part quality and hence roughness of the part needs to be minimum. Literature survey reveals that optimization of processing parameters is suggested by many researchers to improve the surface finish of FDM parts. Bharath et al. proposed that layer thickness and raster width are more significant parameters affecting the part surface roughness when compared with other parameters like build orientation, air gap and model temperature (Bharath et al., 2000). Anitha et al., (2007), also concluded that the effect of layer thickness on FDM part roughness is more significant. Inverse relation between layer thickness and surface roughness was also established through correlation. Horvath et al., (2007), investigated the effect of processing parameters on surface roughness of ABS400 polymer. They suggested that the effect of layer thickness and model temperature were more dominant. Galantucci et al., (2009), recommended that slice height and the raster width are important parameters affecting part surface roughness while the tip diameter has little importance for surfaces running either parallel or perpendicular to the build direction. Nancharaiah et al., (2010), also observed that surface roughness could be improved by using lower value of layer thickness and air gap because it reduced the voids between layers. Mahapatra et al., (2012), attempted to improve the surface roughness of ABS P400 part by combining Bayesian regularization-based levenberg-marquardt neural model with Bacterial Foraging Optimization Algorithm (BFOA). They found that raster width is the most important parameter for improving surface finish at the top face. Part orientation and layer thickness are observed as significant factors for reduction of surface roughness at bottom and side faces. For estimating the roughness value, a 3D roughness profile model was developed by Boschetto et al. (Boschetto et al., 2013). They suggested that roughness was found to be affected by raster orientation. To improve the surface finish of part built by FDM process integration of FDM and BF (barrel finishing) methods is proposed by Boschetto et al. (Boschetto et al., 2015). FDM parts were produced and post processed in a rotating barrel machine. In a BF process FDM part, abrasive media and water are put in a rotating barrel and given specific speed which results in finishing of FDM part. Durgun and Ertan show that parts built by FDM with different part orientations have a strong effect on the surface roughness (Durgun et al., 2014). Reddy et al., (2018), investigated surface texture of FDM parts and concluded that roughness value decreases with increase in build inclination and increases with increase in layer thickness. Literature survey shows that the researchers are mainly focused on studying the effect of processing parameters on the surface roughness of part. Literature also concludes that part orientation and layer thickness was mainly chosen as process parameter for process optimization. Layer thickness and build orientation are geometry specific and their principal influence is on the build time, volume of part material and support structure (Wang et al., 2007). Optimum selection of these parameters varies from part geometry to geometry. Once these parameters are specified surface finish of fabricated part depend upon the layer deposition strategies involved (Lee et al., 2005). Important process parameters influencing layer deposition are raster width, raster angle and air gap. The present work therefore uses raster angle, air gap and raster width as processing parameter for process optimization. Desirability function (DF) method is used as single objective optimization method for optimizing roughness of top, bottom and side faces individually. Industrial requirement however demands single factor setting for minimizing surface roughness of overall part. WPCA based TOPSIS (Weighted principle component analysis based Technique for order of preference by similarity to ideal solution) is used as a multi-objective hybrid method for optimization of overall surface roughness.

METHODOLOGY

The part is fabricated using FDM Vantage SE machine by Stratasys. A well-established design of experiment (DOE) technique known as response surface methodology (RSM) is used for designing experiment. RSM define a polynomial relationship between independent input variables and dependent output response. To reduce the number of levels and number of experimental runs RSM based on face centred central composite design (FCCCD) is adopted in present study (Hattiangadi et al., 2000; Jeff et al., 2002). This design requires only three levels for each factor. FCCCD consists of eight star points, six cube points and six central points (Equbal et al., 2017). FCCCD does not require as many centre runs as other CCD based design. In practice, two or three centre runs are sufficient to provide good variance of prediction throughout the experimental region (Hattiangadi et al., 2000; Jeff et al., 2002). Based on initial trials and exhaustive literature review factors are grouped into fixed factors and control factors (ASTM Designation, 2012; Bharath et al., 2000; Anitha et al., 2001; Nancharaiah et al., 2010; Equbal et al., 2017; Sood et al., 2010; Kumar et al., 2014 Pradhan et al., 2009). Fixed factors as shown in Table 1 are not significant for studied quality measures and their value was not changed during entire experiment runs. Control factors are varied at three levels as given in Table 2. The level of each factor is selected in accordance with previous researches and the permissible minimum and maximum settings recommended by the equipment manufacturer (Sood et al., 2009; Sood et al., 2010).

Control factor levels are coded into +1, 0 and -1 using Eq. 1.

Table 1. Fixed Factors

Factors	Values
Part fill style	Perimeter /Raster
Contour width	0.4064 mm
Orientation	0°
Layer thickness	0.254 mm
X, *Y* and *Z* shrink factor	1.0038
Part interior style	Solid normal
Visible surface	Normal raster
Perimeter to raster air gap	0 mm

Table 2. Control factors and their levels

Factor	Symbol	Levels			Units
		1	2	3	
		low level (-1)	centre level (0)	high level (+1)	
Raster angle	A	0	30	60	degree
Air Gap	B	-0.004	0	0.004	mm
Raster Width	C	0.4064	0.4564	0.5064	mm

$$\xi_{ij} = \left(\frac{x_{ij} - \overline{x_i}}{\Delta x_i} \right) \times 2$$
$$\overline{x_i} = \frac{\sum_{j=1}^{2} x_{ij}}{2}, \qquad \text{and } \Delta x_i = \max(x_{ij}) - \min(x_{ij}) \tag{1}$$

where, $\max(x_i) = \max(x_{ij})$; $\min(x_i) = \min(x_{ij})$; $1 \le i \le k; 1 \le j \le 2$ and ξ_{ij}, x_{ij} are coded and actual value of the jth level of ith factor, $\overline{x_i}$ is mean of values for factor *i*, $\max(x_{ij})$ and $\min(x_{ij})$ is maximum and minimum value of the jth level of ith factor. The FCCCD used in present study is given in Table 3. It consists of 8 axial points, 8 star points and 6 centre runs to get a reasonable estimate of experimental error.

Part Fabrication and Data Collection

CATIAV5 solid modelling software is used to create a solid CAD model of the part shown in Figure 1. CAD model is then converted into machine accepted STL file format and imported into FDM software Insight™. During the fabrication the part is placed such dimension of face that is 30mm is along table *X*-axis and other dimension of this face (10.5mm) is along Y-axis of table. The third dimension (10.5mm) perpendicular to *X*-and *Y*-axis is along part build direction (*Z*-axis). This orientation of the part is 0° orientation.

Factor values are then set in the software. Software slices the STL model, calculate the support structure and generate the tool path for each layer. This information is converted into SML format and send to FDM Vantage SE machine for part fabrication. Machine will fabricate the part layer wise layer based on bottom up approach. Average surface roughness of part faces was

Table 3. Experimental plan for FCCCD runs

Run order	Factors (Coded)		
	Raster angle (*A*)	Air gap (*B*)	Raster width (*C*)
1	-1	-1	-1
2	1	-1	-1
3	-1	1	-1
4	1	1	-1
5	-1	-1	1
6	1	-1	1
7	-1	1	1
8	1	1	1
9	-1	0	0
10	1	0	0
11	0	-1	0
12	0	1	0
13	0	0	-1
14	0	0	1
15	0	0	0
16	0	0	0
17	0	0	0
18	0	0	0
19	0	0	0
20	0	0	0

Figure 1. 3D CAD model of part used

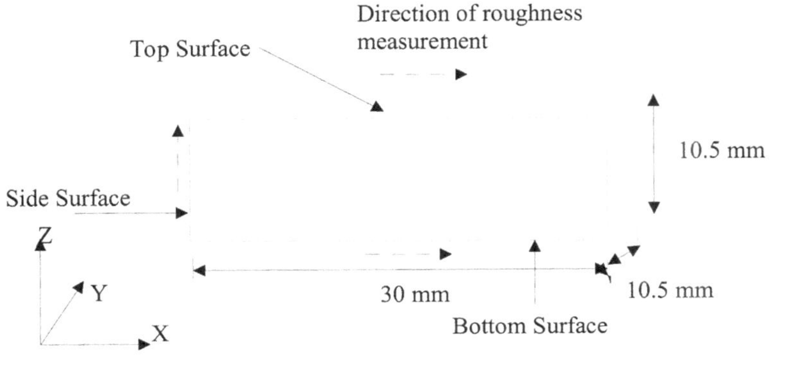

measured using contact type Hommel werke Turbo Wave V7.20 roughness tester. Table 4 shows the roughness measurement conditions. Three readings of average surface roughness on top (Ra^T), bottom (Ra^B) and side (Ra^S) surfaces are taken for fabricated specimen. Mean of these three observations is taken as representative value of respective surface roughness.

Analysis of Data

Analysis of Variance (ANOVA) is used to analyse the outcome of experiments. ANOVA is a statistically based decision-making tool which divides the total variation into accountable sources. It is used to study the effect of individual process parameters and their interaction on the response considered. Assumption in the ANOVA analysis is that given population is normally distributed. This assumption is validated by Anderson darling (AD) plot or normality test. AD plot is plotted to test the effectiveness or suitability of developed model. Here, p value should be greater than 'α'. If the p-value is lower than α, the data do not follow the normal distribution.

Optimization

For Optimization purposes, the present chapter uses three different techniques namely: Desirability, Principal component analysis and TOPSIS are used as optimization techniques in the presented work. Desirability function is used as single objective optimization method for optimization of individual faces roughnesses and WPCA based TOPSIS methos is used as a multi-objective optimization method for optimization of overall part surface roughness.

Table 4. Roughness measuring conditions

Condition	Value
Probe tip radius (TKU 300)	0.005 mm
Measuring range	80μm
Traverse length	4.8 mm
Speed	0.5 mm/s
Filter	ISO11562 [M1]

Desirability approach provides the method of finding out the operating conditions that provide the most desirable response value. For single response, each experimental result is converted into a scale of [0, 1] by calculating their desirability (d), where 1 is highly desirable value (Costa et al., 2011). The maximum value of desirability is then chosen and the factor setting corresponding to that chosen desirability is selected as optimal combination of parameters. The responses are scaled into desirability based on their characteristics namely larger-the-better, smaller-the-better and nominal the better.

- **Larger-The-Better (LTB):** The value of the estimated response is expected to be larger than a lower bound. For this response type, the individual desirability function is defined by Eq. 2:

$$d_i(Y)_i = \begin{cases} 0 & Y < L \\ \left(\dfrac{Y-L}{T-L}\right)^r & L \le Y \le T \\ 1 & Y > T \end{cases} \tag{2}$$

- **Smaller-The-Better (STB):** The value of the estimated response is expected to be smaller than an upper bound. For this response type, the individual desirability function is defined by Eq. 3:

$$d_i(Y)_i = \begin{cases} 1 & Y < T \\ \left(\dfrac{U-Y}{U-T}\right)^r & T \le Y \le U \\ 0 & Y > U \end{cases} \tag{3}$$

- **Nominal-The-Better (NTB):** The value of the estimated response is expected to achieve a particular target value. For this response type, the individual desirability function is defined by Eq. 4:

$$d_i(Y_i) = \begin{cases} 0 & Y < L \\ \left(\dfrac{Y-L}{T-L}\right)^{r_1} & L \le Y \le T \\ \left(\dfrac{U-Y}{U-T}\right)^{r_2} & T \le Y \le U \\ 0 & Y > U \end{cases} \tag{4}$$

where, Y is response, U is upper limit, L is lower limit, T is target value and r, r_1, r_2 are weights.

PCA is a data reduction technique developed in 1933 by Pearson and Hotelling. It is used to resolve the correlation between set of correlated response (Tong et al., 2005; Routara et al., 2010). These uncorrelated responses are called principal components (PCs). In a PCA analysis, responses are represented as cloud of n points in multi-dimensional (k-dimension) space with an axis for each of k response. PCA calculate the centroid of data points. The origin is translated to centroid and axes are orthogonally transformed and called as principal axis or principal component axes. The component of data points is calculated in terms of new reference frame. These axes explain the variation in data such that first principal axis will explain the maximum variation followed by second principal axis and so on but there will be no correlation in data between two axes. Depending upon the percentage of variation explain by these axes dimensions of data space can be reduced by eliminating the axes along which there is least variation of data. The various steps in principal component analysis are given as follows:

1. Collection of response data and normalization of data between [0, 1] as per the objective of response i.e. higher the better, lower the better and nominal the better. Here 1 is the best value and 0 will be worst value.
2. A variance-covariance matrix is constructed between each pair of response. This matrix is a square matrix where diagonal elements are variances and remaining elements are covariances. Variance (Vi) and covariance (Cij) between the responses is calculated using Eq. 5 and Eq. 6.

$$V_i = \frac{1}{n-1}\sum_{i=1}^{n}(X_i - \overline{X_i}) \tag{5}$$

$$C_{ij} = \frac{1}{n-1}\sum_{i=1}^{n}(X_i - \overline{X_i})(X_j - \overline{X_j}) \tag{6}$$

where, $i = 1,2.....n$, $j = 1,2.....k$, $\overline{X_i}$ and $\overline{X_j}$ are mean of ith and j^{th}set of observation.

3. Calculation of the Eigen value $\lambda_1, \lambda_2......\lambda_k$ and the corresponding Eigen vector $\beta_1, \beta_2......\beta_k$ (where, $\beta_i = \beta_{i1}.,.....\beta_{ik}$) from the variance-covariance matrix formed by all the quality characteristics.
4. Calculation of principal components by using Eq. 7.

$$Z_i = \sum_{j=1}^{k} \beta_{ij} Y_{ji} \tag{7}$$

Z_i are principal component and Y_{ji} are normalized data.

The present study uses modified form of PCA which considers all the principal components including the principal components explaining the least variations. This modified approach of PCA is known as WPCA (weighted principal component analysis) (Routara et al., 2010).

TOPSIS (Technique for order of preference by similarity to ideal solution) is a decision- making process that selects the best solution from a set of alternatives, each of which is evaluated against multiple quality characteristic or response (Tong et al., 2005; Lee et al, 2014). The chosen alternative should be the one that is the closest to the ideal alternative and the farthest from the negative-ideal alternative. The various steps used in TOPSIS are discussed below:

Step 1: Construction of decision matrix: A decision matrix $D = \left[X_{ij}\right]_{n\times k}$ is $n \times k$ matrix, where row represent alternatives and column represent quality characteristic or response, X_{ij} is an element corresponding to i^{th} alternative and j^{th} response and $i = 1,2....n; j = 1,2....k$

Step 2: Normalization of decision matrix: Normalize scores for each element of matrix (r_{ij}) is calculated as per Eq. 8.

$$r_{ij} = \frac{x_{ij}}{\sum_{i=1}^{n} x_{ij}^{2}} (for\ i = 1, 2.......n; j = 1, 2....k) \tag{8}$$

Step 3: Weighted normalized decision matrix: Depending upon the importance of each response weights are assigned to them. Each column in normalized decision matrix is then multiplied by its corresponding weight 'w_j' as shown in Eq. 9.

$$V_{ij} = w_j r_{ij} \tag{9}$$

where, V_{ij} is the element of weighted normalized decision matrix.

The weighted normalized decision matrix is represented as:

$$\begin{bmatrix} w_1 r_{11}. & w_2 r_{12}.... & w_j r_{ij} \\ w_1 r_{21} & w_2 r_{22}.... & w_j r_{ij} \\ w_1 r_{n1} & w_2 r_{n2} & w_j r_{nk} \end{bmatrix} = \begin{bmatrix} v_{11} & v_{12}.... & v_{1k} \\ v_{21} & v22.... & v_{2k} \\ v_{n1} & v_{n2}.... & v_{nk} \end{bmatrix}$$

Step 4: The distance of ith alternative is then determined from the ideal and negative-ideal solutions. The distance of the ith alternative from the ideal solution is calculated as Eq. 10 and Eq. 11.

$$d_i^+ = \sqrt{\sum_{j=1}^{k} (v_j^i - v_j^+)^2} \tag{10}$$

$$d_i^- = \sqrt{\sum_{j=1}^{k} (v_j^i - v_j^-)^2} \tag{11}$$

$$\text{where, } v_j^+ v_j^- = \begin{cases} \max(\min)\left\{ v_i^j, for\ i = 1,2...n \right\} \\ \forall v_j^{\,i}(i = 1,2...n;\ \ j = 1,2...k) \\ \min(\max)\left\{ v_j^i, for\ i = 1,2...n \right\} \\ \forall v_j^{\,i}(i = 1,2...n; j = 1,2...k) \end{cases}$$

v_j^+ is used for profit and v_j^- is used for cost.

Step 5: Ranking of the alternative preference order by relative closeness to ideal solution C_i is determined using Eq. 12.

$$C_i = d_i^- / d_i^+ + d_i^- \tag{12}$$

where, $0 < C_i < 1$ and C_i is TOPSIS score. The alternative corresponding to the highest value of C_i is selected as optimized factor setting.

Confirmation Experiment

To verify the experimental conclusions confirmation experiments are conducted by determining the results of test using a specific combination of the factors and levels. Before carrying out the confirmation experiment, a confidence interval (*CI*) is decided. The confidence interval is a maximum and minimum value between which the value should lie at some stated percentage of confidence. *CI* at 95% confidence level for each response is calculated using Eq. 13 (Ross et al., 2005).

$$CI = \left[F_{0.05;1}; V_e \left(\sqrt{\frac{1}{n_{neff}} + \frac{1}{r}} \right) \right] \tag{13}$$

$$n_{eff} = \frac{N}{1 + n^{'}} \tag{14}$$

n_{eff} is number of effective terms calculated using Eq. 14. *F* is *F*- statistic value calculated from *F* table, V_e is degree of freedom of error, $n^{'}$ is number of significant terms, *r* is sample size for confirmation experiment and *N* is the number of experimental trials. It is important to mention that V_e and n_{eff} are noted from pooled ANOVA table. The range for each response is then calculated using Eq. 15.

$$\eta_e^r = \eta_{pre}^r \pm CI^r \tag{15}$$

where, for r^{th} response η_e is range of expected value, η_{pre} is value predicted for respective responses using response surface equations and *CI* is given by Eq. 15 (Ross et al., 2005).

RESULTS

For the part fabricated, Ra^T, Ra^B and Ra^S are measured as per the methodology discussed in section 2. The results of measurement are tabulated in Table 5. Here Exp. No. (experiment number) 1, 2...20 corresponds to Exp. No. mentioned in design matrix given in Table 3 and

-1, 0, 1 is coded values of control factors given in Table 2. Other factors are kept at their fixed values mentioned in Table 1. Run order of experiment is random and for each experiment three specimen is fabricated and their average value corresponding to each response is tabulated in Table 5. Roughness profiles of top, bottom, and side faces are shown in Figure 2. (for a specimen manufactured as per Exp. No. 3). Roughness measurement shows that surface roughness of side face (Ra^S) of the specimen is less as compared to roughness of top face (Ra^T) and bottom face (Ra^B) and this is true for all other experiment run.

Data analysis is performed using statistical software Minitab R16 at 95% of confidence level. ANOVA results corresponding to each studied response is presented from Table 6. In these tables, *SS* corresponds to sum of square, *V* corresponds to variance, *DOF* corresponds to degree of freedom, *Reg* corresponds to regression and *LOF* corresponds to lack of fit. Adequacy of regression and significance of each term is determined using *F*- value given in ANOVA table. Probability of *F* value greater than calculated *F*-value due to noise is indicated by *p*-value.

Table 5. Experimental results for FDM parts

Run Order	Exp. No.	Roughness in μm		
		Ra^T	Ra^B	Ra^S
10	1	4.4220	7.2990	0.1360
15	2	1.4480	9.6175	0.3530
7	3	4.7600	10.0765	0.1960
3	4	2.3000	7.9050	0.4515
5	5	7.8000	7.0890	0.3490
17	6	1.8615	9.6445	0.6253
16	7	6.1965	9.8485	0.2240
2	8	1.9200	9.3040	0.3640
13	9	5.7785	8.7115	0.1550
14	10	1.3970	9.6560	0.4570
20	11	3.2570	6.9230	0.4850
9	12	3.5640	8.3195	0.3986
11	13	3.1640	9.4430	0.2180
1	14	3.7350	9.9055	0.2605
4	15	3.8800	9.0500	0.1645
6	16	3.6500	8.8030	0.2020
8	17	3.7190	8.8690	0.2835
12	18	3.3800	8.9320	0.2520
18	19	3.3400	9.0250	0.1790
19	20	3.6100	9.2775	0.1705

Figure 2. Roughness profiles for FDM part (a) Top face (b) Side face and (c) Bottom face

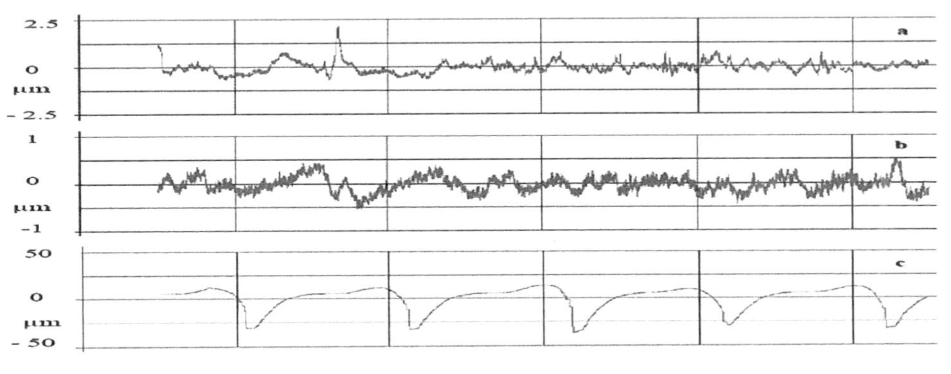

If p-value is less than 0.05, corresponding term is considered as significant or vice-versa. But, for lack of fit, p-value must be greater than 0.05. An insignificant lack of fit is desirable because it indicates any term left out of model is not significant and developed model fits well (Hattiangadi et al. 2000.; Jeff et al., 2002; Montgomery, 2012). ANOVA results show that quadratic model is suitable for predicting the considered response. Based on F-value in ANOVA tables, significant factors and interaction terms for different studied responses are presented Table 7. Figure 3 shows the Anderson darling (A-D) normality plot for surface roughness. Here, p value is greater than 0.05 signifying that residue follow normal distribution and respective models developed by response surface equations (Eq. 16- Eq.18) are suitable for practical engineering applications. Response surface plot for significant interactions of surface roughness is presented in Figure 4.

$$\begin{aligned} Ra^T &= 3.51159 - 2.00305A + 0.54190C \\ &\quad + 0.27200A \times B - 0.59762A \times C - 0.34188B \times C \\ R^2 &= 98.27\ \% \end{aligned} \tag{16}$$

$$\begin{aligned} Ra^B &= 8.9412 + 0.3103A + 0.4880B - 1.24B^2 \\ &\quad + 0.8103C^2 - 0.9487A \times B + 0.2330A \times C \\ R^2 &= 96.22\ \% \end{aligned} \tag{17}$$

$$\begin{aligned} Ra^S &= 0.242406 + 0.119080A + 0.046830C + 0.148659B^2 - 0.068100B \times C \\ R^2 &= 87.17\ \% \end{aligned} \tag{18}$$

R^2 is coefficient of variation which indicates the percentage of total variation explained in the model.

DISCUSSION

Fused deposition modelling is a material extrusion based additive manufacturing process where part is fabricated using fusion bonding principle. In this process, part material (ABS P400) is used in form of wire or filament. Part material is

Table 6. ANOVA Table for surface roughness of Ra^T, Ra^B and Ra^S

Source	DOF	Ra^T in µm				Ra^B in µm				Ra^S in µm			
		SS	V	F	p	SS	V	F	p	SS	V	F	p
A	1	40.1221	40.1221	477.87	0.000	0.9626	0.96255	15.25	0.003	0.141800	0.141800	32.41	0.000
B	1	0.0002	0.0002	0.00	0.959	2.3819	2.38193	37.73	0.000	0.009872	0.009872	2.26	0.164
C	1	2.9366	2.9366	34.98	0.000	0.2104	0.21040	3.33	0.098	0.021930	0.021930	5.01	0.049
A × A	1	0.3340	0.1139	1.36	0.271	0.0182	0.28128	4.46	0.061	0.024304	0.000455	0.10	0.754
B × B	1	0.0082	0.0019	0.02	0.883	2.8204	4.24671	67.27	0.000	0.052798	0.060774	13.89	0.004
C × C	1	0.0117	0.0117	0.14	0.717	1.8057	1.80569	28.60	0.000	0.007987	0.007987	1.83	0.206
A × B	1	0.5919	0.5919	7.05	0.024	7.2010	7.20101	114.06	0.000	0.001196	0.001196	0.27	0.613
A × C	1	2.8572	2.8572	34.03	0.000	0.4343	0.43431	6.88	0.025	0.000395	0.000395	0.09	0.770
B × C	1	0.9350	0.9350	11.14	0.008	0.2292	0.22916	3.63	0.086	0.037101	0.037101	8.48	0.016
Error	10	0.8396	0.0840			0.6313	0.06313			0.043754	0.004375		
Total	19	48.6366				16.6950				0.341138			
Reg	9	47.7970	5.3108	63.25	0.000	16.0637	1.78485	28.27	0.000	0.297384	0.033043	7.55	0.002
LOF	5	0.6285	0.1257	2.98	0.128	0.4909	0.09818	3.50	0.098	0.031944	0.006389	2.70	0.149

Table 7. Significant factors and interactions for studied responses in FDM part fabrication

Significant Factors	Roughness		
	Ra^T	Ra^B	Ra^S
Linear	*A, C*	*A, B*	*A, C*
Interactions	*AXB, AXC, BXC*	*AXB, AXC*	*BXC*

Figure 3. Normality plots for surface roughness (a) Ra^T (b) Ra^B and (c) Ra^S

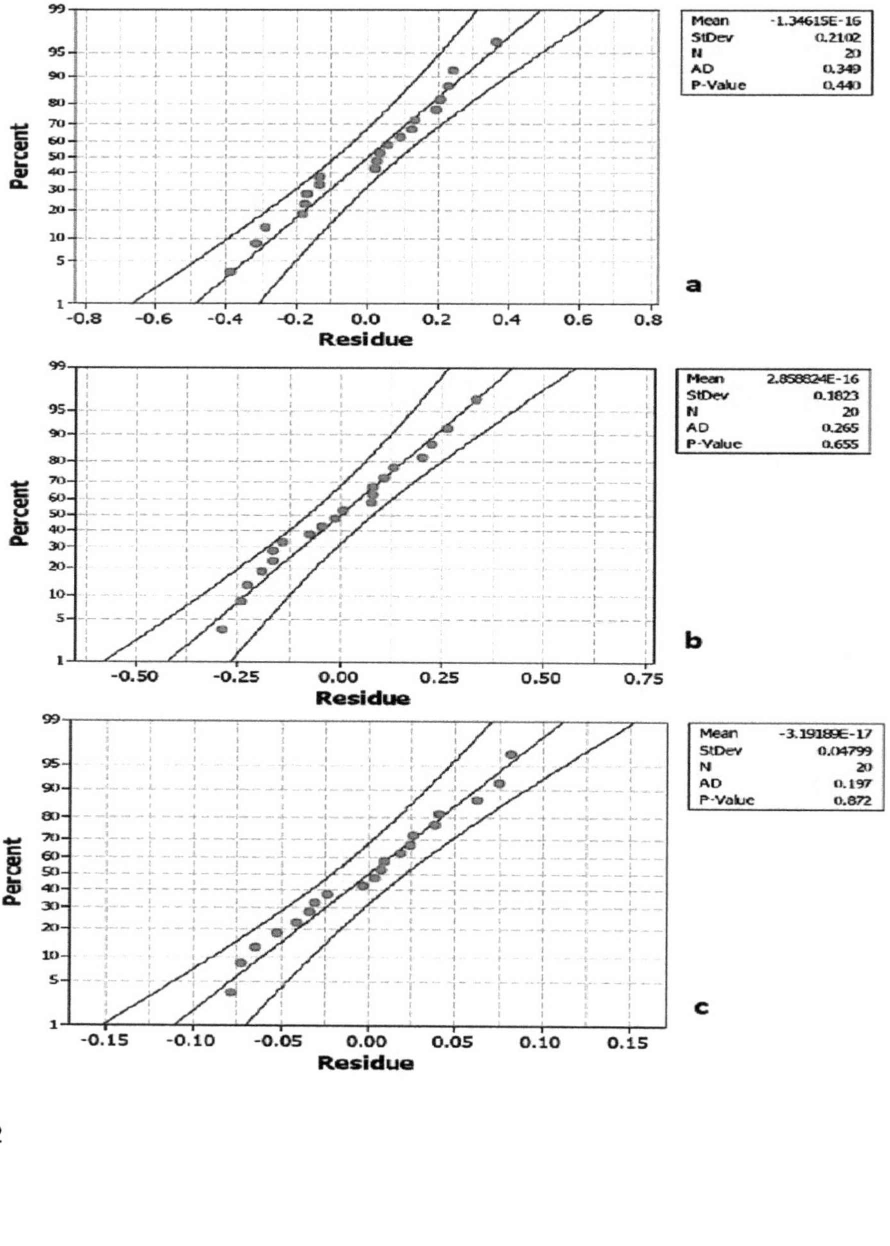

Figure 4. Response surface plots (a) Ra^T, AxB (b) Ra^T, AxC (c) Ra^T, BxC (d) Ra^B, AxB (e) Ra^B, AxC and (f) Ra^S, BxC

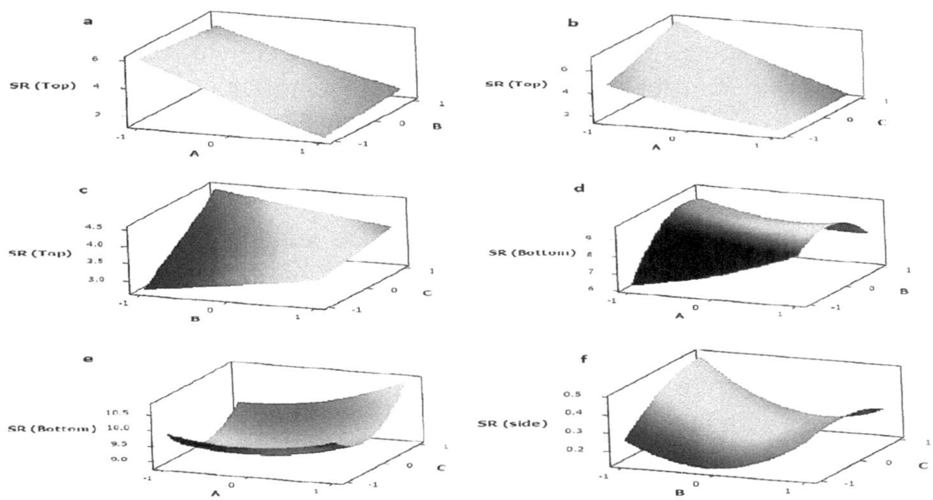

fed through canister provided at the bottom chamber of machine and is guided through roller mechanism into liquifier. When heated above its glass transition temperature (~110°C) the filament is melted to its semi molten state (Gibson et al., 2010). The semi molten material is extruded from the nozzle, expands and deposited in form of continuous beads (known as rasters) on previously deposited material or table placed in temperature-controlled build chamber (~80°C). When the melted material cools its viscosity increases forming it into solid. Heat is transformed to the surrounding in the form of conduction, convection and radiation (Turner et al., 2014). This causes localized heating and melting of previously deposited material which causes fusion bonding between newly deposited material and already deposited materials. For proper fusion bonding, temperature of materials must be above glass transition temperature (T_g) (Bellini et al., 2014). When two adjacent beads diffuse together the type of surface results depend upon the degree of overlap between them. If the overlapping is adequate it will result in a flat surface. Excess of overlapping may result in bump formation and if overlap is less than the critical value it may result in depression. The quality of bonding achieved between adjacent beads is also affected by the amount of overlapping. Bonding is more and better in case of excess overlap in comparison to deficient overlap (Bellehumeur et al., 2008). The profile generated by elliptical cross section of raster deposition and overfilling between contact areas of two raster contributes in variation along part thickness.

The variation of responses with change in input processing parameters is studied using response surface plots. In general, Lower value of raster angle (*A*) produces larger raster lengths whereas the higher raster angle will generate shorter length of raster. The rasters are generated in form of continuous beads. For better surface roughness larger raster angle or shorter raster length is preferred due to less number of deposited beads and less variation is expected. Increase in air gap (*B*) results in larger surface roughness roughness. For a positive air gap, there are spaces on the sides of each deposited raster. When the extruded melt is deposited on previously deposited rasters, it can flow into these spaces in a random manner and can lead to high variation in roughness. At zero or negative air gap the rasters are deposited adjacent to each other with negligible or no spaces between rasters thus reducing the variation in surface roughness. Lower value of raster width (*C*) is preferred for minimizing roughness of FDM part as it reduces the variability of surface roughness because of small height of resulting beads. Increase in further raster width increases the surface roughness of part.

Response surface plots for surface roughness presented in Figure 4 is in clear agreements with the above stated discussions. Figure 4 (a) to Figure 4 (c) shows that surface roughness of top face (Ra^T) decreases with increase of raster angle and increases with increase of air gap and raster width. This is in clear agreement with the reasons explained. Roughness in top surface is mainly contributed by the deposition of raster and also because of process related errors propagating from layers below it, resulting in an accumulated error on it. (Liu et al., 1998). Figure 4 (d) - Figure 4 (e) shows that roughness of bottom face (Ra^B) is also affected by raster angle, air gap and raster width but for bottom face roughness is mainly contributed by support structure. During part fabrication the bottom surface will always remain in contact with support structure and after part fabrication the support is removed away but its impression remains at the face affecting the roughness present in it. Roughness in side face (Ra^S) is because of staircase effect present due to slicing of part. Staircase effect is more dominant in curved profiles however straight profiles are not much affected by it. The presents study uses flat part with side surface having straight profile thus, roughness of side face is less in present study when compared to top and bottom face (Figure 4 (e)).

Optimization

The criterion of desirability function is to select the factor setting with maximum overall desirability. Figure 5 presents the result of desirability for optimization of individual faces roughness.

Figure 5 (a) to Figure 5 (c) presents the desirability plot for minimizing surface roughness in top, bottom and side surface individually. Figure 5 evident desirability of 1 making these models very well suited for minimizing the surface roughness of part.

Desirability method provides individual factor settings for fabricating part with good individual faces roughness. The industrial requirement however demands only one factor setting for fabricating part with good overall part roughness. To achieve the same, WPCA-TOPSIS method has been used. Pearson's correlation coefficient between the normalized responses appears to be non-zero indicates that all response features are correlated to each other. Eigen value, explained variation, and their Eigen vectors are presented in Table 8. Equations for different principal components (PC_S) are given from Eq. 19 to Eq. 21.

Figure 5. Optimized process parameters for minimizing surface roughness at (a) Top (b) Bottom and (c) Side faces

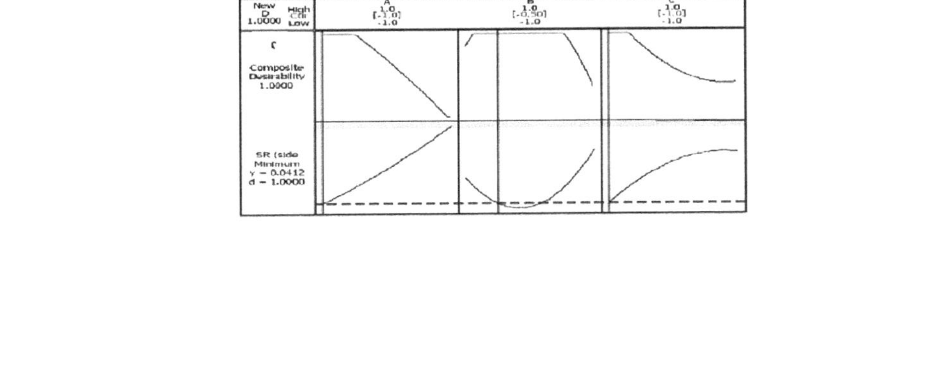

Table 8. Table showing Eigen values and Eigen vectors from PCA analysis

Principle Component	Eigen Value	Explained variation	Cumulative Variation	Eigen Vector
*PC*1	1.5129	1.1152	0.3719	[-0.738, 0.292, 0.609]
*PC*2	0.504	0.372	0.124	[-0.099, 0.845, -0.526]
*PC*3	0.504	0.876	1.000	[0.668, 0.448, 0.594]

$$PC1 = -0.738Ra^{T} + 0.292Ra^{B} + 0.609Ra^{S} \tag{19}$$

$$PC2 = -0.099Ra^{T} + 0.845Ra^{B} - 0.526Ra^{S} \tag{20}$$

$$PC3 = 0.668Ra^{T} + 0.448Ra^{B} + 0.594Ra^{S} \tag{21}$$

where, *PC1, PC2,* PC3 are three principal components.

TOPSIS uses principal components as response and explained variation is used as weights. TOPSIS Score and their desirability value are presented in Table 9.

Desirability of 1 (Table 9) corresponds to factor setting 5 [-1, -1, 1] in coded unit as per FCCCD design given in Table 3. Thus, at factor setting of 0 ° raster angles, -0.004 mm air gap and 0.5064 mm raster width fabricated FDM part will have minimum surface roughness.

Confirmation Experiment

For confirming the results of analysis, confirmation experiments are conducted for each response at optimum factor levels of [-1, -1, 1]. The results are presented in Table 10. The resulting model seems to be capable of predicting Ra^{T}, $Ra^{B}$and $Ra^{S}$to a reasonable accuracy.

Table 9. TOPSIS score and their desirability values

Exp. No.	d_i^+	d_i^-	$d_i^+ + d_i^-$	C_i	Desirability
1.	0.13127	0.435759	0.567029	0.7685	0.715273
2.	0.417073	0.122433	0.539506	0.2269	0
3.	0.337199	0.293432	0.63063	0.4653	0.23117
4.	0.307973	0.264378	0.572352	0.4619	0.15948
5.	*0.035643*	*0.521825*	*0.557468*	*0.9361*	*1*
6.	0.475684	0.1407	0.616384	0.2283	0
7.	0.288138	0.348494	0.636632	0.5474	0.319187
8.	0.381379	0.157957	0.539336	0.2929	0
9.	0.216245	0.407856	0.624102	0.6535	0.504959
10.	0.445482	0.105819	0.551301	0.1919	0
11.	0.240096	0.364711	0.604807	0.603	0.480476
12.	0.265251	0.277041	0.542292	0.5109	0.313442
13.	0.329369	0.24375	0.573119	0.4253	0.136097
14.	0.350389	0.23038	0.580769	0.3967	0.154545
15.	0.278669	0.31121	0.589879	0.5276	0.237429
16.	0.266067	0.300552	0.566619	0.5304	0.237429
17.	0.275431	0.273576	0.549008	0.4983	0.237429
18.	0.287332	0.266373	0.553704	0.4811	0.237429
19.	0.290764	0.285059	0.575822	0.495	0.237429
20.	0.303026	0.287631	0.590658	0.487	0.237429

Table 10. Confirmation test results for FDM fart fabrication

Response	Optimal parameters combination			Experimental value	Predicted Value	Range
	A	B	C			
Ra^T	-1	-1	1	7.80000	8.22600	[7.25873, 8.34127]
Ra^B	-1	-1	1	7.08900	7.42000	[6.65540, 7.62401]
Ra^S	-1	-1	1	0.34900	0.23200	[0.20340, 0.49410]

CONCLUSION

Parts are fabricated by FDM Vantage SE machine using combination of process variables: raster angle (*A*), air gap (*B*) and raster width (*C*) as shown in Table 3 the value for which is given in Table 2. Effect of process variables on surface roughness is studied by response surface plots. Surface roughness of FDM part is measured by measuring roughness of top, bottom and side faces. ANOVA analysis is performed for evaluating the effect of process variables on defined quality characteristics. Variation in surface roughness of top, bottom and side faces with change of process parameters is expressed through response surface plots. A mathematical model is proposed in terms of response surface equations to correlate part quality with studied parameters. Suitability of model is tested with Normality plots. Significance of factors and their significant level are determined but actual part fabrication is done at one factor setting and thus optimization is performed. Desirability function is used for single objective optimization and WPCA based TOPSIS hybrid approach is used as for multi-objective optimization. Important conclusions drawn are:

1. Roughness in top face is contributed by raster pattern. Higher roughness of bottom face is mainly because of impression of support structure. Side face experiences roughness because of slicing or staircase effect and profile of contour. In the present study surface roughness of side surface is low as part is flat and staircase effect is not present.
2. Lower value of raster angles (*A*) forms larger raster length and larger length has more ridges which increases roughness of part. Increasing the raster angle thus reduces roughness by shortening the raster length.
3. Larger air gap (*B*) increases the surface roughness as there are spaces on the sides of each deposited raster. During material deposition melt can flow into these spaces in a random manner and can lead to high surface roughness. Slight negative or no air gap eliminates the presence of these voids thus reducing the surface roughness.
4. Small raster width has less variation in roughness because of presence of smaller ridges whereas roughness values continuously increase with increasing raster width (*C*).
5. Result of optimization shows that lower raster angle, lower air gap and larger raster width is preferred for reducing the overall part roughness.

REFERENCES

Anitha, R., Arunachalam, S., & Radhakrishnan, P. (2001). Critical parameters influencing the quality of prototypes in fused deposition modelling. *Journal of Materials Processing Technology, 118*(1-3), 385-388. doi:(01)00980-3. doi:10.1016/S0924-0136

Bellehumeur, C. T., Gu, P., Sun, Q., & Rizvi, G. M. (2008). Effect of processing conditions on the bonding quality of FDM polymer filaments. *Rapid Prototyping Journal*, *14*(2), 72–80. doi:10.1108/13552540810862028

Bellini, A., Güçeri, S., & Bertoldi, M. (2004). Liquefier dynamics in fused deposition. *Journal of Manufacturing Science and Engineering*, *126*(2), 237–246. doi:10.1115/1.1688377

Bharath, V., Prakash, N. D., Anshuman, R., & Henderson, M. (2000). Sensitivity of RP surface finish to process parameter variation. In *Solid free form fabrication proceedings*. The University of Texas.

Boschetto, A., & Bottini, L. (2015). Surface improvement of fused deposition modelling parts by barrel finishing. *Rapid Prototyping Journal*, *21*(6), 686–696. doi:10.1108/RPJ-10-2013-0105

Boschetto, A., Giordano, V., & Veniali, F. (2013). 3D roughness profile model in fused deposition modelling. *Rapid Prototyping Journal*, *19*(4), 240–252. doi:10.1108/13552541311323254

Costa, N. R., Lourenço, J., & Pereira, Z. L. (2011). Desirability function approach: A review and performance evaluation in adverse conditions. *Chemometrics and Intelligent Laboratory Systems*, *107*(2), 234–244. doi:10.1016/j.chemolab.2011.04.004

Durgun, I., & Ertan, R. (2014). Experimental Investigation of FDM Process for improvement of mechanical properties and production cost. *Rapid Prototyping Journal*, *20*(3), 228–235. doi:10.1108/RPJ-10-2012-0091

Equbal, A., Dixit, N. K., & Sood, A. K. (2015). Electroless metallisation of ABS plastic: A comparative study. *International Journal of Rapid Manufacturing*, *5*(3-4), 255–275. doi:10.1504/IJRAPIDM.2015.074806

Equbal, A., Sood, A.K., Ansari, A. K., & Equbal, A. (2017). Optimization of process parameters of FDM part for minimiizing its dimensional inaccuracy. *International Journal of Mechanical and Production Engineering Research and Development, 7*(2), 57-65.

Galantucci, L.M., Lavecchia, F. Percoco, G. (2009). Experimental study aiming to enhance the surface finish of fused deposition modelled parts. *CIRP Annals - Manufacturing Technology, 58*(1), 189-192. doi:.2009.03.071 doi:10.1016/j.cirp

Gibson, I., Rosen, D. W., & Stucker, B. (2010). Additive Manufacturing Technologies Rapid prototyping to direct digital manufacturing. Springer. doi:10.1007/978-1-4419-1120-9

Hattiangadi, A., & Bandyopadhyay, A. (2000). Modeling of Multiple Pore Ceramic Materials Fabricated via Fused Deposition Process. *Scripta Materialia, 42*(6), 581–588. doi:10.1016/S1359-6462(99)00370-X

Horvath, D., Noorani, R., & Mendelson, M. (2007). Improvement of surface roughness on ABS 400 polymer using design of experiments (DOE). *Material Science Forum, 561*, 2389-2392. Retrieved from www.scientific.net/MSF.561-565.2389

Jeff, C. F. W., & Hamada, M. (2002). *Experiments: Planning, Analysis, and Parameter Design Optimization*. New Delhi, India: John Wiley & Sons.

Kim, M. K., Lee, I. H., & Kim, H. C. (2018). Effect of fabrication parameters on surface roughness of FDM parts. *International Journal of Precision Engineering and Manufacturing, 19*(1), 137–142. doi:10.100712541-018-0016-0

Kumar, S.D., Kannan, V.N., & Sankaranarayanan, G. (2014). Parameter optimization of ABS-M30i parts produced by fused deposition modelling for minimum surface roughness. *International Journal of Current Engineering and Technology, 3*, 93-97.

Lee, B. H., Abdullah, J., & Khan, Z. A. (2005). Optimization of rapid prototyping parameters for production of flexible ABS object. *Journal of Materials Processing Technology, 169*(1), 54–61. doi:10.1016/j.jmatprotec.2005.02.259

Lee, K. K., Lee, K. H., Woo, E. T., & Han, S. H. (2014). Optimization Process for Concept Design of tactical missiles by Using Pareto Front and TOPSIS. *International Journal of Precision Engineering and Manufacturing*, *15*(7), 1371–1376. doi:10.100712541-014-0478-7

Liu, W., Li, L., & Kochhar, A. K. (1998). A Method for Assessing Geometrical Errors in Layered Manufacturing. Part 1: Error Interaction and Transfer Mechanisms. *International Journal of Advanced Manufacturing Technology*, *14*(9), 637–643. doi:10.1007/BF01192283

Mahapatra, S. S., & Sood, A. K. (2012). Bayesian regularization-based Levenberg-Marquardt neural model combined with BFOA for improving surface finish of FDM processed part. *International Journal of Advanced Manufacturing Technology*, *60*(9-12), 1223–1235. doi:10.100700170-011-3675-x

Mohamed, O. A., Masood, S. H., Bhowmik, J. L., Mostafa, N. M., & Azadmanjiri, J. (2016). Effect of process parameters on dynamic mechanical performance of FDM PC/ABS printed parts through design of experiment. *Journal of Materials Engineering and Performance*, *25*(7), 2922–2935. doi:10.100711665-016-2157-6

Montgomery, D. C. (2012). *Design and Analysis of Experiments*. Singapore: John Wiley & Sons.

Nancharaiah, T., Raju, D.R., & Raju, V.R. (2010). An experimental investigation on surface quality and dimensional accuracy of FDM components. *International Journal of Emerging Technology, 1*(2), 106-111.

Pradhan, M.K., & Biswas, C.K. (2009). Modelling and analysis of process parameters on surface roughness in EDM of AISI D2 tool Steel by RSM Approach. *International Journal of Mechanical and Mechatronics Engineering, 3*(9), 1132-1137.

Reddy, V. F. O., Chaparala, A., Berrimi, C. E., Amogh, V., & Rosen, B. G. (2018). Study on surface texture of Fused Deposition Modeling. *Procedia Manufacturing*, *25*, 389–396. doi:10.1016/j.promfg.2018.06.108

Ross, P. J. (2005). Taguchi techniques for quality engineering (2nd ed.). Tata McGraw-Hill Publishing Company Limited.

Routara, B. C., Mohanty, S. D., Datta, S., Bandyopadhyay, A., & Mahapatra, S. S. (2010). Combined quality loss (CQL) concept in WPCA-based Taguchi philosophy for optimization of multiple surface quality characteristics of UNS C34000 brass in cylindrical grinding. *International Journal of Advanced Manufacturing Technology*, *51*(1-4), 135–143. doi:10.100700170-010-2599-1

Sood, A. K., Ohdar, R. K., & Mahapatra, S. S. (2009). Improving dimensional accuracy of fused deposition modelling processed part using grey Taguchi method. *Materials & Design*, *30*(10), 4243–4252. doi:10.1016/j.matdes.2009.04.030

Sood, A. K., Ohdar, R. K., & Mahapatra, S. S. (2010). Parametric appraisal of mechanical property of fused deposition modelling processed parts. *Materials & Design*, *31*(1), 287–295. doi:10.1016/j.matdes.2009.06.016

Standard terminology for additive manufacturing technologies. (2012). *ASTM international Designation: F2792-12a*. doi:10.1520/F2792-12A

Tong, L. I., Wang, C. H., & Chen, H. C. (2005). Optimization of multiple responses using principal component analysis and technique for order preference by similarity to ideal solution. *International Journal of Advanced Manufacturing Technology*, *27*(3-4), 407–414. doi:10.100700170-004-2157-9

Turner, B. N., Strong, R., & Gold, S. A. (2014). A review of melt extrusion additive manufacturing processes: I. Process design and modelling. *Rapid Prototyping Journal*, *20*(3), 192–204. doi:10.1108/RPJ-01-2013-0012

Wang, C. C., Lin, T. W., & Hu, S. S. (2007). Optimizing the rapid prototyping process by integrating the Taguchi method with the gray relational analysis. *Rapid Prototyping Journal*, *13*(5), 304–315. doi:10.1108/13552540710824814

Compilation of References

(2013). Science and society: Experts warn against bans on 3D printing. *Science*, *342*(6157), 439. PMID:24163835

(ISO) International Organization for Standardization. (2015). Additive manufacturing - General principles. ISO - International Organization for Standardization.

(ISO) International Organization for Standardization. (2016). Specification for Additive Manufacturing File Format (AMF) Version 1.2. *International Organization for Standardization*. ISO.

A.M. UK. (2017). *Additive Manufacturing UK: National Strategy, 2018–2025*. Author.

Abdulrahman, K. O., Akinlabi, E. T., & Mahamood, R. M. (2018). Manufacturing of aluminium composite materials: A review. In K. Kumar & P. Davim (Eds.), *Hierarchical composite materials: Materials, Manufacturing and Engineering*. Berlin: Witer de Gruyter. doi:10.1515/9783110545104-002

AbouHashem, Y., Dayal, M., Savanah, S., & Strkali, G. (2015). The application of 3D printing in anatomy education. *Medical Education Online*, *20*(1), 29847. doi:10.3402/meo.v20.29847 PMID:26478143

Adam, G. A. O., & Zimmer, D. (2014). Design for Additive Manufacturing—Element transitions and aggregated structures. *CIRP Journal of Manufacturing Science and Technology*, *7*(1), 20–28. doi:10.1016/j.cirpj.2013.10.001

Aguado, B. A., Mulyasasmita, W., Su, J., Lampe, K. J., & Heilshorn, S. C. (2011). Improving viability of stem cells during syringe needle flow through the design of hydrogel cell carriers. *Tissue Engineering. Part A*, *18*(7-8), 806–815. doi:10.1089/ten.tea.2011.0391 PMID:22011213

Ahmed, E. M. (2015). Hydrogel: Preparation, characterization, and applications. *Journal of Advanced Research*, *6*(2), 105–121. doi:10.1016/j.jare.2013.07.006 PMID:25750745

Ahn, S., Montero, M., Odell, D., Roundy, S., & Wright, P. (2002). Anisotropic material properties of fused deposition modeling ABS. *Rapid Prototyping Journal*, *8*(4), 248–257. doi:10.1108/13552540210441166

Ahsan, M. N. (2011). *Modelling and analysis of laser direct metal deposition of Ti-6Al-4V alloy* (PhD thesis). Manchester, UK: The University of Manchester.

Ahuja, B., Schaubb, A., Karga, M., Lechnerb, M., & Merkleinb, M. (2014). Developing LBM process parameters for Ti-6Al-4V thin wall structures and determining the corresponding mechanical characteristics. *8th International Conference on Photonic Technologies LANE 2014*. 10.1016/j.phpro.2014.08.102

Akinlabi, E. T., & Akinlabi, S. A. (2016). Advanced Coating : Laser Metal Deposition of Aluminium Powder on Titanium Substrate. In *Proceedings of the World Congress on Engineering* (*Vol. 2*). London: Academic Press.

Akinlabi, E. T. (2015). *Laser deposition of Titanium Carbide on Titanium alloy grade 5*. University of Johannesburg.

Alabort, E., Barba, D., & Reed, R. C. (2019). Design of metallic bone by additive manufacturing. *Scripta Materialia*, *164*, 110–114. doi:10.1016/j.scriptamat.2019.01.022

Alafaghani, A., Qattawi, A., Alrawi, B., & Guzman, A. (2017). Experimental Optimization of Fused Deposition Modelling Processing Parameters: A Design for Manufacturing Approach, 45th SME NAMRC conference. *Procedia Manufacturing*, *10*, 791 – 803.

Aliakbari, M. (2012). *Additive Manufacturing: State-of-the-Art, Capabilities, and Sample Applications with Cost Analysis*. KTH. Retrieved from https://www.diva-portal.org/smash/get/diva2:560827/FULLTEXT02.pdf

Al-Mouh, N., Al-Khalifa, H. S., Al-Ghamdi, S. A., Al-Onaizy, N., Al-Rajhi, N., Al-Ateeq, W., & Al-Habeeb, B. (2016). A professional development workshop on advanced computing technologies for high and middle school teachers. *2016 15th Int. Conf. Inf. Technol. Based High. Educ. Train*, 4–7. .2016.776069610.1109/ITHET

American Dye Source. (2002). *Water soluble thiophene polymer*. American Dye Source Inc. Retrieved from http://www.adsdyes.com/products/pdf/polythiophene/ADS2000P_DATA.pdf

Anitha, R., Arunachalam, S., & Radhakrishnan, P. (2001). Critical parameters influencing the quality of prototypes in fused deposition modelling. *Journal of Materials Processing Technology, 118*(1-3), 385-388. doi:(01)00980-3. doi:10.1016/S0924-0136

Anitha, R., Arunachalam, S., & Radhakrishnan, P. (2001). Critical parameters influencing the quality of prototypes in fused deposition modelling. *Journal of Materials Processing Technology*, *118*(1-3), 385–388. doi:10.1016/S0924-0136(01)00980-3

Anoop, K. S., Ohdar, R. K., & Siba, S. M. (2009). Improving dimensional accuracy of fused deposition modelling processed part using grey Taguchi method. *4th International Conference on Materials Processing and Characterization Proceedings*, *2*, 4243 – 4252.

Arısoy, Y., Criales, L., & Özel, T. (2019). Modeling and simulation of thermal field and solidification in laser powder bed fusion of nickel alloy IN625. *Optics & Laser Technology*, *109*, 278–292. doi:10.1016/j.optlastec.2018.08.016

ASTM International. (n.d.). *The Global Leader in Additive Manufacturing Standards*. ASTM International. Retrieved November 17, 2018, from https://www.astm.org/ABOUT/OverviewsforWeb2014/Additive-Manufacturing.pdf

ASTM. (2010). *F2792-10e1 Standard terminology for additive manufacturing technologies*. ASTM International. Retrieved from http://enterprise.astm.org/filtrexx40.cgi?+REDLINE_PAGES/F2792.htm

Atala, A., Bauer, S. B., Soker, S., Yoo, J. J., & Retik, A. B. (2006). Tissue-engineered autologous bladders for patients needing cystoplasty. *Lancet*, *367*(9518), 1241–1246. doi:10.1016/S0140-6736(06)68438-9 PMID:16631879

Bagley, J. R., & Galpin, A. J. (2015). Three-dimensional printing of human skeletal muscle cells: An interdisciplinary approach for studying biological systems. *Biochemistry and Molecular Biology Education*, *43*(6), 403–407. doi:10.1002/bmb.20891 PMID:26345697

Bal, B., Gumus, B., Gerstein, G., Canadinc, D., & Maier, H. (2015). On the micro-deformation mechanisms active in high-manganese austenitic steels under impact loading. *Materials Science and Engineering A*, *632*, 29–34. doi:10.1016/j.msea.2015.02.054

Balla, V. K., Das, M., Mohammad, A., & Al-Ahmari, A. M. (2016). Additive manufacturing of γ-TiAl: Processing, Microstructure, and Properties. *Advanced Engineering Materials*, *18*(7), 1208–1215. doi:10.1002/adem.201500588

Baloyi, N. M., Popoola, A. P., & Pityana, S. L. (2014). Laser coating of Zirconium and ZrO2 composites on Ti6Al4V forbiomedical applications. *South African Journal of Industrial Engineering*, *25*(1), 62–70. doi:10.7166/25-1-661

Banks, J. (2013). Adding value in additive manufacturing: Researchers in the United Kingdom and Europe look to 3D printing for customization. *IEEE Pulse*, *4*(6), 22–26. doi:10.1109/MPUL.2013.2279617 PMID:24233187

Barr, C., Sun, S. D., Easton, M., Orchowski, N., Matthews, N., & Brandt, M. (2018). Influence of macrosegregation on solidification cracking in laser clad ultra-high strength steels. *Surface and Coatings Technology*, *340*, 126–136. doi:10.1016/j.surfcoat.2018.02.052

Barron, J., Spargo, B., & Ringeisen, B. (2004). Biological laser printing of threedimensional cellular structures. *Applied Physics. A, Materials Science & Processing*, *79*(4-6), 1027–1030. doi:10.100700339-004-2620-3

Bartolo, P. J., Almeida, H., & Laoui, T. (2009). Rapid prototyping and manufacturing for tissue engineering scaffolds. *International Journal of Computer Applications in Technology*, *36*(1), 1–9. doi:10.1504/IJCAT.2009.026664

Barzilai. (1997). Deriving weights from pairwise comparison matrices. *Journal of Operational Research Society*, *48*, 1226–1232.

Basavaraj, C.K., & Vishwas, M. (2017). Studies on Effect of Fused Deposition Modelling Process Parameters on Ultimate Tensile Strength and Dimensional Accuracy of Nylon. *Journal of Procedia Manufacturing*, *10*, 791-801.

Baumers, M., Tuck, C., Bourell, D. L., Sreenivasan, R., & Hague, R. (2011). Sustainability of additive manufacturing: Measuring the energy consumption of the laser sintering process. *IMechE Part B: J Eng Manuf*, *225*(12), 2228–2239. doi:10.1177/0954405411406044

Bayode, A., Akinlabi, E. T., & Pityana, S. (2018). Fabrication of stainless steel-based FGM by laser metal deposition. In K. Kumar & P. Davim (Eds.), *Hierarchical composite materials: Materials, Manufacturing and Engineering*. Berlin: Walter de Gruyter.

Behzadnasab, M., & Yousefi, A. (2016). Effects of 3D printer nozzle head temperature on the physical and mechanical properties of PLA based product. *12th International Seminar on Polymer Science and Technology*.

Bellehumeur, C. T., Gu, P., Sun, Q., & Rizvi, G. M. (2008). Effect of processing conditions on the bonding quality of FDM polymer filaments. *Rapid Prototyping Journal, 14*(2), 72–80. doi:10.1108/13552540810862028

Bellini, A., Güçeri, S., & Bertoldi, M. (2004). Liquefier dynamics in fused deposition. *Journal of Manufacturing Science and Engineering, 126*(2), 237–246. doi:10.1115/1.1688377

Bengtson, J., & Bunnett, B. (2012). Across the table: Competing perspectives for managing technology in a library setting. *Journal of Library Administration, 52*(8), 699–715. doi:10.1080/01930826.2012.746877

Benson, C. L., Triulzi, G., & Magee, C. L. (2018). Is There a Moore's Law for 3D Printing? *3D Printing and Additive Manufacturing, 5*(1), 53-62. doi:10.1089/3dp.2017.0041

Bertol, L. S., Júnior, W. K., Silva, F. P., & Aumund-Kopp, C. (2010). Medical design: Direct Metal Laser Sintering of Ti–6Al–4V. *Materials & Design, 31*(8), 3982–3988. doi:10.1016/j.matdes.2010.02.050

Bharath, V., Prakash, N. D., Anshuman, R., & Henderson, M. (2000). Sensitivity of RP surface finish to process parameter variation. In *Solid free form fabrication proceedings*. The University of Texas.

Bharti, N., Gonzalez, S., & Buhler, A. (2015). 3D Technology in Libraries: Applications for Teaching and Research. *4th Int. Symp. Emerg. Trends Technol. Libr. Inf. Serv.*, 161–166. 10.1109/ETTLIS.2015.7048191

Bharti, N., & Singh, S. (2017). Three-dimensional (3D) printers in libraries: Perspective and preliminary safety analysis. *Journal of Chemical Education, 94*(7), 879–885. doi:10.1021/acs.jchemed.6b00745

Bhowmik, S., & Jagadish. (2017). Multi-criteria decision making for optimization of product development under green manufacturing environment. In *Design and Optimization of Mechanical Engineering Products*. Hershey, PA: IGI Global.

Bilen, S. G., Wheeler, T. F., & Bock, R. G. (2015). MAKER: applying 3D printing to model rocketry to enhance learning in undergraduate engineering design projects. *ASEE Annu. Conf. Expo*. 10.18260/p.24448

Blauch, D. N., & Carroll, F. A. (2014). 3D printers can provide an added dimension for teaching structure-energy relationships. *Journal of Chemical Education*, *91*(8), 1254–1256. doi:10.1021/ed4007259

Borah, A., & Jagadish. (2018) A Multi Criteria Decision Making Approach for Rapid Prototyping Process Selection. *Proceedings of IRF International Conference*, 1-5.

Boruslawski, P. (2015). *Danit Peleg 3D Prints Entire Graduate Fashion Collection at Home*. Retrieved from https://www.designboom.com/technology/danit-peleg-3d-prints-fashion-collection07-27-2015/

Boschetto, A. (2017). Surface Characterization in Fused Deposition Modeling. *3D Printing: Breakthroughs in Research and Practice*, 2-47. doi:10.4018/978-1-5225-1677-4.ch002

Boschetto, A., & Bottini, L. (2015). Surface improvement of fused deposition modeling parts by barrel finishing. *Rapid Prototyping Journal*, *21*(6), 686–696. doi:10.1108/RPJ-10-2013-0105

Boschetto, A., Giordano, V., & Veniali, F. (2013). 3D roughness profile model in fused deposition modelling. *Rapid Prototyping Journal*, *19*(4), 240–252. doi:10.1108/13552541311323254

Bourell, D., Kruth, J., Leu, M., Levy, G., Rosen, D., Beese, A., & Clare, A. (2017). Materials for additive manufacturing. *CIRP Annals*, *66*(2), 659–681. doi:10.1016/j.cirp.2017.05.009

Bremen, S., Meiners, W., & Diatlov, A. (2012). Selective Laser Melting A manufacturing technology for the future? *Laser-Technik-Journal*. doi:10.1002/latj.201290018

Brenner, D. J., & Hall, E. J. (2007). Computed tomography—An increasing source of radiation exposure [Review]. *The New England Journal of Medicine*, *357*(22), 2277–2284. doi:10.1056/NEJMra072149 PMID:18046031

Brueckner, F., Finaske, T., Willner, R., Seidel, A., Nowotny, S., Leyens, C., & Beyer, E. (2015). Laser additive manufacturing with crack-sensitive materials. *Laser-Technik-Journal*, *12*(2), 28–30. doi:10.1002/latj.201500015

Buehler, E., Comrie, N., Hofmann, M., McDonald, S., & Hurst, A. (2019). Investigating the implications of 3D printing in special education. *ACM Trans. Access. Comput.*, *8*. . doi:10.1145/2870640

Buehler, E., Hurst, A., & Hofmann, M. (2014). Coming to Grips: 3D Printing for Accessibility. *ASSETS' 14 Proc. 16th Int. ACM SIGACCESS Conf. Comput. Access.*, 291–292. 10.1145/2661334.2661345

Buehler, E., Kane, S. K., & Hurst, A. (2014). ABC and 3D: opportunities and obstacles to 3D printing in Special education environments. *ASSETS' 14 Proc. 16th Int. ACM SIGACCESS Conf. Comput. Access.*, 107–114, 10.1145/2661334.2661365

Bull, G., Chiu, J., Berry, R., Lipson, H., & Xie, C. (2014). Advancing children's engineering through desktop manufacturing. In Handb. Res. Educ. Commun. Technol. (4th ed.). Springer Science +Business Media. doi:10.1007/978-1-4614-3185-5_54

Bull, G., Haj-Hariri, H., Atkins, R., & Moran, P. (2015). An educational framework for digital manufacturing in schools, 3D print. *Addit. Manuf.*, *2*, 42–49. doi:10.1089/3dp.2015.0009

Callister, W. D. (2007). Classification of Materials. In Materials Science and Enmgineering - An introduction (7th ed.; pp. 5-8, 38-44, 80-105, 207-245). John Wiley and Sons Inc.

Caminero, M., Chacón, J., García-Moreno, I., & Rodríguez, G. (2018). Impact damage resistance of 3D printed continuous fibre reinforced thermoplastic composites using fused deposition modelling. *Composites. Part B, Engineering*, *148*, 93–103. doi:10.1016/j.compositesb.2018.04.054

Campbell, I., Bourell, D., & Gibson, I. (2012). Additive Manufacturing: Rapid Prototyping Comes of Age. *Rapid Prototyping Journal, 18*(4), 255–258. doi:10.1108/13552541211231563

Carabello, B. A., & Grossman, W. (2006). Calculation of stenotic valve orifice area. In D. S. Baim (Ed.), *Grossman's Cardiac Catheterization, Angiography, and Intervention* (7th ed.; pp. 173–183). Philadelphia: Lippincott Williams & Wilkins.

Castro, G., Rodríguez, J., Montealegre, M., Arias, J., Yañez, A., Panedas, S., & Rey, L. (2015). Laser Additive Manufacturing of High Added Value Pieces. *Procedia Engineering, 132*, 102–109. doi:10.1016/j.proeng.2015.12.485

Cavanaugh, T., & Eastham, N. (2017). The 3D printer as assistive technology. *Soc. Inf. Technol. Teach. Educ. Int. Conf.*, 95–102. Retrieved from https://www.learntechlib.org/p/ 177280/

Celani, G. (2012). Digital fabrication laboratories: Pedagogy and impacts on architectural education. *Nexus Network Journal, 14*(3), 469–482. doi:10.100700004-012-0120-x

Chacon, J. M., Caminero, M. A., Garcia-Plaza, E., & Nunez, P. J. (2017). Additive manufacturing of PLA structures using fused deposition modelling: Effect of process parameters on mechanical properties and their optimal selection. *Materials & Design, 124*, 143–157. doi:10.1016/j.matdes.2017.03.065

Chambers, J. (2014, October). Prosthetic heart valves. *International Journal of Clinical Practice, 68*(10), 1227–1230. doi:10.1111/ijcp.12309 PMID:24423099

Chan, J. R., & Enimil, S. A. (2015). Copyright Considerations for Providing 3D Printing Services in the Library. *Bulletin of the American Society for Information Science and Technology, 42*, 26–31. doi:10.1002/bul2.2015.1720420109

Chen, M., Zhang, Y., & Zhang, Y. (2014). Effects of a 3D printing course on mental rotation ability among 10-year-old primary students. *International Journal of Psychophysiology, 94*(2), 240. doi:10.1016/j.ijpsycho.2014.08.925

Chen, Y., Zhang, K., Huang, J., Hosseini, S. R., & Li, Z. (2016). Characterization of heat affected zone liquation cracking in laser additive manufacturing of Inconel 718. *Materials & Design*, *90*, 586–594. doi:10.1016/j.matdes.2015.10.155

Chery, D., Mburu, S., Ward, J., & Fontecchio, A. (2015). Integration of the arts and technology in GK-12 science courses. *2015 IEEE Front. Educ. Conf.*, *1–4*. doi:10.1109/FIE.2015.7344165

Chien, K. B., Makridakis, E., & Shah, R. N. (2013). Three-dimensional printing ofsoy protein scaffolds for tissue regeneration. *Tissue Engineering. Part C, Methods*, *19*(6), 417–426. doi:10.1089/ten.tec.2012.0383 PMID:23102234

Chiu, P. H. P., Lai, K. W. C., Fan, T. K. F., & Cheng, S. H. (2015). A pedagogical model for introducing 3D printing technology in a freshman level course based on a classic instructional design theory. *2015 IEEE Front. Educ. Conf.*, 1–6. 10.1109/FIE.2015.7344287

Choi, W. S., Ha, D., Park, S., & Kim, T. (2011). Synthetic multicellular cell-to-cell communication in inkjet printed bacterial cell systems. *Biomaterials*, *32*(10), 2500–2507. doi:10.1016/j.biomaterials.2010.12.014 PMID:21208654

Chong, L., Ramakrishna, S., & Singh, S. (2018). A review of digital manufacturing-based hybrid additive manufacturing processes. *International Journal of Advanced Manufacturing Technology*, *95*(5-8), 2281–2300. doi:10.100700170-017-1345-3

Christiyan, K. J., Chandrasekhar, U., & Venkateswarlu, K. (2016). A study on the influence of process parameters on the mechanical properties of 3D printed ABS composite. *Materials Science and Engineering*, *114*.

Chung, J. H. Y., Naficy, S., Yue, Z. L., Kapsa, R., Quigley, A., Moulton, S. E., & Wallace, G. G. (2013). Bio-ink properties and printability for extrusion printing living cells. *Biomaterials Science*, *1*(7), 763–773. doi:10.1039/c3bm00012e

Cicala, G., Giordano, D., Tosto, C., Filippone, G., Recca, A., & Blanco, I. (2018). Polylactide (PLA) Filaments a Biobased Solution for Additive Manufacturing: Correlating Rheology and Thermomechanical Properties with Printing Quality. *Materials (Basel)*, *11*(7), 1191. doi:10.3390/ma11071191 PMID:29997365

Coogan, T. J., & Kazmer, D. O. (2017). Bond and Part Strength in Fused Deposition Modeling. *Rapid Prototyping Journal*, *23*(2), 414–422. doi:10.1108/RPJ-03-2016-0050

Cook, K. L., Bush, S. B., & Cox, R. (2015). Creating a prosthetic hand: 3D printers innovate and inspire and maker movement. *Sci. Child*, *53*, 80–86. Retrieved from http://stats.lib.pdx.edu/ proxy.php?url=http://search.ebscohost.com/login.aspx?direct=true&db=ehh&AN=111061979&site=ehost-live

Cook, K. L., Bush, S. B., & Cox, R. (2015). Engineering encounters: Creating a prosthetic hand. *Science and Children*, *53*(4), 80–86. doi:10.2505/4c15_053_04_80

Cordes, J. (2019). *Mass Finishing for FDM Parts*. Retrieved from http://articles.stratasys.com/finishing-processes/mass-finishing-for-fdm-parts

Corum, K., & Garofalo, J. (2015). Using digital fabrication to support student learning, 3D Print. *Addit. Manuf.*, *2*, 50–55. doi:10.1089/3dp.2015.0008

Costa, N. R., Lourenço, J., & Pereira, Z. L. (2011). Desirability function approach: A review and performance evaluation in adverse conditions. *Chemometrics and Intelligent Laboratory Systems*, *107*(2), 234–244. doi:10.1016/j.chemolab.2011.04.004

Cottam, R. (2012, December 12). *INTECH Open Science Open Minds.* Retrieved April 6, 2015, from www.intechopen.com/...corrosion.../laser-materials-processing-for-impro

Dahle, R., & Rasel, R. (2016). 3-d printing as an effective educational tool for MEMS design and fabrication. *IEEE Transactions on Education*, *59*(3), 210–215. doi:10.1109/TE.2016.2515071

Dasi, L. P., Simon, H. A., Sucosky, P., & Yoganathan, A. P. (2009, February). Fluid mechanics of artificial heart valves. *Clinical and Experimental Pharmacology & Physiology*, *36*(2), 225–237. doi:10.1111/j.1440-1681.2008.05099.x PMID:19220329

Das, P., Jayaganthan, R., & Singh, I. (2011). Tensile and impact-toughness behaviour of cryorolled Al 7075 alloy. *Materials & Design*, *32*(3), 1298–1305. doi:10.1016/j.matdes.2010.09.026

Datta, S., Barua, R., Sarkar, R., Barui, A., Roy Chowdhury, A., & Datta, P. (2018). Design and development of alginate: Poly-l-lysine scaffolds by 3D bio printing and studying their mechanical, structural and cell viability properties. *IOP Conf. Series: Materials Science and Engineering, 402*.

Datta, S., Sarkar, R., Vyas, V., Bhutoria, S., Barui, A., Roy Chowdhury, A., & Datta, P. (2018). Alginate-honey bioinks with improved cell responses for applications as bioprinted tissue engineered constructs. *Journal of Materials Research*, 1–11.

Daver, F., Lee, K., Brandt, M., & Shanks, R. (2018). Cork–PLA composite filaments for fused deposition modelling. *Composites Science and Technology*, *168*, 230–237. doi:10.1016/j.compscitech.2018.10.008

de Sampaio, C. P. de O., Spinosa, R.M., Tsukahara, D.Y., da Silva, J.C., Borghi, S.L.S., Rostirolla, F., & Vicentin, J. (2013). 3D printing in graphic design education: educational experiences using fused deposition modeling (FDM) in a Brazilian university. *Proc. 6th Int. Conf. Adv. Res. Virtual Rapid Prototyp.*, 25–30.

Delaere, P. R., & Hermans, R. (2009). Clinical transplantation of a tissue-engineered airway. *Lancet*, *373*(9665), 717–718, author reply 718–771. doi:10.1016/S0140-6736(09)60429-3 PMID:19249622

Delfs, P., Tows, M., & Schmid, H. J. (2016). Optimized Build Orientation of Additive Manufactured Parts for Improved Surface Quality and Build Time. *Additive Manufacturing*, *12*, 314–320. doi:10.1016/j.addma.2016.06.003

Despeisse, M., Baumers, M., Brown, P., Charnley, F., Ford, S.J., Garmulewicz, A., … Rowley, J. (2017). Unlocking value for a circular economy through 3D printing: a research agenda. *Technol. Forecast. Soc. Change, 115*, 75–84, .2016.09.021 doi:10.1016/j.techfore

Detsch, R., Sarker, B., Grigore, A., & Boccaccini, A. R. (2013). Alginate and gelatine blending for bone cell printing and biofabrication. In *IASTED International Conference Biomedical Engineering*. Innsbruck, Austria: ACTA Press. 10.2316/P.2013.791-177

DeWall, R. A., Qasim, N., & Carr, L. (2000, May). Evolution of mechanical heart valves. *The Annals of Thoracic Surgery*, *69*(5), 1612–1621. doi:10.1016/S0003-4975(00)01231-5 PMID:10881865

Dexter, R., Anami, K., Albretcht, P., Brakke, B., Bucak, O., Connor, R., . . . Fisher, J. (2013). Manual for Repair and Retrofit of fatigue Cracks in Steel Bridges. New York: U.S. Department of Transportation Federal Highway Administration (FHWA).

Di Cesare, E. (2001). *MRI of the cardiomyopathies*. Academic Press.

Di Cesare, E., Cademartiri, F., Carbone, I., Carriero, A., Centonze, M., De Cobelli, F., ... Sardanelli, F. (2013). Natale. Clinical indications for the use of cardiac MRI. By the SIRM Study Group on Cardiac Imaging. *La Radiologia Medica*, *118*(5), 752–798. doi:10.100711547-012-0899-2 PMID:23184241

Di Cesare, E., Carbone, I., Carriero, A., Centonze, M., De Cobelli, F., De Rosa, R., ... Cademartiri, F. (2012). Clinical indications for cardiac computed tomography. From the working group of the Cardiac Radiology Section of the Italian Society of Medical Radiology (SIRM). *La Radiologia Medica*, *117*(6), 901–938. doi:10.100711547-012-0814-x PMID:22466874

Di Cesare, E., Splendiani, A., Barile, A., Squillaci, E., Di Cesare, A., Brunese, L., & Masciocchi, C. (2016). Carlo Masciocchi CT and MR imaging of the thoracic aorta. *Open Medicine: a Peer-Reviewed, Independent, Open-Access Journal*, *11*(1), 143–151. doi:10.1515/med-2016-0028 PMID:28352783

Dinda, G. P., Dasgupta, A. K., & Mazumder, J. (2009). Laser aided direct metal deposition of iconel 625 superalloy: Microstructural evolution and thermal stability. *Materials Science and Engineering A*, *509*(1), 98–104. doi:10.1016/j.msea.2009.01.009

Dizon, J. R., Espera, A. H. Jr, Chen, Q., & Advincula, R. C. (2018). Mechanical characterization of 3D-printed polymers. *Additive Manufacturing*, *20*, 44–67. doi:10.1016/j.addma.2017.12.002

Drizo, A., & Pegna, J. (2006). Environmental impacts of rapid prototyping: An overview of research to date. *Rapid Prototyping Journal*, *12*(2), 64–71. doi:10.1108/13552540610652393

Duan, B., Hockaday, L. A., Kang, K. H., & Butcher, J. T. (2013). 3D bioprinting ofheterogeneous aortic valve conduits with alginate/gelatin hydrogels. *Journal of Biomedical Materials Research. Part A*, *101*(5), 1255–1264. doi:10.1002/jbm.a.34420 PMID:23015540

Durgun, I., & Ertan, R. (2014). Experimental Investigation of FDM Process for improvement of mechanical properties and production cost. *Rapid Prototyping Journal*, *20*(3), 228–235. doi:10.1108/RPJ-10-2012-0091

Easley, W., Buehler, E., Salib, G., & Hurst, A. (2017) Fabricating engagement: using 3D printing to engage underrepresented students in STEM learning. *ASEE Annu. Conf. Expo.* 10.18260/1-2--28347

Eisenberg, M. (2013). 3D printing for children: What to build next? Int. J. Child-Comput. *Interact.*, *1*, 7–13. doi:10.1016/j.ijcci.2012.08.004

Elrod, R. E. (2016). Classroom innovation through 3D printing. *Library Hi Tech News*, *33*(3), 5–7. doi:10.1108/LHTN-12-2015-0085

Equbal, A., Sood, A.K., Ansari, A. K., & Equbal, A. (2017). Optimization of process parameters of FDM part for minimiizing its dimensional inaccuracy. *International Journal of Mechanical and Production Engineering Research and Development, 7*(2), 57-65.

Equbal, A., Dixit, N. K., & Sood, A. K. (2015). Electroless metallisation of ABS plastic: A comparative study. *International Journal of Rapid Manufacturing*, *5*(3-4), 255–275. doi:10.1504/IJRAPIDM.2015.074806

Estevez, M. E., Lindgren, K. A., & Bergethon, P. R. (2010). A novel three-dimensional tool for teaching human neuroanatomy. *Anatomical Sciences Education*, *3*(6), 309–317. doi:10.1002/ase.186 PMID:20939033

European Commission. (2014). *Additive Manufacturing in FP7 and Horizon 2020: Report from the EC Workshop on Additive Manufacturing*. Author.

European Powder Metallurgy Association. (2015). *Introduction to Additive Manufacturing Technology*. European Powder Metallurgy Association. Retrieved November 18, 2018, from http://www.epma.com/am

Fang, Y., Frampton, J. P., Raghavan, S., Sabahi-Kaviani, R., Luker, G., Deng, C. X., & Takayama, S. (2012). Rapid generation of multiplexed cell cocultures using acoustic droplet ejection followed by aqueous two-phase exclusion patterning. *Tissue Engineering. Part C, Methods*, *18*(9), 647–657. doi:10.1089/ten.tec.2011.0709 PMID:22356298

Farayibi, P. K., Abioye, T. E., Murray, J. W., Kinnell, P. K., & Clare, A. T. (2015). Surface improvement of laser clad Ti–6Al–4V using plain waterjet and pulsed electron beam irradiation. *Journal of Materials Processing Technology*, *218*, 1–11. doi:10.1016/j.jmatprotec.2014.11.035

Farooqi, K. M., & Sengupta, P. P. (2015). Echocardiography and three-dimensional printing: Sound ideas to touch a heart. *Journal of the American Society of Echocardiography*, *28*(4), 398–403. doi:10.1016/j.echo.2015.02.005 PMID:25839152

Fendrick, J. (2016). 3D Printing is Passing the Aerospace Test. *Manufacturing Engineering*, 37.

Fernandes, S. C. F., & Simoes, R. (2016). Collaborative use of different learning styles through 3D printing. *2016 2nd Int. Conf. Port. Soc. Eng. Educ.*

Finley, T. K. (2016). The impact of 3D printing services on library stakeholders: A case study. *Public Services Quarterly*, *12*(2), 152–163. doi:10.1080/15228959.2016.1160808

Gagné, R. M. (1985). *The Conditions of Learning and Theory of Instruction* (4th ed.). New York, NY: Holt, Rinehart & Winston.

Galantucci, L.M., Lavecchia, F. Percoco, G. (2009). Experimental study aiming to enhance the surface finish of fused deposition modelled parts. *CIRP Annals - Manufacturing Technology, 58*(1), 189-192. doi:.2009.03.071 doi:10.1016/j.cirp

Galbally, J., & Satta, R. (2016). Three-dimensional and two-and-a-half-dimensional face recognition spoofing using three-dimensional printed models. *IET Biometrics*, *5*(2), 83–91. doi:10.1049/iet-bmt.2014.0075

Gatto, A., Bassoli, E., Denti, L., Iuliano, L., & Minetola, P. (2015). Multi-disciplinary approach in engineering education: Learning with additive manufacturing and reverse engineering. *Rapid Prototyping Journal*, *21*(5), 598–603. doi:10.1108/RPJ-09-2014-0134

Gibson, I., Rosen, D. W., & Stucker, B. (2010). *Additive Manufacturing Technologies*. New York, NY: Springer. doi:10.1007/978-1-4419-1120-9

Gibson, I., Rosen, D., & Stucker, B. (2015). *Additive Manufacturing Technologies: 3D Printing, Rapid Prototyping, and Direct Digital Manufacturing* (2nd ed.). New York: Springer. doi:10.1007/978-1-4939-2113-3

Goitein, O., Salem, Y., & Jacobson, J. (2014). The role of cardiac computed tomography in infants with congenital heart disease. *The Israel Medical Association Journal*, *16*(3), 147–152. PMID:24761701

Go, J., & Hart, A. J. (2016). A framework for teaching the fundamentals of additive manufacturing and enabling rapid innovation. *Addit. Manuf.*, *10*, 76–87. doi:10.1016/j.addma.2016.03.001

Golub, M., Guo, X., Jung, M., & Zhang, J. (2016). *3D Printed ABS and Carbon Fiber Reinforced Polymer Specimens for Engineering Education. In REWAS 2016 Towar. Mater. Resour. Sustain* (pp. 281–285). Cham: Springer. doi:10.1007/978-3-319-48768-7_43

Gonzalez, S. R., & Bennett, D. B. (2014). Planning and implementing a 3d printing service in an academic library. *Issues Sci. Technol. Librariansh.*, *78*, 1–11. doi:10.5062/F4M043CC

Gora, W. S., Tian, Y., Cabo, A. P., Ardron, M., Maier, R. R. J., Prangnell, P., ... Hand, D. P. (2016). Enhancing Surface Finish of Additively Manufactured Titanium and Cobalt Chrome Elements Using Laser Based Finishing. *Physics Procedia*, *83*, 258–263. doi:10.1016/j.phpro.2016.08.021

Gosnell, J., Pietila, T., Samuel, B. P., Kurup, H. K. N., Haw, M. P., & Vettukattil, J. J. (2016). Integration of computed tomography and three-dimensional echocardiography for hybrid three-dimensional printing in congenital heart disease. *Journal of Digital Imaging*, *29*(6), 665–669. doi:10.100710278-016-9879-8 PMID:27072399

Graf, B., Gumenyuka, A., & Rethmeiera, M. (2012). Laser metal deposition as repair technology for stainless steel and titanium alloys. *Physics Procedia*, *39*(39), 376–381. doi:10.1016/j.phpro.2012.10.051

Grant, C. A., MacFadden, B. J., Antonenko, P., & Perez, V. J. (2016). 3-d fossils for K–12 education: A case example using the giant extinct shark carcharocles megalodon. *Paleontol. Soc. Pap., 22*, 197–209. doi:10.1017cs.2017.15

Greenfield, A. (2017). *Radical Technologies: The Design of Everyday Life*. London: Verso.

Greil, G. F., Wolf, I., Kuettner, A., Fenchel, M., Miller, S., Martirosian, P., ... Sieverding, L. (2007). Stereolithographic reproduction of complex cardiac morphology based on high resolution imaging. *Clinical Research in Cardiology; Official Journal of the German Cardiac Society*, *96*(3), 176–185. doi:10.100700392-007-0482-3 PMID:17225916

Griffey, J. (2012). Absolutely Fab-ulous. *Library Technology Reports*, *48*(3), 21–24.

Griffith, K. M., de Cataldo, R., & Fogarty, K. H. (2016). Do-It-Yourself: 3D models of hydrogenic orbitals through 3D printing. *Journal of Chemical Education*, *93*(9), 1586–1590. doi:10.1021/acs.jchemed.6b00293

Groenendyk, M., & Gallant, R. (2013). 3D printing and scanning at the Dalhousie University Libraries: A pilot project. *Library Hi Tech*, *31*(1), 34–41. doi:10.1108/07378831311303912

Gross, B. C., Erkal, J. L., Lockwood, S. Y., Chen, C., & Spence, D. M. (2014). Evaluation of 3D printing and its potential impact on biotechnology and the chemical sciences. *Analytical Chemistry*, *86*(7), 3240–3253. doi:10.1021/ac403397r PMID:24432804

Group, U. A. M. S. (2016). *Additive Manufacturing*. Leading Additive Manufacturing in the UK.

Grunkemeier, G. L., & Anderson, W. N. (1998). Clinical evaluation and analysis of heart valve substitutes. *The Journal of Heart Valve Disease*, (7): 163–169. PMID:9587856

Guillot, A., Champely, S., Batier, C., Thiriet, P., & Collet, C. (2007). Relationship between spatial abilities, mental rotation, and functional anatomy learning. *Advances in Health Sciences Education: Theory and Practice*, *12*(4), 491–507. doi:10.100710459-006-9021-7 PMID:16847728

Hacene, A., & Mekki, A. (2010). Bio-CAD Reverse Engineering of free-form surfaces by planar contours. *Computer-Aided Design and Applications, 7*(S1). doi:10.1080/16864360.2010.10738809

Haleem, A., Javaid, M., & Saxena, A. (2018). Additive manufacturing applications in cardiology: A review. *The Egyptian Heart Journal, 70*(4), 433–441. doi:10.1016/j.ehj.2018.09.008 PMID:30591768

Hall, S., Grant, G., Arora, D., Karaksha, A., McFarland, A., Lohning, A., & Dukie, S. (2017). A pilot study assessing the value of 3D printed molecular modelling tools for pharmacy student education. *Curr. Pharm. Teach. Learn., 9*(4), 723–728. doi:10.1016/j.cptl.2017.03.029 PMID:29233449

Hanzl, P., Zetek, M., Baksa, T., & Kroupa, T. (2015). The influence of processing parameters on the mechanical properties of SLM parts. *Procedia Engineering, 100*, 1405–1413. doi:10.1016/j.proeng.2015.01.510

Hart, P. (1999). Weld Metal Hydrogen Cracking in Pipeline Girth Welds. *1st International Conference.*

Hashemite-University. (n.d.). *The Hashemite University NDT Centre: Defectology.* Retrieved November 14, 2018, from https://eis.hu.edu.jo/ACUploads/10526/Defectology.pdf

Hassan, W., Dong, Y., & Wang, W. (2013). Encapsulation and 3D culture of humanadipose-derived stem cells in an in-situ crosslinked hybrid hydrogelcomposed of peg-based hyperbranched copolymer and hyaluronicacid. *Stem Cell Research & Therapy, 4*(2), 32. doi:10.1186crt182 PMID:23517589

Hattiangadi, A., & Bandyopadhyay, A. (2000). Modeling of Multiple Pore Ceramic Materials Fabricated via Fused Deposition Process. *Scripta Materialia, 42*(6), 581–588. doi:10.1016/S1359-6462(99)00370-X

He, H., Yang, Y., & Pan, Y. (2019). Machine learning for continuous liquid interface production: Printing speed modelling. *Journal of Manufacturing Systems, 50*, 236–246. doi:10.1016/j.jmsy.2019.01.004

Heng, Z., Qingbin, L., & Yaole, X. (2015). The microstructure and texture analysis of Ti-6Al-4V alloy through linear friction welding. *International Conference on Materials, Environmental and Biological Engineering (MEBE 2015).*

Herderick, E. (2011). Additive Manufacturing of Metals : A Review. *Materials Science and Technology*, (176252), 1413–1425.

Hillis, L. D., Lange, R. A., Winniford, M. D., & Page, R. L. (1995). *Manual of Clinical Problems in Cardiology*. Little, Brown and Company.

Horejsi, M. (2014). Teaching STEM with a 3D printer. *Science Teacher (Normal, Ill.)*, *10*. doi:10.1126cience.1153539

Hornbeck, L. J. (1997). Digital light processing for high-brightness high-resolution applications. In *Proceedings of Electronic Imaging '97*. International Society for Optics and Photonics. doi:10.1117/12.273880

Horowitz, S. S., & Schultz, P. H. (2014). Printing space: Using 3D printing of digital terrain models in geosciences education and research. *Journal of Geoscience Education*, *62*(1), 138–145. doi:10.5408/13-031.1

Horvath, D., Noorani, R., & Mendelson, M. (2007). Improvement of surface roughness on ABS 400 polymer using design of experiments (DOE). *Material Science Forum, 561*, 2389-2392. Retrieved from www.scientific.net/MSF.561-565.2389

Hoyek, N., Collect, C., & Rastello, O. (2009). Enhancement of mental rotation abilities and its effect on anatomy learning. *Teaching and Learning in Medicine*, *21*(3), 201–206. doi:10.1080/10401330903014178 PMID:20183339

Hoy, M. B. (2013). 3D printing: Making things at the library. *Medical Reference Services Quarterly*, *32*(1), 94–99. doi:10.1080/02763869.2013.749139 PMID:23394423

Huang, B., & Singamneni, S. B. (2015). Curved Layer Adaptive Slicing (CLAS) for Fused Deposition Modelling. *Rapid Prototyping Journal*, *21*(4), 354–367. doi:10.1108/RPJ-06-2013-0059

Huang, S. H., Liu, P., Mokasdar, A., & Hou, L. (2013). Additive manufacturing and its societal impact: A literature review. *International Journal of Advanced Manufacturing Technology*, *67*(5-8), 1191–1203. doi:10.100700170-012-4558-5

Huleihil, M. (2017). 3D printing technology as innovative tool for math and geometry teaching applications. *IOP Conf. Ser.: Mater. Sci. Eng., 164*.

Hull, C. (1988). Stereolithography: Plastic prototype from CAD data without tooling. *Modern Casting*, *78*, 38.

Hu, Y., Wang, H., Ning, F., & Cong, W. (2016). Laser Engineered Net Shaping of Commercially Pure Titanium : Effects of Fabricating Variables. *Proceedings of the ASME 2016 International Manufacturing Science and Engineering Conference*. 10.1115/MSEC2016-8812

i Materialise. (2018). *ABS: Design Guide*. Retrieved from https://i.materialise.com/3d-printing-materials/abs/design-guide

Inoue, T., Kimura, Y., & Ochiai, S. (2012). Shape effect of ultrafine-grained structure on static fracture toughness in low-alloy steel. *Science and Technology of Advanced Materials*, *13*(3), 035005. doi:10.1088/1468-6996/13/3/035005 PMID:27877493

Irwin, J. L., Pearce, J. M., Anzalone, G., & Oppliger, D. E. (2014). The RepRap 3-D printer revolution in STEM education. *ASEE Annu. Conf. Expo.*, 24.1242.1- 24.1242.13.

Ishengoma, F. R., & Mtaho, A. B. (2014). 3D printing: Developing countries perspectives. *International Journal of Computers and Applications*, *104*, 30–34. doi:10.5120/18249-9329

Jacobs, S., Grunert, R., & Mohr, F.W. (2008). 3D-Imaging of cardiac structures using 3D heart models for planning in heart surgery: a preliminary study. *Interact CardioVasc Thorac Surg.*, *7*(1), 6–9. doi:10.1510/icvts.2007.156588.

Jacobs, S., Schull, J., White, P., Lehrer, R., Vishwakarma, A., & Bertucci, A. (2016). e-NABLING education: curricula and models for teaching students to print Hands. *2016 IEEE Front. Educ. Conf.* 10.1109/FIE.2016.7757460

Jagadish & Ray, A. (2014). Green cutting fluid selection using MOOSRA method. *International Journal of Research in Engineering and Technology*, *3*(3), 559–563.

Jaksic, N. I. (2014). New inexpensive 3D printers open doors to novel experiential learning practices in engineering education. *ASEE Annu. Conf. Expo.,* 24.932.1-24.932.23. Retrieved from https://peer.asee.org/22865

Jani, M. (2018). *Print detailed objects faster using adaptive layers in Ultimaker Cura*. Retrieved from https://ultimaker.com/en/blog/52520-print-detailed-objects-faster-using-adaptive-layers-in-ultimaker-cura

Janković, N. Z., Slijepčević, M. Z., Čantrak, D. S., & Gađanski, I. I. (2016). Application of 3D printing in M.Sc. Studies – axial turbocompressors. *Int. Conf. Multidiscip. Eng. Des. Optim*, 96–99. 10.1109/MEDO.2016.7746545

Javaid, M., & Haleem, A. (2018). Additive manufacturing applications in orthopaedics: A review. *Journal of Clinical Orthopaedics and Trauma*, *9*(3), 202–206. doi:10.1016/j.jcot.2018.04.008 PMID:30202149

Jaymes, C. (2012). Adverse health effects of oil mist in machine tool industries. *SD Editorials*. Retrieved from http://www.streetdirectory.com/travel_guide/159029/health/adverse_health_effects_of_oil_mist_in_machine_tool_industries.html

Jeff, C. F. W., & Hamada, M. (2002). *Experiments: Planning, Analysis, and Parameter Design Optimization*. New Delhi, India: John Wiley & Sons.

Jones, B. M. (2015). 3D printing in libraries: a view from within the american library association: privacy, intellectual freedom and ethical policy framework. *Bulletin of the American Society for Information Science and Technology*, *42*, 36–41. Retrieved from https://search.proquest.com/docview/1812277903?accountid=10297%0Ahttp://resolver.ebscohost.com/openurl?ctx_ver=Z39.88-2004 &ctx_enc=info:ofi/enc:UTF-8&rfr_id=info:sid/ProQ%3Aabiglobal&rft_val_fmt=info:ofi/fmt:kev:mtx:journal

Jong, J. P., & Bruijn, E. D. (2013). *Innovation Lessons From 3-D Printing*. Retrieved from https://sloanreview.mit.edu/article/innovation-lessons-from-3-d-printing/

Jo, W. (2016). Introduction of 3d printing technology in the classroom for visually impaired students. *Journal of Visual Impairment & Blindness*, *110*(2), 115–121. doi:10.1177/0145482X1611000205

Junk, S., & Matt, R. (2015). New approach to introduction of 3D digital technologies in design education. *Procedia Cirp*, *36*, 35–40. .2015.01.045 doi:10.1016/j.procir

Junk, S., & Matt, R. (2015). Workshop rapid prototyping - a new approach to introduce digital manufacturing in engineering education. *2015 Int. Conf. Inf. Technol. Based High. Educ. Train.*, 1–6. 10.1109/ITHET.2015.7217965

Kai, C. C., & Fai, L. K. (1997). Rapid Prototyping: Principles and Applications in Manufacturing. Singapore: John Willey & Sons (Asia) Pte Ltd.

Kamrani, A. K., & Nasr, E. A. (2010). *Engineering design and rapid prototyping*. New York: Springer. doi:10.1007/978-0-387-95863-7

Kandemir, V., Dogan, O., & Yaman, U. (2018). Topology optimization of 2.5D parts using the SIMP method with a variable thickness approach. *Procedia Manufacturing*, *17*, 29–36. doi:10.1016/j.promfg.2018.10.009

Kane, S. K., & Bigham, J. P. (2014). Tracking @stemxcomet: Teaching Programming to Blind Students via 3D Printing, Crisis Management, and Twitter. *SIGCSE' 14 Proc. 45th ACM Tech. Symp. Comput. Sci. Educ.*, 247–252, 10.1145/2538862.2538975

Karsh, P. K., & Singh, H. (2018). Multi-Characteristic Optimization in Wire Electrical Discharge Machining of Inconel-625 by Using Taguchi-Grey Relational Analysis (GRA) Approach: Optimization of an Existing Component/ Product for Better Quality at a Lower Cost. In Design and Optimization of Mechanical Engineering Products (pp. 283–303). Hershey, PA: IGI Global. doi:10.4018/978-1-5225-3401-3.ch014

Kayfi, R., Ragab, D., & Tutunji, T. A. (2015). Mechatronic system design project: a 3D printer case study. *2015 IEEE Jordan Conf. Appl. Electr. Eng. Comput. Technol*, 1–6. 10.1109/AEECT.2015.7360570

Kim, H., Zhao, Y., & Zhao, L. (2016). *Process-level modeling and simulation for HP's Multi Jet Fusion 3D printing technology*. Paper presented at the 2016 1st International Workshop on Cyber-Physical Production Systems (CPPS), Vienna, Austria. 10.1109/CPPS.2016.7483916

Kim, M. K., Lee, I. H., & Kim, H. C. (2018). Effect of fabrication parameters on surface roughness of FDM parts. *International Journal of Precision Engineering and Manufacturing*, *19*(1), 137–142. doi:10.100712541-018-0016-0

Klahn, C., Leutenecker, B., & Meboldt, M. (2014). Design for Additive Manufacturing – Supporting the substitution of components in series products. *24th CIRP Design Conference*. 10.1016/j.procir.2014.03.145

Kobryn, P. A., Ontko, N. R., Perkins, L. P., & Tiley, J. S. (2006). Additive Manufacturing of Aerospace Alloys for Aircraft Structures. In *Meeting Proceedings RTO-AVT-139* (pp. 3-1-3–14). Neuilly-sur-Seine, France: Academic Press. Retrieved from http://www.rto.nato.int/abstracts.asp

Kobryn, P. A., & Semiatin, S. L. (2002). *Mechanical Properties of Laser-Deposited Ti-6AI-4V*. Wright-Patterson Air Force Base.

Kocovic, P. (2017). *From Modeling to 3D Printing, IGI Global. 3D Printing and Its Impact on the Production of Fully Functional Components: Emerging Research and Opportunities*. doi:10.4018/978-1-5225-2289-8.ch002

Kolitsky, M. (2014). 3D printed tactile learning objects: Proof of concept. *J. Blind. Innov. Res., 4*. doi:10.5241/4-51

Kostakis, V., Niaros, V., & Giotitsas, C. (2015). Open source 3D printing as a means of learning: An educational experiment in two high schools in Greece. *Telematics and Informatics, 32*(1), 118–128. doi:10.1016/j.tele.2014.05.001

Kou, S. (1987). *Welding Metallurgy*. John Wiley & Sons, Inc.

Krassenstein, B. (2014). *The Moore's Law of 3D Printing... Yes it Does Exist, and Could Have Staggering Implications*. Retrieved from http://3dprint.com/7543/3d-printing-moores-law/

Krassenstein, E. (2014). *Andreas Bastian Creates Incredible Bendable 3D Printed Mesostructured Material*. Retrieved from https://3dprint.com/2739/bastian-mesostructured/

Krassenstein, E. (2015). *Danit Peleg Creates First 3D Printed Fashion Collection Printed Entirely at Home*. Retrieved from https://3dprint.com/83423/danit-peleg-3d-printed-fashion/

Kröger, E., Dekiff, M., & Dirksen, D. (2016). (n.d.). 3D printed simulation models based on real patient situations for hands-on practice. *European Journal of Dental Education*, 1–7. doi:10.1111/eje.12229 PMID:27470072

Kroll, E., & Artzi, D. (2011). Enhancing aerospace engineering students' learning with 3D printing wind-tunnel models. *Rapid Prototyping Journal, 17*(5), 393–402. doi:10.1108/13552541111156522

Kruth, J. P., Leu, M. C., & Nakagawa, T. (1998). Progress in additive manufacturing and rapid prototyping. *CIRP Ann-Manuf Techn, 47*(2), 525–540. doi:10.1016/S0007-8506(07)63240-5

Kumar, N., Kumar, H., & Khurmi, J.S. (2016). Experimental Investigation of process parameters for rapid prototyping technique (Selective Laser Sintering) to enhance the part quality of prototype by Taguchi method. *3rd ICIAME Procedia Technology*, *23*, 352 – 360.

Kumar, S.D., Kannan, V.N., & Sankaranarayanan, G. (2014). Parameter optimization of ABS-M30i parts produced by fused deposition modelling for minimum surface roughness. *International Journal of Current Engineering and Technology, 3*, 93-97.

Kumar, P. (2009). Crack Detection through Non-Destructive Testing. In *Elements of Fracture Mechanics*. McGraw-Hill.

Kumar, S., & Choudhary, A. K. S., & Rakesh. (2014). Effect of the Process Parameters on Geometrical Characteristics of the Parts in Direct Metal Deposition: A Review. *International Journal of Mechanical Engineering and Technology*, *5*(4), 116–122.

Kumar, S., Kannan, V. N., & Sankaranarayanan, G. (2014). Parameter Optimization of ABS-M30i Parts Produced by Fused Deposition Modeling for Minimum Surface Roughness. *International Journal of Current Engineering and Technology*, *3*, 93–97.

Kumke, M., Watschke, H., & Vietor, T. (2016). A new methodological framework for design for additive manufacturing. *Virtual and Physical Prototyping*, *11*(1), 3–19. doi:10.1080/17452759.2016.1139377

Kurup, H. K. N., Samuel, B. P., & Vettukattil, J. J. (2015). Hybrid 3D printing: A game-changer in personalized cardiac medicine? *Expert Review of Cardiovascular Therapy*, *13*(12), 1281–1284. doi:10.1586/14779072.2015.1100076 PMID:26465262

Lalehpour, A., & Barari, A. (2016). Post processing for Fused Deposition Modeling Parts with Acetone Vapour Bath. *IFAC, 49*(31), 42-48. doi:10.1016/j.ifacol.2016.12.159

Langer, R., & Vacanti, J. P. (2013). Tissue engineering. Science, 1993, 260:920–926Trappmann B, Chen CS. How cells sense extracellular matrix stiffness: a material's perspective. *Current Opinion in Biotechnology, 24*, 948–953. PMID:23611564

Lantada & Morgado. (2012). Rapid Prototyping for Biomedical Engineering: Current Capabilities and Challenges. *Annual Review of Biomedical Engineering, 14* . doi:10.1146/annurev-bioeng-071811-150112 PMID:22524389

Laoui, T., & Shaik, S. K. (2003). Rapid prototyping techniques used to produce medical models/implants. In *Proceedings of the 4th national conference on rapid and virtual prototyping and applications* (pp. 23-32). Buckinghamshire, UK: Chilterns University College.

Lapin, J. (2009). TiAl-based alloys: Present status and future. *Hradec Nad Moravici*, 1–12. Retrieved from http://metal2013.tanger.cz/files/proceedings/metal_09/Lists/Papers/077.pdf

Laschinger, J. C., Vannier, M. W., & Gutierrez, E. (1974). Preoperative three-dimensional reconstruction of the heart and great vessels in patients with congenital heart disease. *Computers and Biomedical Research, an International Journal*, *7*(6), 544–553. doi:10.1016/0010-4809(74)90031-7 PMID:4457270

Lee, B. H., Abdullah, J., & Khan, Z. A. (2005). Optimization of rapid prototyping parameters for production of flexible ABS object||. *Journal of Materials Processing Technology*, *169*(1), 54–61. doi:10.1016/j.jmatprotec.2005.02.259

Lee, C. Y., & Liu, C. Y. (2019). The influence of forced-air cooling on a 3D printed PLA part manufactured by fused filament fabrication. *Additive Manufacturing*, *25*, 196–203. doi:10.1016/j.addma.2018.11.012

Lee, K. K., Lee, K. H., Woo, E. T., & Han, S. H. (2014). Optimization Process for Concept Design of tactical missiles by Using Pareto Front and TOPSIS. *International Journal of Precision Engineering and Manufacturing*, *15*(7), 1371–1376. doi:10.100712541-014-0478-7

Lefrak, E. A., & Starr, A. (1979). Starr-Edwards ball valve. In *Cardiac valve prostheses* (pp. 67–117). New York: Appleton-Century-Crofts.

Le, H. P. (1998). Progress and trends in ink-jet print technology. *The Journal of Imaging Science and Technology*, *42*, 49–62.

León-Cabezas, M., Martínez-García, A., & Varela-Gandía, F. (2017). Innovative advances in additive manufactured moulds for short plastic injection series. *Procedia Manufacturing*, *13*, 732–737. doi:10.1016/j.promfg.2017.09.124

Letnikova, G., & Xu, N. (2017). Academic library innovation through 3D printing services. *Library Management*, *38*(4/5), 208–218. doi:10.1108/LM-12-2016-0094

Levy, G. N., Schindel, R., & Kruth, J. P. (2003). Rapid manufacturing and rapid tooling with layer manufacturing (LM) technologies: State of the art and future perspectives. *CIRPAnn-Manuf Techn*, *52*(2), 589–609. doi:10.1016/S0007-8506(07)60206-6

Lin, F., Zhang, L., Zhang, T., Wang, J., & Zhang, R. (2012). Innovative education in additive manufacturing in China. *23rd Annu. Int. Solid Free. Fabr. Symp.*, 14–44. Retrieved from http://www.scopus.com/inward/record.url?eid=2-s2.0-84898487345&partnerID=40&md5=669226e0bb4642ee5e7d2d578c02fb63

Liou, F. W., Leu, M. C., & Landers, R. G. (2012). Interactions of an Additive Manufacturing Program with Society. *23rd Annu. Int. Solid Free. Fabr. Symp.*, 45–61.

Lipson, H. (2012). Design in the Age of 3D Printing. *Mechanical Engineering (New York, N.Y.)*, *134*(10), 30–35. doi:10.1115/1.2012-JAN-1

Lipson, H., & Kurman, M. (2013). *Fabricated: The New World of 3D Printing*. Somerset, NJ: Wiley.

Liu, W., & Dupont, J. N. (2004). Fabrication of Carbide-Particle-Reinforced Titanium Aluminide – Matrix Composites by Laser-Engineered Net Shaping. *Metallurgical and Materials Transactions. A, Physical Metallurgy and Materials Science*, *35A*(March), 1133–1140. doi:10.100711661-004-0039-2

Liu, W., Li, L., & Kochhar, A. K. (1998). A Method for Assessing Geometrical Errors in Layered Manufacturing. Part 1: Error Interaction and Transfer Mechanisms. *International Journal of Advanced Manufacturing Technology*, *14*(9), 637–643. doi:10.1007/BF01192283

Li, Y.-J., Chen, C.-H., Hoe, Z.-Y., & Yin, Z.-X. (2016). Design a Stretchable Elbow Brace by the Use of 3D Printed Mesostructure. In R. Goonetilleke & W. Karwowski (Eds.), *Advances in Physical Ergonomics and Human Factors* (Vol. 489, pp. 739–750). Cham: Springer International Publishing. doi:10.1007/978-3-319-41694-6_71

Li, Y.-y., Li, L., & Li, B. (2015). Direct write printing of three-dimensional ZrO2 biological scaffolds. *Materials & Design*, *72*, 16–20. doi:10.1016/j.matdes.2015.02.018

Locknstitch-Inc. (n.d.). *Aluminum Crack Repair Without Welding - Lock-N-Stitch.* Retrieved November 17, 2018, from http://www.locknstitch.com/RepairExamples.htm

Lolur, P., & Dawes, R. (2014). 3D Printing of Molecular Potential Energy Surface Models. *Journal of Chemical Education.*

Loucachvsky, N. (2017). *Impression 3D: Application actuelle en odontologie et perspectives* (Thèse du doctorat). Université de Nantes.

Lou, X., Andresen, P. L., & Rebak, R. B. (2018). Oxide inclusions in laser additive manufactured stainless steel and their effects on impact toughness and stress corrosion cracking behavior. *Journal of Nuclear Materials*, *499*, 182–190. doi:10.1016/j.jnucmat.2017.11.036

Loy, J. (2014). eLearning and eMaking: 3D Printing blurring the Digital and the Physical. *Education in Science*, *4*(1), 108–121. doi:10.3390/educsci4010108

Luo, Y. C., Ji, Z. M., Leu, M. C., & Caudill, R. (1999). *Environmental performance analysis of solid freeform fabrication processes. In The 1999 IEEE Int Symp on Electron and the Environ* (pp. 1–6). IEEE.

Lutjering, G., & Williams, J. C. (2003). *Titanium*. New York: Springer-Verlag. doi:10.1007/978-3-540-71398-2

Mahamood, R. M., Akinlabi, E. T., Shukla, M., & Pityana, S. (2013). *Proceedings of the International MultiConference of Engineers and Computer Scientists 2013* (*Vol. 2*). Hong Kong: Academic Press.

Mahamood, R. M., Akinlabi, E. T., Shukla, M., & Pityana, S. (2014). *Proceedings of the International MultiConference of Engineers and Computer Scientists 2014* (*Vol. 2*). Hong Kong: Academic Press.

Mahamood, R.M. (2017). Laser Metal Deposition Process. *3D Printing: Breakthroughs in Research and Practice*, 172-182. doi:10.4018/978-1-5225-1677-4.ch009

Mahamood, R.M., & Akinlabi, E.T. (2017). Laser Additive Manufacturing. *3D Printing: Breakthroughs in Research and Practice*, 154-171. doi:10.4018/978-1-5225-1677-4.ch008

Mahamood, R. M., & Akinlabi, E. T. (2017). *Functionally graded materials* (C. P. Bergmann, Ed.). Springer Nature. doi:10.1007/978-3-319-53756-6

Mahamood, R. M., Akinlabi, E. T., Shukla, M., & Pityana, S. (2013, March 28). Scanning velocity influence on microstructure, microhardness and wear resistance performance of laser deposited Ti6Al4V/TiC composite. *Materials & Design*, *50*, 656–666. doi:10.1016/j.matdes.2013.03.049

Mahapatra, S. S., & Sood, A. K. (2012). Bayesian regularization-based Levenberg-Marquardt neural model combined with BFOA for improving surface finish of FDM processed part. *International Journal of Advanced Manufacturing Technology*, *60*(9-12), 1223–1235. doi:10.100700170-011-3675-x

Majumdar, J. D. (2011). Laser Gas Alloying of Ti-6Al-4V. *Physics Procedia 12. LiM*, *12*, 472–477.

Makino, M., Suzuki, K., Takamatsu, K., Shiratori, A., Saito, A., Sakai, K., & Furukawa, H. (2017). 3D printing of police whistles for STEM education. *Microsystem Technologies*, *1–4*. doi:10.100700542-017-3393-x

Maloy, R., Trust, T., Kommers, S., Malinowski, A., & LaRoche, I. (2017). 3D modeling and printing in History/Social studies classrooms: Initial lessons and insights. *Contemporary Issues in Technology & Teacher Education*, *17*, 229–249. Retrieved from https://citejournal.s3. amazonaws.com/wp-content/uploads/v17i2socialstudies1.pdf

Mann, A. (2011). Forensic Engineering: Cracks in Steel Structures. *Proceedings - Institution of Civil Engineers*, *164*(FE1), 15–23.

Marazani, T., Madyira, D. M., & Akinlabi, E. T. (2017). Repair of Cracks in Metals. *Procedia Manufacturing*, *8*, 673–679. doi:10.1016/j.promfg.2017.02.086

Marks, D. (2011). 3D printing advantages for prototyping applications. *Articles Base*. Retrieved from http://www.articlesbase.com/technologyarticles/3d-printing-advantages-for-prototyping-applications- 1843958.html

Martinez, M. O., Morimoto, T. K., Taylor, A. T., Barron, A. C., Pultorak, J. D. A., Wang, J., . . . Okamura, A. M. (2016). 3-D Printed Haptic Devices for Educational Applications. *2016 IEEE Haptics Symp.*, 126–133. 10.1109/HAPTICS.2016.7463166

Matsumoto, K., Ishiduka, T., Yamada, H., Yonehara, Y., Arai, Y., & Honda, K. (2014). Clinical use of three-dimensional models of the temporomandibularjoint established by rapid prototyping based on cone-beam computed tomography imaging data. *Oral Radiology*, *30*(1), 98–104. doi:10.100711282-013-0127-3

Matthews, A. M. (1998). The development of the Starr-Edwards heart valve. *Texas Heart Institute Journal*, *25*(4), 282–293. PMID:9885105

Mazumder, J., Schifferer, A., & Choi, J. (1999). Direct materials deposition: Designed macro and microstructure. *Materials Research Innovations*, *3*(3), 118–131. doi:10.1007100190050137

McGahern, P., Bosch, F., & Poli, D. (2015). Enhancing learning using 3D printing: An alternative to traditional student project methods. *The American Biology Teacher*, *77*(5), 376–377. doi:10.1525/abt.2015.77.5.9

McMenamin, P. G., Quayle, M. R., McHenry, C. R., & Adams, J. W. (2014). The production of anatomical teaching resources using three-dimensional (3D) printing technology. *Anatomical Sciences Education*, *7*(6), 479–486. doi:10.1002/ase.1475 PMID:24976019

Melchels, F. P. W., Feijen, J., & Grijpma, D. W. (2010). A review on stereolithography and its applications in biomedical engineering. *Biomaterials*, *31*(24), 6121–6130. doi:10.1016/j.biomaterials.2010.04.050 PMID:20478613

Mercuri, R., & Meredith, K. (2014). An educational venture into 3D printing. *2014 IEEE Integr. STEM Educ. Conf*, 1–6. 10.1109/ISECon.2014.6891037

Mertz, L. (2013). Dream it, design it, print it in 3-D: What can 3-D printing do for you? *IEEE Pulse*, *4*(6), 15–21. doi:10.1109/MPUL.2013.2279616 PMID:24233186

Michna, S., Wu, W., & Lewis, J. A. (2005). Concentrated hydroxyapatite inks for direct-write assembly of 3-D periodic scaffolds. *Biomaterials*, *26*(28), 5632–5639. doi:10.1016/j.biomaterials.2005.02.040 PMID:15878368

Minetola, P., Iuliano, L., Bassoli, E., & Gatto, A. (2015). Impact of additive manufacturing on engineering education – evidence from Italy. *Rapid Prototyping Journal*, *21*(5), 535–555. doi:10.1108/RPJ-09-2014-0123

Mironov, V., Prestwich, G., & Forgacs, G. (2007). Bio-printing living structures. *Journal of Materials Chemistry*, *17*(20), 2054–2060. doi:10.1039/b617903g

Mironov, V., Trusk, T., Kasyanov, V., Little, S., Swaja, R., & Markwald, R. (2009). Biofabrication: A 21st century manufacturing paradigm. *Biofabrication*, *1*(2), 022001. doi:10.1088/1758-5082/1/2/022001 PMID:20811099

Miyanaji, H., Zhang, S., Lassell, A., Zandinejad, A. A., & Yang, L. (2016). Optimal process parameters for 3D printing of porcelain structures. *Procedia Manufacturing*, *5*, 870–887. doi:10.1016/j.promfg.2016.08.074

Mogali, S. R., Yeong, W. J., Kuan, H., Tan, J., Jit, G., Tan, S., . . . Ferenczi, M. A. (2017). Evaluation by medical students of the educational value of multi-material and multi-colored three-dimensional printed models of the upper limb for anatomical education. *Anat. Sci. Educ.*

Mohamed, O.A., Masood, S.H., & Bhowmik, J.L. (2016). Mathematical modeling and FDM process parameters optimization using response surface methodology based on Q-optimal design. *Journal of Applied Mathematical Modelling*, *40*, 10052–10073.

Mohamed, O. A., Masood, S. H., Bhowmik, J. L., Mostafa, N. M., & Azadmanjiri, J. (2016). Effect of process parameters on dynamic mechanical performance of FDM PC/ABS printed parts through design of experiment. *Journal of Materials Engineering and Performance*, *25*(7), 2922–2935. doi:10.100711665-016-2157-6

Montgomery, D. C. (2012). *Design and Analysis of Experiments*. Singapore: John Wiley & Sons.

Moorefield-Lang, H. M. (2014). Makers in the library: Case studies of 3D printers and maker spaces in library settings. *Library Hi Tech*, *32*(4), 583–593. doi:10.1108/LHT-06-2014-0056

Morales-Planas, S., Minguella-Canela, J., Lluma-Fuentes, J., Travieso-Rodriguez, J., & Garcia-Granada, A.-A. (2018). Multi Jet Fusion PA12 Manufacturing Parameters for Watertightness, Strength and Tolerances. *Materials (Basel)*, *11*(8), 1472. doi:10.3390/ma11081472 PMID:30126216

MSC Additive Manufacturing, Computer Program, Simufact Volume VII. (2017). Retrieved from http://www.mscsoftware.com/product/simufact-additive

Mueller, B., & Kochan, D. (1999). Laminated object manufacturing for rapidtooling and patternmaking in foundry industry. *Computers in Industry*, *39*(1), 47–53. doi:10.1016/S0166-3615(98)00127-4

Murphy, S. V., & Atala, A. (2014). 3D bio-printing of tissues and organs. *Nature Biotechnology*, *32*(8), 773–785. doi:10.1038/nbt.2958 PMID:25093879

Nakamura, M., Kobayashi, A., Takagi, F., Watanabe, A., Hiruma, Y., Ohuchi, K., ... Takatani, S. (2005). Biocompatible inkjet printing technique for designed seeding of individual living cells. *Tissue Engineering*, *11*(11-12), 1658–1666. doi:10.1089/ten.2005.11.1658 PMID:16411811

Nancharaiah, T., Raju, D.R., & Raju, V.R. (2010). An experimental investigation on surface quality and dimensional accuracy of FDM components. *International Journal of Emerging Technology, 1*(2), 106-111.

Narra, S. P., Mittwede, P. N., Wolf, S. D., & Urish, K. L. (2019). Additive Manufacturing in Total Joint Arthroplasty. *The Orthopedic Clinics of North America*, *50*(1), 13–20. doi:10.1016/j.ocl.2018.08.009 PMID:30477702

Nemorin, S. (2016). The frustrations of digital fabrication: An auto/ethnographic exploration of "3D Making" in school. *International Journal of Technology and Design Education*. doi:10.100710798-016-9366-z

Nemorin, S., & Selwyn, N. (2016). Making the best of it? Exploring the realities of 3D printing in school. *Research Papers in Education*, *32*(5), 578–595. doi:10.1080/02671522.2016.1225802

Nevarez, H. E. L., Pitcher, M. T., Perez, O. A., Gomez, H., Espinoza, P. A., Hemmitt, H., & Anaya, R. H. (2016). Work in progress: designing a university 3D printer open lab 3D model. *ASEE Annu. Conf. Expo*. 10.18260/p.27219

Ngo, T., Kashani, A., Imbalzano, G., Nguyen, K., & Hui, D. (2018). Additive manufacturing (3D printing): A review of materials, methods, applications and challenges. *Composites. Part B, Engineering*, *143*, 172–196. doi:10.1016/j.compositesb.2018.02.012

Niaki, M.K., & Nonino, F. (2017). Additive manufacturing management: a review and future research agenda. *Int. J. Prod. Res., 7543*.

Nidagundi, V. B., Keshavamurthy, R., & Prakash, C. P. S. (2015). Studies on Parametric Optimization for Fused Deposition Modelling Process. *4th International Conference on Materials Processing and Characterization Proceedings*, *2*, 1691 – 1699. 10.1016/j.matpr.2015.07.097

NIOSH. (2011). *Noise and hearing loss prevention*. Centers for Disease Control and Prevention. Retrieved from http://www.cdc.gov/niosh/topics/ noise/

Novak, J. I. (2015). *A Study of Bicycle Frame Customisation Through the use of Additive Manufacturing Technology*. Paper presented at the RAPID 2015, Long Beach, CA. Retrieved from https://www.sme.org/globalassets/sme.org/about/awards/a-study-of-bicycle-frame-customization-through-the-use-of-additive-manufacturing-technology.pdf

Novak, J. I. (2016). *Mesostructure*. Retrieved from https://edditiveblog.wordpress.com/2016/03/07/mesostructure/

Novak, J. I. (2018). Re-educating the Educators: Collaborative 3D Printing Education. In I. M. Santos, N. Ali, & S. Areepattamannil (Eds.), *Interdisciplinary and International Perspectives on 3D Printing in Education* (pp. 28–49). Hershey, PA: IGI Global.

Nowlan, G. A. (2015). Developing and implementing 3D printing services in an academic library. *Library Hi Tech*, *33*(4), 472–479. doi:10.1108/LHT-05-2015-0049

Nurul Amin, A. K. M., & Shah Alam, M. (2012). *High Speed Machining of Titanium Alloy, Ti6Al4V: to Achieve Nano Level Surface Roughness*. LAP Lambert Academic Publishing.

O'Connor, H. J., Dickson, A. N., & Dowling, D. P. (2018). Evaluation of the mechanical performance of polymer parts fabricated using a production scale multi jet fusion printing process. *Additive Manufacturing*, *22*, 381–387. doi:10.1016/j.addma.2018.05.035

O'Reilly, M. K., Reese, S., Herlihy, T., Geoghegan, T., Cantwell, C. P., Feeney, R. N. M., & Jones, J. F. X. (2016). Fabrication and assessment of 3D printed anatomical models of the lower limb for anatomical teaching and femoral vessel access training in medicine. *Anatomical Sciences Education*, *9*(1), 71–79. doi:10.1002/ase.1538 PMID:26109268

Odde, D. J., & Renn, M. J. (1999). Laser-guided direct writing for applications inbiotechnology. *Trends in Biotechnology*, *17*(10), 385–389. doi:10.1016/S0167-7799(99)01355-4 PMID:10481169

Olivieri, L. J., Krieger, A., Loke, Y. H., Nath, D. S., Kim, P. C. W., & Sable, C. A. (2015). Three-dimensional printing of intracardiac defects from threedimensional echocardiographic images: Feasibility and relative accuracy. *Journal of the American Society of Echocardiography*, *28*(4), 392–397. doi:10.1016/j.echo.2014.12.016 PMID:25660668

Olivieri, L., Krieger, A., Chen, M. Y., Kim, P., & Kanter, J. P. (2014). 3D heart model guides complex stent angioplasty of pulmonary venous baffle obstruction in a mustard repair of D-TGA. *International Journal of Cardiology*, *172*(2), e297–e298. doi:10.1016/j.ijcard.2013.12.192 PMID:24447757

Oregon Health and Science University. (n.d.). Retrieved March 2016 from http://www.ohsu.edu/xd/education/library/about/collections/historical-collections-archives/exhibits/miles-lowell-edwards.cfm

Oxman, R., & Oxman, R. (2010). The new structuralism design, engineering and architectural technologies. *Archit. Des., 80*, 15–25.

Page, H., Flood, P., & Reynaud, E. G. (2013). Three-dimensional tissue cultures: Current trends and beyond. *Cell and Tissue Research, 352*(1), 123–131. doi:10.100700441-012-1441-5 PMID:22729488

Paio, E., Eloy, S., Rato, V. M., Resende, R., & de Oliveira, M. J. (2012). Prototyping Vitruvius, New Challenges: Digital Education, Research and Practice. *Nexus Network Journal, 14*(3), 409–429. doi:10.100700004-012-0124-6

Palma, T., Munther, M., Damasus, P., Salari, S., Beheshti, A., & Davami, K. (2019). Multiscale mechanical and tribological characterizations of additively manufactured polyamide 12 parts with different print orientations. *Journal of Manufacturing Processes, 40*, 76–83. doi:10.1016/j.jmapro.2019.03.004

Panda, S.K., Padhee, S., Anoop K.S., & Mahapatra, S.S. (2009). Optimization of fused deposition modeling (FDM) process parameters using Bacterial Foraging Technique. *Journal of Intelligent Information Management, 1*, 89.

Pati, F., Jang, J., Ha, D. H., Won Kim, S., Rhie, J. W., Shim, J. H., ... Cho, D. W. (2014). Printing three-dimensional tissue analogues with decellularizedextracellular matrix bioink. *Nature Communications, 5*(1), 3935. doi:10.1038/ncomms4935 PMID:24887553

Paudel, A. M., & Kalla, D. K. (2016). *Direct digital manufacturing course into mechanical engineering technology curriculum. ASEE Annu. Conf. Expo.* doi:10.18260/p.26848

Payne, B. R. (2015). Using 3D printers in a computer graphics survey course. *Journal of Computing Sciences in Colleges, 31*, 44–251. doi:10.1017/CBO9781107415324.004

Pepper, M. E., Parzel, C. A., Burg, T., Boland, T., Burg, K. J. L., & Groff, R. E. (2009). Designand implementation of a two-dimensional inkjet bio-printer. *Proceedings of Annual International Conference of the IEEE Engineering in Medicine and Biology Society*, 6001–6005. 10.1109/IEMBS.2009.5332513

Peters, M., Hemptenmacher, J., Kumpfert, J., & Leyens, C. (2003a). Structure and Properties of Titanium Alloys. In C. Leyens & M. Peters (Eds.), *Titanium and Titanium Alloys, Fundamentals and Applications*. Weinheim: WILEY-VCH. doi:10.1002/3527602119.ch1

Petrick, I. J., & Simpson, T. W. (2013). 3D printing disrupts manufacturing. *Research Technology Management*, *56*(6), 12–16. doi:10.5437/08956308X5606193

Pham, D. T., & Gault, R. S. (1998). A comparison of rapid prototyping technologies. *International Journal of Machine Tools & Manufacture*, *38*(10-11), 1257–1287. doi:10.1016/S0890-6955(97)00137-5

Phatak, A. M., & Pandee, S. S. (2012). Optimum part orientation in Rapid Prototyping using generic algorithm. *Journal of Manufacturing Systems*, *31*(4), 395–402. doi:10.1016/j.jmsy.2012.07.001

Pieterse, F. F., & Nel, A. L. (2016). The advantages of 3D printing in undergraduate mechanical engineering research. *2016 IEEE Glob. Eng. Educ. Conf.*, 25–31. 10.1109/EDUCON.2016.7474526

Pinkerton, A. J., Wang, W., & Li, L. (2008). *Component repair using laser direct metal deposition. In Engineering Manufacture* (Vol. 222, pp. 827–836). Academic Press.

Pityana, S., Mahamood, R. M., Akinlabi, E. T., & Shukla, M. (2013). Gas Flow Rate and Powder Flow Rate Effect on Properties of Laser Metal Deposited Ti6Al4V. *Proceedings of the International MultiConference of Engineers and Computer Scientists 2013 (Vol 2)*. Hong Kong: Academic Press.

Plemmons, A. (2014). Building a culture of creation. *Teach. Libr., 41*, 12–16. Retrieved from http:// search.proquest.com/docview/1548229289?accountid=8194%5Cnhttp://primo.unilinc.edu.au/openurl/ACU/ACU_SERVICES_PAGE?url_ver=Z39.88-2004&rft_val_fmt=info:ofi/fmt:kev:mtx:journal&genre=article&sid=ProQ:ProQ%3Aeducation&atitle=Building+a+Culture+of+Cre.

Pradel, P., Zhu, Z., Bibb, R. J., & Moultrie, J. (2017). *Complexity Is Not For Free: The Impact of Component Complexity on Additive Manufacturing Build Time*. Paper presented at the Rapid Design, Prototyping & Manufacturing (RDPM2017), Newcastle, UK.

Pradel, P., Zhu, Z., Bibb, R., & Moultrie, J. (2018). A framework for mapping design for additive manufacturing knowledge for industrial and product design. *Journal of Engineering Design*, *29*(6), 291–326. doi:10.1080/09544828.2018.1483011

Pradhan, M.K., & Biswas, C.K. (2009). Modelling and analysis of process parameters on surface roughness in EDM of AISI D2 tool Steel by RSM Approach. *International Journal of Mechanical and Mechatronics Engineering, 3*(9), 1132-1137.

Pritish, S., Arnab, S., & Teg, C. (2016). The influence of layer thickness on mechanical properties of the 3D printed ABS polymer by fused deposition modeling. *Key Engineering Materials*, *706*, 63–67. doi:10.4028/www.scientific.net/KEM.706.63

Quinlan, H. E., Hasan, T., Jaddou, J., & Hart, A. J. (2017). Industrial and Consumer Uses of Additive Manufacturing: A Discussion of Capabilities, Trajectories, and Challenges. *Journal of Industrial Ecology*, *21*(S1), S15–S20. doi:10.1111/jiec.12609

Radharamanan, R. (2017). Additive manufacturing in manufacturing education: a new course development and implementation. *ASEE Annu. Conf. Expo.*

Raju, B.S., Shekar, U.C., Venkateswarlu, K., & Drakashayani, D.N. (2014). Establishment of Process model for rapid prototyping technique (Stereolithography) to enhance the part quality by Taguchi method. *2nd ICIAME 2014 Procedia Technology*, *14*, 380 – 389.

Rankouhi, B., Javadpour, S., Delfanian, F., & Letcher, T. (2016). Failure analysis and mechanical characterization of 3D printed ABS with respect to layer thickness and orientation. *Journal of Failure Analysis and Prevention*, *16*(: 3), 467–481. doi:10.100711668-016-0113-2

Ratcliffe, J. H., Hunneyball, I. M., Smith, A., Wilson, C. G., & Davis, S. S. (1984). Preparation and evaluation of biodegradable polymeric systems for the intra-articular delivery of drugs. *The Journal of Pharmacy and Pharmacology*, *36*(7), 431–436. doi:10.1111/j.2042-7158.1984.tb04419.x PMID:6146685

Reddy, V. F. O., Chaparala, A., Berrimi, C. E., Amogh, V., & Rosen, B. G. (2018). Study on surface texture of Fused Deposition Modeling. *Procedia Manufacturing, 25*, 389–396. doi:10.1016/j.promfg.2018.06.108

Redwood, B., Schöffer, F., & Garret, B. (2017). *The 3D Printing Handbook*. 3D Hubs.

Reggia, E., Calabro, K. M., & Albrecht, J. (2015). A scalable instructional method to introduce first-year engineering students to design and manufacturing processes by coupling 3D printing with CAD assignments. *ASEE Annu. Conf. Expo*. 10.18260/p.23447

Rengier, F., Mehndiratta, A., von Tengg-Kobligk, H., Zechmann, C. M., Unterhinninghofen, R., Kauczor, H. U., & Giesel, F. L. (2010). 3D printing based on imaging data: review of medical applications. *International Journal of Computer Assisted Radiology and Surgery, 5*(4), 335-341.

Rengier, F., von Tengg-Kobligk, H., Zechmann, C. M., Kauczor, H. U., & Giesel, F. L. (2008). Beyond the eye – Medical applications of 3D rapid prototyping objects. *European Medical Imaging Review, 1*, 76-80.

Ricci, J. L., Clark, E. A., Murriky, A., & Smay, J. E. (2012). Three-dimensional printing of bone repair and replacement materials: Impact on craniofacialsurgery. *The Journal of Craniofacial Surgery, 23*(1), 304–308. doi:10.1097/SCS.0b013e318241dc6e PMID:22337431

Roscoe, J. F., Fearn, S., & Posey, E. (2014) Teaching computational thinking by playing games and building robots. *2014 Int. Conf. Interact. Technol. Games*, 9–12. 10.1109/iTAG.2014.15

Ross, P. J. (2005). Taguchi techniques for quality engineering (2nd ed.). Tata McGraw-Hill Publishing Company Limited.

Rottwinkel, B., Nölke, C., Kaierle, S., & Wesling, V. (2014). Crack repair of single crystal turbine blades using laser cladding technology. *Procedia CIRP 22 (2014): 3rd International Conference on Through-life Engineering Services: Session: Recent Progress in Jet-Engine Regeneration.*

Routara, B. C., Mohanty, S. D., Datta, S., Bandyopadhyay, A., & Mahapatra, S. S. (2010). Combined quality loss (CQL) concept in WPCA-based Taguchi philosophy for optimization of multiple surface quality characteristics of UNS C34000 brass in cylindrical grinding. *International Journal of Advanced Manufacturing Technology, 51*(1-4), 135–143. doi:10.100700170-010-2599-1

Saboori, A., Gallo, D., Biamino, S., Fino, P., & Lombardi, M. (2017). An Overview of Additive Manufacturing of Titanium Components by Directed Energy Deposition: Microstructure and Mechanical Properties. *Applied Sciences, 7*(9), 883. doi:10.3390/app7090883

Sahu, R., Mahapatra, S., & Sood, A. (2013). A Study on Dimensional Accuracy of Fused Deposition Modeling (FDM) Processed Parts using Fuzzy Logic. *Journal for Manufacturing Science & Production, 13*(3), 183–197. doi:10.1515/jmsp-2013-0010

Samuel, B. P., Pinto, C., Pietila, T., & Vettukattil, J. J. (2015). Ultrasound-derived three-dimensional printing in congenital heart disease. *Journal of Digital Imaging, 28*(4), 459–461. doi:10.100710278-014-9761-5 PMID:25537458

Scalfani, V. F., & Sahib, J. (2013). A model for managing 3D printing services in academic libraries. *Issues Sci. Technol. Librariansh., 72*, 1–9.

Schade, C. T., Murphy, T. F., & Walton, C. (2014). Development of atomized powders for additive manufacturing. *World Congress on Powder Metallurgy and Particulate Materials*, 215–226. Retrieved from https://pdfs.semanticscholar.org/164c/56f1e3a3e525162cc28a65d975e2dd39e357.pdf

Schelly, C., Anzalone, G., Wijnen, B., & Pearce, J. M. (2015). Open-source 3-D printing technologies for education: Bringing additive manufacturing to the classroom. *Journal of Visual Languages and Computing, 28*, 226–237. doi:10.1016/j.jvlc.2015.01.004

Schubert, C., van Langeveld, M. C., & Donoso, L. A. (2014). Innovations in 3D printing: A 3D overview from optics to organs. *The British Journal of Ophthalmology, 98*(2), 159–161. doi:10.1136/bjophthalmol-2013-304446 PMID:24288392

Schuurman, W., Levett, P. A., Pot, M. W., van Weeren, P. R., Dhert, W. J., Hutmacher, D. W., ... Malda, J. (2013). Gelatinmethacrylamidehydrogels as potential biomaterials for fabrication of tissue-engineered cartilage constructs. *Macromolecular Bioscience*, *13*(5), 551–561. doi:10.1002/mabi.201200471 PMID:23420700

Seepersad, C. C. (2014). Challenges and Opportunities in Design for Additive Manufacturing. *3D Printing and Additive Manufacturing, 1*(1), 10-13. doi:10.1089/3dp.2013.0006

Seitz, H., Rieder, W., Irsen, S., Leukers, B., & Tille, C. (2005). Three-dimensional printing of porous ceramic scaffolds for bone tissue engineering. *Journal of Biomedical Materials Research. Part B, Applied Biomaterials*, *74*(2), 782–788. doi:10.1002/jbm.b.30291 PMID:15981173

Sekine, H., Shimizu, T., Sakaguchi, K., Dobashi, I., Wada, M., Yamato, M., ... Okano, T. (2013). In vitro fabrication of functional three-dimensional tissues with perfusable blood vessels. *Nature Communications*, *4*(1), 1399. doi:10.1038/ncomms2406 PMID:23360990

Seuring, S., & Muller, M. (2008). From a literature review to a conceptual framework for sustainable supply chain management. *Journal of Cleaner Production*, *16*(15), 1699–1710. doi:10.1016/j.jclepro.2008.04.020

Shah, K., Pinkerton, A. J., Salman, A., & Li, L. (2010). Effects of Melt Pool Variables and Process Parameters in Laser Direct Metal Deposition of Aerospace Alloys. *Materials and Manufacturing Processes*, *25*(12), 1372–1380. doi:10.1080/10426914.2010.480999

Shapeways 3d Printing News and Innovation. (2011). *3D Printing Bone on a Budget!* Retrieved from http://www.shapeways.com/blog/archives/995-3DPrinting- Bone-on-a-budget!html

Shapeways. (2018). *PLA Material Information*. Retrieved from https://www.shapeways.com/materials/pla

Sharman, A. R. C., Hughes, J. I., & Ridgway, K. (2018). Characterisation of titanium aluminide components manufactured by laser metal deposition. *Intermetallics*, *93*, 89–92. doi:10.1016/j.intermet.2017.11.013

Shukla, M., Mahamood, R. M., Akinlabi, E. T., & Pityana, S. (2012). Effect of Laser Power and Powder Flow Rate on Properties of Laser Metal Deposited Ti6Al4V. *World Academy of Science and Technology*, *6*, 44–48.

Siba, S. M., & Narayan, P. B. (2013). Benchmarking of rapid prototyping systems using grey relational analysis. *International Journal of Services and Operations Management*, *16*(4), 460–477. doi:10.1504/IJSOM.2013.057509

Simpson, T. W., Williams, C. B., & Hripko, M. (2017). Preparing industry for additive manufacturing and its applications: summary & recommendations from a National Science Foundation workshop. *Addit. Manuf.*, *13*, 166–178. 10.1016/j.addma.2016.08.002

Singh, S. C., Zeng, H., Guo, C., & Cai, W. (2012). Lasers: Fundamentals, Types, and Operations. In S. C. Singh, H. Zeng, C. Guo, & W. Cai (Eds.), *Nanomaterials: Processing and Characterization with Lasers*. Wiley-VCH Verlag GmbH & Co. KGaA. Retrieved from https://application.wiley-vch.de/books/sample/3527327150_c01.pdf

Singh, M., Haverinen, H. M., Dhagat, P., & Jabbour, G. E. (2010). Inkjet printing: Process and its applications. *Advanced Materials*, *22*(6), 673–685. doi:10.1002/adma.200901141 PMID:20217769

Singh, R. (2011). Process capability study of polyjet printing for plastic components. *J MechSciTechnol*, *25*, 1011–1015.

Singh, S., & Ramakrishna, S. (2017). Biomedical applications of additive manufacturing: Present and future. *Current Opinion in Biomedical Engineering*, *2*, 105–115. doi:10.1016/j.cobme.2017.05.006

Sinha, N. (n.d.). *Additive Manufacturing*. Retrieved November 18, 2018, from http://home.iitk.ac.in/~nsinha/Additive_Manufacturing%20I.pdf

Sirringhaus, H., Kawase, T., Friend, R. H., Shimoda, T., Inbasekaran, M., Wu, W., & Woo, E. P. (2000). High-resolution inkjet printing of all-polymer transistor circuits. *Science*, *290*(5499), 2123–2126. doi:10.1126cience.290.5499.2123 PMID:11118142

Snyder, T. J., Andrews, M., Weislogel, M., Moeck, P., Sundberg, J. S., Birkes, D., ... Graft, J. (2014). 3D systems' technology overview and new applications in manufacturing, engineering, science, and education, 3D print. *Addit. Manuf.*, *1*, 169–177. doi:10.1089/3dp.2014.1502 PMID:28473997

Sobiyi, K., Akinlabi, E., & Akinlabi, S. (2017). The influence of scanning speed on the laser metal deposition of Ti / TiC powders. *Materials Technology*, *51*(2), 345–351.

Sood, A.K., Ohdar, R.K., & Mahapatra, S.S. (2010). Parametric appraisal of mechanical property of fused deposition modelling processed parts. *Journal of Materials and Design*, 287-295.

Sood, A. K., Ohdar, R. K., & Mahapatra, S. S. (2009). Improving dimensional accuracy of fused deposition modelling processed part using grey Taguchi method. *Materials & Design*, *30*(10), 4243–4252. doi:10.1016/j.matdes.2009.04.030

Sood, A. K., Ohdar, R. K., & Mahapatra, S. S. (2010). Parametric appraisal of mechanical property of fused deposition modelling processed parts. *Materials & Design*, *31*(1), 287–295. doi:10.1016/j.matdes.2009.06.016

Sood, A. K., Ohdar, R. K., & Mahapatra, S. S. (2014). Parametric appraisal of fused deposition modelling process using the grey Taguchi method. *Proceedings of the Institution of Mechanical Engineers. Part B, Journal of Engineering Manufacture*, *224*(1), 135–145. doi:10.1243/09544054JEM1565

Spoerk, M., Savandaiah, C., Arbeiter, F., Traxler, G., Cardon, L., Holzer, C., & Sapkota, J. (2018). Anisotropic properties of oriented short carbon fibre filled polypropylene parts fabricated by extrusion-based additive manufacturing. *Composites. Part A, Applied Science and Manufacturing*, *113*, 95–104. doi:10.1016/j.compositesa.2018.06.018

Srivastava, D. U., Chang, I. T. H., & Loretto, M. H. (2000). The optimisation of processing parameters and characterisation of microstructure of direct laser fabricated TiAl alloy components. *Materials & Design*, *21*(4), 425–433. doi:10.1016/S0261-3069(99)00091-6

Standard terminology for additive manufacturing technologies. (2012). *ASTM international Designation: F2792-12a*. doi:10.1520/F2792-12A

Stansell, A., & Tyler-Wood, T. (2016). Digital fabrication for STEM projects: a Middle school example. *IEEE 16th Int. Conf. Adv. Learn. Technol.*, 483–485. 10.1109/ICALT.2016.44

Stanton, R. A., & Billmire, D. A. (2002). Skin resurfacing for the burned patient. *Clinics in Plastic Surgery*, *29*(1), 29–51. doi:10.1016/S0094-1298(03)00085-3 PMID:11827368

Stein, A. (2012). Disadvantages of 3D printers. *eHow TECH*. Retrieved from http://www.ehow.com/facts_7652991_disadvantages-3d-printers.html

Steven, R. (2014). *A2 & New North Press' 3D-printed letterpress font*. Retrieved from https://www.creativereview.co.uk/a2-new-north-press-3d-printed-letterpress-font/

Stier, K., & Brown, R. (2000). Integrating rapid prototyping technology into the curriculum. *J. Ind. Technol., 17*, 1–6. Retrieved from http://www.scopus.com/inward/record.url?eid=2-s2.0-3242746763&partnerID=40&md5=6ee1529a2624ddb9053d19f9f18949d2

Stone-Sundberg, J., Kaminsky, W., Snyder, T., & Moeck, P. (2015). 3D printed models of small and large molecules, structures and morphologies of crystals, as well as their anisotropic physical properties. *Crystal Research and Technology*, *50*(6), 432–441. doi:10.1002/crat.201400469

Struers. (2015). *The Professional Metallographer Short Version*. Struers.

Stucker, B. (2015). *Additive Manufacturing Technologies*. Louisville, KY: University of Louisville. Retrieved November 12, 2018, from http://www.uta.fi/yky/en/studies/disciplines/northamericanstudies/fulbright/Stucker%20America%20in%20Living%20Color%20presentation.pdf

Sullivan, P., & McCartney, H. (2017). Integrating 3D printing into an early childhood teacher preparation course: Reflections on practice. *Journal of Early Childhood Teacher Education*, *38*(1), 39–51. doi:10.1080/10901027.2016.1274694

Sun, C., Fang, N., Wu, D.M., & Zhang, X. (2005). Projection micro-stereolithography using digital micro-mirror dynamic mask. *Sensor Actuat a-Phys, 121*, 113–120.

Sun, G. F., Shen, X. T., Wang, Z. D., Zhang, M. J., Yao, S., Zhou, R., & Ni, Z. (2019). Laser metal deposition as a repair technology 316 stainless steel: Influence of feeding powder compositions on microstructure and mechanical properties. *Optics & Laser Technology*, *109*, 71–83. doi:10.1016/j.optlastec.2018.07.051

Suresh, G., & Narayana, K. L. (2017). A Review on Fabricating Procedures in Rapid Prototyping. *3D Printing: Breakthroughs in Research and Practice*, 1-21. doi:10.4018/978-1-5225-1677-4.ch001

Svensson, M., Sabbadini, S., Pelissero, F., Gennaro, P., Filippini, M., Beretta, S., ... Badini, C. (2010). Additive Manufacturing of Gamma Titanium Aluminide Parts by Electron Beam Melting (EBM ®). In Titanium 2010, Orlando, FL.

Tang, L., Ruan, J., Landers, R. G., & Liou, F. (2007). *Variable Powder Flow Rate Control in Laser Metal Deposition Processes*. Retrieved from https://sffsymposium.engr.utexas.edu/Manuscripts/2007/2007-03-Tang.pdf

Tay, B. Y., Evans, J. R. G., & Edirisinghe, M. J. (2003). Solid freeform fabrication of ceramics. *International Materials Reviews*, *48*(6), 341–370. doi:10.1179/095066003225010263

Taylor, B. (2015). *Metallographic preparation of titanium Application Notes*. Struers Ltd. Retrieved from http://cdnstruersproduction.azureedge.net/-/media/Struers-media-library/Materials/Application-reports/Application_Note_Titanium_2015_ENG.pdf

Taylor, B., & Weidmann, E. (2015). *Application Notes: Metallographic Preparation of Titanium*. Struers.

Technical white paper, HP Multi Jet Fusion technology: A disruptive 3D printing technology for a new era of manufacturing. (2017, May). Retrieved from https://www8.hp.com/us/en/printers/3d-printers/resource/3dtechpaper.html

Thilmany, J. (2012). Printed Life. *Mechanical Engineering (New York, N.Y.)*, *134*(1), 44–47. doi:10.1115/1.2012-JAN-5

Thompson, M. K., Moroni, G., Vaneker, T., Fadel, G., Campbell, R. I., Gibson, I., ... Martina, F. (2016). Design for Additive Manufacturing: Trends, Opportunities, Considerations, and Constraints. *CIRP Annals*, *65*(2), 737–760. doi:10.1016/j.cirp.2016.05.004

Titanium Company. (2014, February 18). *Titanium alloy guide*. Retrieved April 6, 2015, from https://www.scribd.com/doc/207688992/Titanium-Alloy-Guide

Tlotleng, M., Masina, B., & Pityana, S. (2016). Characteristics of laser In-situ alloyed titanium aluminides coatings. *Procedia Manufacturing*, *7*, 39–45. doi:10.1016/j.promfg.2016.12.013

Toker, S., Canadinc, D., Taube, A., Gerstein, G., & Maier, H. (2014). On the role of slip–twin interactions on the impact behavior of high-manganese austenitic steels. *Materials Science and Engineering A*, *593*, 120–126. doi:10.1016/j.msea.2013.11.033

Tong, L. I., Wang, C. H., & Chen, H. C. (2005). Optimization of multiple responses using principal component analysis and technique for order preference by similarity to ideal solution. *International Journal of Advanced Manufacturing Technology*, *27*(3-4), 407–414. doi:10.100700170-004-2157-9

Tschumperlin, D. J., Liu, F., & Tager, A. M. (2013). Biomechanical regulation ofmesenchymal cell function. *Current Opinion in Rheumatology*, *25*(1), 92–100. doi:10.1097/BOR.0b013e32835b13cd PMID:23114589

Tumbleston, J. R., Shirvanyants, D., Ermoshkin, N., Janusziewicz, R., Johnson, A. R., Kelly, D., ... DeSimone, J. M. (2015). Continuous liquid interface production of 3D objects. *Science*, *347*(6228), 1349–1352. doi:10.1126cience.aaa2397 PMID:25780246

Turner, B. N., Strong, R., & Gold, S. A. (2014). A review of melt extrusion additive manufacturing processes: I. Process design and modelling. *Rapid Prototyping Journal*, *20*(3), 192–204. doi:10.1108/RPJ-01-2013-0012

Upadhyay, A., Prakash, V., & Sharma, V. (2018). Optimizing Material Removal Rate Using Artificial Neural Network for Micro-EDM. In Design and optimization of Mechanical Engineering Products (pp. 209–233). India: IGI global.

US DOE. (2012). *Additive manufacturing: Pursuing the promise*. US Department of Energy.

Utela, B., Storti, D., Anderson, R., & Ganter, M. (2008). A review of process developmentsteps for new material systems in three dimensional printing (3DP). *Journal of Manufacturing Processes*, *10*(2), 96–104. doi:10.1016/j.jmapro.2009.03.002

Vacanti, C. A. (2006). The history of tissue engineering. *Journal of Cellular and Molecular Medicine*, *10*(3), 569–576. doi:10.1111/j.1582-4934.2006.tb00421.x PMID:16989721

Vaezi, M., & Chua, C. K. (2011). Effects of layer thickness and binder saturation level parameters on 3D printing process. *International Journal of Advanced Manufacturing Technology*, *53*(1-4), 275–284. doi:10.100700170-010-2821-1

Valero-Gomez, A., Gónzalez-Gómez, J., González-Pacheco, V., & Salichs, M. A. (2012). Printable creativity in plastic valley UC3M. *Glob. Eng. Educ. Conf. (EDUCON), 2012 IEEE*, 1–9. 10.1109/EDUCON.2012.6201151

Van Den Bulcke, A. I., Bogdanov, B., De Rooze, N., Schacht, E. H., Cornelissen, M., & Berghmans, H. (2000). Structural and rheological properties of methacrylamide modified gelatin hydrogels. *Biomacromolecules*, *1*(1), 31–38. doi:10.1021/bm990017d PMID:11709840

Van Epps, A., Huston, D., Sherrill, J., Alvar, A., & Bowen, A. (2015). How 3D printers support teaching in engineering, technology and beyond. *Bulletin of the American Society for Information Science and Technology*, *42*, 16–20. doi:10.1002/bul2.2015.1720420107

Vasilescu, M. D., & Groza, I. V. (2017). Influence of Technological Parameters on the Roughness and Dimension of Flat Parts Generated by FDM 3D Printing. *Revista de Tehnologii Neconventionale*, *21*(3), 18–23.

Villalpando, L., Eiliat, H., & Urbanic, R. J. (2014). An optimization approach for components built by fused deposition modeling with parametric internal structures, Product Services Systems and Value Creation. *Proceedings of the 6th CIRP Conference on Industrial Product-Service Systems*, 800-805.

Vinodh, S., Anesh, R. R., & Gautham, S. G. (2011). Application of fuzzy analytic network process for supplier selection in a manufacturing organization. *Expert Systems with Applications*, *38*(1), 272–280. doi:10.1016/j.eswa.2010.06.057

Waldo, C. (2012). 10 Ways 3D Printing Is Changing the Medical World. *3D Printer*. Retrieved from http://www.3dprinter.net/10-ways-3d-printing-is-changing-themedical-world

Wang, C. C., Lin, T. W., & Hu, S. S. (2007). Optimizing the rapid prototyping process by integrating the Taguchi method with the gray relational analysis. *Rapid Prototyping Journal*, *13*(5), 304–315. doi:10.1108/13552540710824814

Weishiet, A., Mordike, B. L., Smarsly, W., & Richter, K.-H. (2000). Laser surface remelting and laser surface gas alloying of an intermetallic TiAl alloy. *Lasers in Engineering*, *10*, 63–81.

Wei, Y., Tay, D., Panda, B., Paul, S. C., Mohamed, N. A. N., Tan, M. J., & Leong, K. F. (2017). 3D printing trends in building and construction industry: A review. *Virtual and Physical Prototyping*, *12*(3), 261–276. doi:10.1080/17452759.2017.1326724

Weng, B., Liu, X., Shepherd, R., & Wallace, G. G. (2012). Inkjet printed polypyrrole/collagen scaffold: A combination of spatial control and electrical stimulation of PC12 cells. *Synthetic Metals*, *162*(15-16), 1375–1380. doi:10.1016/j.synthmet.2012.05.022

Wieting, D. W. (1996, August). The Björk-Shiley Delrin tilting disc heart valve: Historical perspective, design and need for scientific analyses after 25 years. *The Journal of Heart Valve Disease*, *5*(2S), 157–168. PMID:8905516

Williams, B. F., & Folkman, M. (2017). Librarians as makers. *Journal of Library Administration*, *57*(1), 17–35. doi:10.1080/01930826.2016.1215676

Wilson, W. C., & Boland, T. (2003). Cell and organ printing 1: Protein and cellprinters. *The Anatomical Record. Part A, Discoveries in Molecular, Cellular, and Evolutionary Biology*, *272*(2), 491–496. doi:10.1002/ar.a.10057 PMID:12740942

Wu, T., & Cheung, E. H. M. (2006). Enhanced STL. *International Journal of Advanced Manufacturing Technology*, *29*(11), 1143–1150. doi:10.100700170-005-0001-5

Wyatt, M. C. (2015). Custom 3D-printed acetabular implants in hip surgery - innovative breakthrough or expensive bespoke upgrade? *Hip International*, *25*(4), 375–379. doi:10.5301/hipint.5000294 PMID:26351112

Yan, L., Chen, X., Zhang, Y., Newkirk, J. W., & Liou, F. (2017). Fabrication of functionally graded Ti and γ-TiAl by laser metal deposition. *JOM: The Materials. Metals & Materials Society, 69*(12), 2756–2761. doi:10.100711837-017-2582-5

Young, W. C., Budynas, R. G., & Sadegh, A. M. (2012). Fatigue and Fracture. In *Roark's Formulas for Stress and Strain* (8th ed.). McGraw-Hill.

Yu, J., Choi, Y., Shim, D., & Park, S. (2018). Repairing cast part using laser assisted metal-layer deposition and its mechanical properties. *Optics & Laser Technology, 106*, 87–93. doi:10.1016/j.optlastec.2018.04.007

Yu, J., Rombouts, M., Maes, G., & Motmans, F. (2012). Material properties of Ti6Al4V parts produced by laser metal deposition. *Physics Procedia, 39*, 416–424. doi:10.1016/j.phpro.2012.10.056

Yvonni-Effrosyni, D. (2014). *Direct Laser Metal Deposition- [DLMD] of Titanium Matrix Composites : Analysis of Microstructure and Mechanical*. National Technical University of Athens. Retrieved from https://higherlogicdownload.s3.amazonaws.com/SNAME/a09ed13c-b8c0-4897-9e87-eb86f500359b/UploadedImages/DamianidouY_SNAME_ThesisPresentation_23jan2014 (no videos)_(EN).pdf

Zareiyan, B., & Khoshnevis, B. (2017). Interlayer adhesion and strength of structures in Contour Crafting - Effects of aggregate size, extrusion rate, and layer thickness. *Automation in Construction, 81*, 112–121. doi:10.1016/j.autcon.2017.06.013

Zelený, P., Safka, J., & Elkina, I. (2014). The Mechanical Characteristics of 3D Printed Parts According to the Build Orientation. *Applied Mechanics and Materials, 474*, 381-386. Retrieved from www.scientific.net/AMM.474.381

Zghair, Y. A., & Lachmayer, R. (2017). Additive repair design approach: Case study to repair aluminium base components. In *21st International Conference On Engineering Design, ICED17* (pp. 141-150). Design to X.

Zhang, K., Chou, C. K., Xia, X., Hung, M. C., & Qin, L. (2014). Block-cell-printing for live single-cell printing. *Proceedings of the National Academy of Sciences of the United States of America, 111*(8), 2948–2953. doi:10.1073/pnas.1313661111 PMID:24516129

Zha, W., & Anand, S. (2015). Geometric Approaches to Input File Modification for Part Quality Improvement in Additive Manufacturing. *Journal of Manufacturing Processes*, *20*, 465–477. doi:10.1016/j.jmapro.2015.06.021

Zhu, K., Dancu, A., & Zhao, S. (2016). *FusePrint: A DIY 2.5D Printing Technique Embracing Everyday Artifacts.* Paper presented at the 2016 ACM Conference on Designing Interactive Systems, Brisbane, QLD, Australia. 10.1145/2901790.2901792

Ziaeefard, S., Ribeiro, G. A., & Mahmoudian, N. (2015). GUPPIE, underwater 3D printed robot a game changer in control design education. *2015 Am. Control Conf.*, 2789–2794. 10.1109/ACC.2015.7171157

Zotti, A., Zuppolini, S., Tábi, T., Grasso, M., Ren, G., Borriello, A., & Zarrelli, M. (2018). Effects of 1D and 2D nanofillers in basalt/poly(lactic acid) composites for additive manufacturing. *Composites. Part B, Engineering*, *153*, 364–375. doi:10.1016/j.compositesb.2018.08.128

Zuberbier, D. P., Agarwala, R., Sanders, M. M., & Chin, R. A. (2016). An academic library's role in improving accessibility to 3-D printing. *ASEE Annu. Conf. Expo.* 10.18260/p.26551

Zyl, I., Van, & Yadroitsava, I., & Yadroitsev. (2016). Residual stress in ti6al4v objects produced by direct metal laser sintering. *South African Journal of Industrial Engineering*, *27*(December), 134–141.

Related References

To continue our tradition of advancing academic research, we have compiled a list of recommended IGI Global readings. These references will provide additional information and guidance to further enrich your knowledge and assist you with your own research and future publications.

Abed, S., Khir, T., & Ben Brahim, A. (2016). Thermodynamic and Energy Study of a Regenerator in Gas Turbine Cycle and Optimization of Performances. *International Journal of Energy Optimization and Engineering*, *5*(2), 25–44. doi:10.4018/IJEOE.2016040102

Abu Bakar, W. A., Abdullah, W. N., Ali, R., & Mokhtar, W. N. (2016). Polymolybdate Supported Nano Catalyst for Desulfurization of Diesel. In T. Saleh (Ed.), *Applying Nanotechnology to the Desulfurization Process in Petroleum Engineering* (pp. 263–280). Hershey, PA: IGI Global. doi:10.4018/978-1-4666-9545-0.ch009

Addo-Tenkorang, R., Helo, P., & Kantola, J. (2016). Engineer-To-Order Product Development: A Communication Network Analysis for Supply-Chain's Sustainable Competitive Advantage. In R. Addo-Tenkorang, J. Kantola, P. Helo, & A. Shamsuzzoha (Eds.), *Supply Chain Strategies and the Engineer-to-Order Approach* (pp. 43–59). Hershey, PA: IGI Global. doi:10.4018/978-1-5225-0021-6.ch003

Adebiyi, I. D., Popoola, P. A., & Pityana, S. (2016). Mitigation of Wear Damage by Laser Surface Alloying Technique. In E. Akinlabi, R. Mahamood, & S. Akinlabi (Eds.), *Advanced Manufacturing Techniques Using Laser Material Processing* (pp. 172–196). Hershey, PA: IGI Global. doi:10.4018/978-1-5225-0329-3.ch007

Ahmad, W. (2016). Sulfur in Petroleum: Petroleum Desulfurization Techniques. In T. Saleh (Ed.), *Applying Nanotechnology to the Desulfurization Process in Petroleum Engineering* (pp. 1–52). Hershey, PA: IGI Global. doi:10.4018/978-1-4666-9545-0.ch001

Ahmed, I., Ahmad, N., Mehmood, I., Haq, I. U., Hassan, M., & Khan, M. U. (2016). Applications of Nanotechnology in Transportation Engineering. In A. Khitab & W. Anwar (Eds.), *Advanced Research on Nanotechnology for Civil Engineering Applications* (pp. 180–207). Hershey, PA: IGI Global. doi:10.4018/978-1-5225-0344-6.ch006

Aikhuele, D. (2018). A Study of Product Development Engineering and Design Reliability Concerns. *International Journal of Applied Industrial Engineering*, *5*(1), 79–89. doi:10.4018/IJAIE.2018010105

Al-Najar, B. T., & Bououdina, M. (2016). Bioinspired Nanoparticles for Efficient Drug Delivery System. In M. Bououdina (Ed.), *Emerging Research on Bioinspired Materials Engineering* (pp. 69–103). Hershey, PA: IGI Global. doi:10.4018/978-1-4666-9811-6.ch003

Al-Shebeeb, O. A., Rangaswamy, S., Gopalakrishan, B., & Devaru, D. G. (2017). Evaluation and Indexing of Process Plans Based on Electrical Demand and Energy Consumption. *International Journal of Manufacturing, Materials, and Mechanical Engineering*, *7*(3), 1–19. doi:10.4018/IJMMME.2017070101

Alexakis, H., & Makris, N. (2016). Validation of the Discrete Element Method for the Limit Stability Analysis of Masonry Arches. In V. Sarhosis, K. Bagi, J. Lemos, & G. Milani (Eds.), *Computational Modeling of Masonry Structures Using the Discrete Element Method* (pp. 292–325). Hershey, PA: IGI Global. doi:10.4018/978-1-5225-0231-9.ch012

AlMegren, H. A., Gonzalez-Cortes, S., Huang, Y., Chen, H., Qian, Y., Alkinany, M., ... Xiao, T. (2016). Preparation of Deep Hydrodesulfurzation Catalysts for Diesel Fuel using Organic Matrix Decomposition Method. In H. Al-Megren & T. Xiao (Eds.), *Petrochemical Catalyst Materials, Processes, and Emerging Technologies* (pp. 216–253). Hershey, PA: IGI Global. doi:10.4018/978-1-4666-9975-5.ch009

Alshammari, A., Kalevaru, V. N., Bagabas, A., & Martin, A. (2016). Production of Ethylene and its Commercial Importance in the Global Market. In H. Al-Megren & T. Xiao (Eds.), *Petrochemical Catalyst Materials, Processes, and Emerging Technologies* (pp. 82–115). Hershey, PA: IGI Global. doi:10.4018/978-1-4666-9975-5.ch004

Amel, M. (2016). Synthesis, Characterizations, and Biological Effects Study of Some Quinoline Family. In M. Bououdina (Ed.), *Emerging Research on Bioinspired Materials Engineering* (pp. 160–196). Hershey, PA: IGI Global. doi:10.4018/978-1-4666-9811-6.ch006

Amna, T., Haasan, M. S., Khil, M., & Hwang, I. (2016). Impact of Electrospun Biomimetic Extracellular Environment on Proliferation and Intercellular Communication of Muscle Precursor Cells: An Overview – Intercellular Communication of Muscle Precursor Cells with Extracellular Environment. In M. Bououdina (Ed.), *Emerging Research on Bioinspired Materials Engineering* (pp. 247–265). Hershey, PA: IGI Global. doi:10.4018/978-1-4666-9811-6.ch009

Amuda, M. O., Lawal, T. F., & Akinlabi, E. T. (2017). Research Progress on Rheological Behavior of AA7075 Aluminum Alloy During Hot Deformation. *International Journal of Materials Forming and Machining Processes, 4*(1), 53–96. doi:10.4018/IJMFMP.2017010104

An, M., & Qin, Y. (2016). Challenges of Railway Safety Risk Assessment and Maintenance Decision Making. In B. Rai (Ed.), *Handbook of Research on Emerging Innovations in Rail Transportation Engineering* (pp. 173–211). Hershey, PA: IGI Global. doi:10.4018/978-1-5225-0084-1.ch009

Anil, M., Ayyildiz-Tamis, D., Tasdemir, S., Sendemir-Urkmez, A., & Gulce-Iz, S. (2016). Bioinspired Materials and Biocompatibility. In M. Bououdina (Ed.), *Emerging Research on Bioinspired Materials Engineering* (pp. 294–322). Hershey, PA: IGI Global. doi:10.4018/978-1-4666-9811-6.ch011

Armutlu, H. (2018). Intelligent Biomedical Engineering Operations by Cloud Computing Technologies. In U. Kose, G. Guraksin, & O. Deperlioglu (Eds.), *Nature-Inspired Intelligent Techniques for Solving Biomedical Engineering Problems* (pp. 297–317). Hershey, PA: IGI Global. doi:10.4018/978-1-5225-4769-3.ch015

Arokiyaraj, S., Saravanan, M., Bharanidharan, R., Islam, V. I., Bououdina, M., & Vincent, S. (2016). Green Synthesis of Metallic Nanoparticles Using Plant Compounds and Their Applications: Metallic Nanoparticles Synthesis Using Plants. In M. Bououdina (Ed.), *Emerging Research on Bioinspired Materials Engineering* (pp. 1–34). Hershey, PA: IGI Global. doi:10.4018/978-1-4666-9811-6.ch001

Atik, M., Sadek, M., & Shahrour, I. (2017). Single-Run Adaptive Pushover Procedure for Shear Wall Structures. In V. Plevris, G. Kremmyda, & Y. Fahjan (Eds.), *Performance-Based Seismic Design of Concrete Structures and Infrastructures* (pp. 59–83). Hershey, PA: IGI Global. doi:10.4018/978-1-5225-2089-4.ch003

Aydin, A., Akyol, E., Gungor, M., Kaya, A., & Tasdelen, S. (2018). Geophysical Surveys in Engineering Geology Investigations With Field Examples. In N. Ceryan (Ed.), *Handbook of Research on Trends and Digital Advances in Engineering Geology* (pp. 257–280). Hershey, PA: IGI Global. doi:10.4018/978-1-5225-2709-1.ch007

Azevedo, N. M., Lemos, J. V., & Rocha de Almeida, J. (2016). Discrete Element Particle Modelling of Stone Masonry. In V. Sarhosis, K. Bagi, J. Lemos, & G. Milani (Eds.), *Computational Modeling of Masonry Structures Using the Discrete Element Method* (pp. 146–170). Hershey, PA: IGI Global. doi:10.4018/978-1-5225-0231-9.ch007

Bamufleh, H. S., Noureldin, M. M., & El-Halwagi, M. M. (2016). Sustainable Process Integration in the Petrochemical Industries. In H. Al-Megren & T. Xiao (Eds.), *Petrochemical Catalyst Materials, Processes, and Emerging Technologies* (pp. 150–163). Hershey, PA: IGI Global. doi:10.4018/978-1-4666-9975-5.ch006

Banerjee, S., Gautam, R. K., Gautam, P. K., Jaiswal, A., & Chattopadhyaya, M. C. (2016). Recent Trends and Advancement in Nanotechnology for Water and Wastewater Treatment: Nanotechnological Approach for Water Purification. In A. Khitab & W. Anwar (Eds.), *Advanced Research on Nanotechnology for Civil Engineering Applications* (pp. 208–252). Hershey, PA: IGI Global. doi:10.4018/978-1-5225-0344-6.ch007

Bas, T. G. (2017). Nutraceutical Industry with the Collaboration of Biotechnology and Nutrigenomics Engineering: The Significance of Intellectual Property in the Entrepreneurship and Scientific Research Ecosystems. In T. Bas & J. Zhao (Eds.), *Comparative Approaches to Biotechnology Development and Use in Developed and Emerging Nations* (pp. 1–17). Hershey, PA: IGI Global. doi:10.4018/978-1-5225-1040-6.ch001

Beale, R., & André, J. (2017). *Design Solutions and Innovations in Temporary Structures*. Hershey, PA: IGI Global. doi:10.4018/978-1-5225-2199-0

Behnam, B. (2017). Simulating Post-Earthquake Fire Loading in Conventional RC Structures. In P. Samui, S. Chakraborty, & D. Kim (Eds.), *Modeling and Simulation Techniques in Structural Engineering* (pp. 425–444). Hershey, PA: IGI Global. doi:10.4018/978-1-5225-0588-4.ch015

Ben Hamida, I., Salah, S. B., Msahli, F., & Mimouni, M. F. (2018). Distribution Network Reconfiguration Using SPEA2 for Power Loss Minimization and Reliability Improvement. *International Journal of Energy Optimization and Engineering*, *7*(1), 50–65. doi:10.4018/IJEOE.2018010103

Benjamin, S. R., de Lima, F., & Rathoure, A. K. (2016). Genetically Engineered Microorganisms for Bioremediation Processes: GEMs for Bioremediaton. In A. Rathoure & V. Dhatwalia (Eds.), *Toxicity and Waste Management Using Bioremediation* (pp. 113–140). Hershey, PA: IGI Global. doi:10.4018/978-1-4666-9734-8.ch006

Bhaskar, S. V., & Kudal, H. N. (2017). Effect of TiCN and AlCrN Coating on Tribological Behaviour of Plasma-nitrided AISI 4140 Steel. *International Journal of Surface Engineering and Interdisciplinary Materials Science*, *5*(2), 1–17. doi:10.4018/IJSEIMS.2017070101

Bhowmik, S., Sahoo, P., Acharyya, S. K., Dhar, S., & Chattopadhyay, J. (2016). Effect of Microstructure Degradation on Fracture Toughness of 20MnMoNi55 Steel in DBT Region. *International Journal of Manufacturing, Materials, and Mechanical Engineering*, *6*(3), 11–27. doi:10.4018/IJMMME.2016070102

Bhutto, A. W., Abro, R., Abbas, T., Yu, G., & Chen, X. (2016). Desulphurization of Fuel Oils Using Ionic Liquids. In H. Al-Megren & T. Xiao (Eds.), *Petrochemical Catalyst Materials, Processes, and Emerging Technologies* (pp. 254–284). Hershey, PA: IGI Global. doi:10.4018/978-1-4666-9975-5.ch010

Bhuyan, D. (2018). Designing of a Twin Tube Shock Absorber: A Study in Reverse Engineering. In K. Kumar & J. Davim (Eds.), *Design and Optimization of Mechanical Engineering Products* (pp. 83–104). Hershey, PA: IGI Global. doi:10.4018/978-1-5225-3401-3.ch005

Bouloudenine, M., & Bououdina, M. (2016). Toxic Effects of Engineered Nanoparticles on Living Cells. In M. Bououdina (Ed.), *Emerging Research on Bioinspired Materials Engineering* (pp. 35–68). Hershey, PA: IGI Global. doi:10.4018/978-1-4666-9811-6.ch002

Brunetti, A., Sellaro, M., Drioli, E., & Barbieri, G. (2016). Membrane Engineering and its Role in Oil Refining and Petrochemical Industry. In H. Al-Megren & T. Xiao (Eds.), *Petrochemical Catalyst Materials, Processes, and Emerging Technologies* (pp. 116–149). Hershey, PA: IGI Global. doi:10.4018/978-1-4666-9975-5.ch005

Bügler, M., & Borrmann, A. (2016). Simulation Based Construction Project Schedule Optimization: An Overview on the State-of-the-Art. In F. Miranda & C. Abreu (Eds.), *Handbook of Research on Computational Simulation and Modeling in Engineering* (pp. 482–507). Hershey, PA: IGI Global. doi:10.4018/978-1-4666-8823-0.ch016

Calderon, F. A., Giolo, E. G., Frau, C. D., Rengel, M. G., Rodriguez, H., Tornello, M., ... Gallucci, R. (2018). Seismic Microzonation and Site Effects Detection Through Microtremors Measures: A Review. In N. Ceryan (Ed.), *Handbook of Research on Trends and Digital Advances in Engineering Geology* (pp. 326–349). Hershey, PA: IGI Global. doi:10.4018/978-1-5225-2709-1.ch009

Carmona-Murillo, J., & Valenzuela-Valdés, J. F. (2016). Motivation on Problem Based Learning. In D. Fonseca & E. Redondo (Eds.), *Handbook of Research on Applied E-Learning in Engineering and Architecture Education* (pp. 179–203). Hershey, PA: IGI Global. doi:10.4018/978-1-4666-8803-2.ch009

Ceryan, N. (2016). A Review of Soft Computing Methods Application in Rock Mechanic Engineering. In P. Samui (Ed.), *Handbook of Research on Advanced Computational Techniques for Simulation-Based Engineering* (pp. 1–70). Hershey, PA: IGI Global. doi:10.4018/978-1-4666-9479-8.ch001

Ceryan, N., & Can, N. K. (2018). Prediction of The Uniaxial Compressive Strength of Rocks Materials. In N. Ceryan (Ed.), *Handbook of Research on Trends and Digital Advances in Engineering Geology* (pp. 31–96). Hershey, PA: IGI Global. doi:10.4018/978-1-5225-2709-1.ch002

Ceryan, S. (2018). Weathering Indices Used in Evaluation of the Weathering State of Rock Material. In N. Ceryan (Ed.), *Handbook of Research on Trends and Digital Advances in Engineering Geology* (pp. 132–186). Hershey, PA: IGI Global. doi:10.4018/978-1-5225-2709-1.ch004

Chandrasekaran, S., Silva, B., Patil, A., Oo, A. M., & Campbell, M. (2016). Evaluating Engineering Students' Perceptions: The Impact of Team-Based Learning Practices in Engineering Education. *International Journal of Quality Assurance in Engineering and Technology Education*, *5*(4), 42–59. doi:10.4018/IJQAETE.2016100103

Chen, H., Padilla, R. V., & Besarati, S. (2017). Supercritical Fluids and Their Applications in Power Generation. In L. Chen & Y. Iwamoto (Eds.), *Advanced Applications of Supercritical Fluids in Energy Systems* (pp. 369–402). Hershey, PA: IGI Global. doi:10.4018/978-1-5225-2047-4.ch012

Chen, L. (2017). Principles, Experiments, and Numerical Studies of Supercritical Fluid Natural Circulation System. In L. Chen & Y. Iwamoto (Eds.), *Advanced Applications of Supercritical Fluids in Energy Systems* (pp. 136–187). Hershey, PA: IGI Global. doi:10.4018/978-1-5225-2047-4.ch005

Clementi, F., Di Sciascio, G., Di Sciascio, S., & Lenci, S. (2017). Influence of the Shear-Bending Interaction on the Global Capacity of Reinforced Concrete Frames: A Brief Overview of the New Perspectives. In V. Plevris, G. Kremmyda, & Y. Fahjan (Eds.), *Performance-Based Seismic Design of Concrete Structures and Infrastructures* (pp. 84–111). Hershey, PA: IGI Global. doi:10.4018/978-1-5225-2089-4.ch004

Cortés-Polo, D., Calle-Cancho, J., Carmona-Murillo, J., & González-Sánchez, J. (2017). Future Trends in Mobile-Fixed Integration for Next Generation Networks: Classification and Analysis. *International Journal of Vehicular Telematics and Infotainment Systems*, *1*(1), 33–53. doi:10.4018/IJVTIS.2017010103

Cui, X., Zeng, S., Li, Z., Zheng, Q., Yu, X., & Han, B. (2018). Advanced Composites for Civil Engineering Infrastructures. In K. Kumar & J. Davim (Eds.), *Composites and Advanced Materials for Industrial Applications* (pp. 212–248). Hershey, PA: IGI Global. doi:10.4018/978-1-5225-5216-1.ch010

Dalgıç, S., & Kuşku, İ. (2018). Geological and Geotechnical Investigations in Tunneling. In N. Ceryan (Ed.), *Handbook of Research on Trends and Digital Advances in Engineering Geology* (pp. 482–529). Hershey, PA: IGI Global. doi:10.4018/978-1-5225-2709-1.ch014

de la Varga, D., Soto, M., Arias, C. A., van Oirschot, D., Kilian, R., Pascual, A., & Álvarez, J. A. (2017). Constructed Wetlands for Industrial Wastewater Treatment and Removal of Nutrients. In Á. Val del Río, J. Campos Gómez, & A. Mosquera Corral (Eds.), *Technologies for the Treatment and Recovery of Nutrients from Industrial Wastewater* (pp. 202–230). Hershey, PA: IGI Global. doi:10.4018/978-1-5225-1037-6.ch008

del Valle-Zermeño, R., Chimenos, J. M., & Formosa, J. (2016). Flue Gas Desulfurization: Processes and Technologies. In T. Saleh (Ed.), *Applying Nanotechnology to the Desulfurization Process in Petroleum Engineering* (pp. 337–377). Hershey, PA: IGI Global. doi:10.4018/978-1-4666-9545-0.ch011

Delgado, J. M., Henriques, A. A., & Delgado, R. M. (2016). Structural Non-Linear Models and Simulation Techniques: An Efficient Combination for Safety Evaluation of RC Structures. In F. Miranda & C. Abreu (Eds.), *Handbook of Research on Computational Simulation and Modeling in Engineering* (pp. 540–584). Hershey, PA: IGI Global. doi:10.4018/978-1-4666-8823-0.ch018

Delgado, P. S., Arêde, A., Pouca, N. V., & Costa, A. (2016). Numerical Modeling of RC Bridges for Seismic Risk Analysis. In F. Miranda & C. Abreu (Eds.), *Handbook of Research on Computational Simulation and Modeling in Engineering* (pp. 457–481). Hershey, PA: IGI Global. doi:10.4018/978-1-4666-8823-0.ch015

Deng, Y., & Liu, S. (2016). Catalysis with Room Temperature Ionic Liquids Mediated Metal Nanoparticles. In H. Al-Megren & T. Xiao (Eds.), *Petrochemical Catalyst Materials, Processes, and Emerging Technologies* (pp. 285–329). Hershey, PA: IGI Global. doi:10.4018/978-1-4666-9975-5.ch011

Deperlioglu, O. (2018). Intelligent Techniques Inspired by Nature and Used in Biomedical Engineering. In U. Kose, G. Guraksin, & O. Deperlioglu (Eds.), *Nature-Inspired Intelligent Techniques for Solving Biomedical Engineering Problems* (pp. 51–77). Hershey, PA: IGI Global. doi:10.4018/978-1-5225-4769-3.ch003

Dias, G. L., Magalhães, R. R., Ferreira, D. D., & Vitoriano, F. A. (2016). The Use of a Robotic Arm for Displacement Measurements in a Cantilever beam. *International Journal of Manufacturing, Materials, and Mechanical Engineering*, *6*(3), 45–57. doi:10.4018/IJMMME.2016070104

Dimitratos, N., Villa, A., Chan-Thaw, C. E., Hammond, C., & Prati, L. (2016). Valorisation of Glycerol to Fine Chemicals and Fuels. In H. Al-Megren & T. Xiao (Eds.), *Petrochemical Catalyst Materials, Processes, and Emerging Technologies* (pp. 352–384). Hershey, PA: IGI Global. doi:10.4018/978-1-4666-9975-5.ch013

Dixit, A. (2018). Application of Silica-Gel-Reinforced Aluminium Composite on the Piston of Internal Combustion Engine: Comparative Study of Silica-Gel-Reinforced Aluminium Composite Piston With Aluminium Alloy Piston. In K. Kumar & J. Davim (Eds.), *Composites and Advanced Materials for Industrial Applications* (pp. 63–98). Hershey, PA: IGI Global. doi:10.4018/978-1-5225-5216-1.ch004

Drei, A., Milani, G., & Sincraian, G. (2016). Application of DEM to Historic Masonries, Two Case-Studies in Portugal and Italy: Aguas Livres Aqueduct and Arch-Tympana of a Church. In V. Sarhosis, K. Bagi, J. Lemos, & G. Milani (Eds.), *Computational Modeling of Masonry Structures Using the Discrete Element Method* (pp. 326–366). Hershey, PA: IGI Global. doi:10.4018/978-1-5225-0231-9.ch013

Dutta, S., Roy, P. K., & Nandi, D. (2016). Optimal Allocation of Static Synchronous Series Compensator Controllers using Chemical Reaction Optimization for Reactive Power Dispatch. *International Journal of Energy Optimization and Engineering*, *5*(3), 43–62. doi:10.4018/IJEOE.2016070103

Dutta, S., Roy, P. K., & Nandi, D. (2016). Quasi Oppositional Teaching-Learning based Optimization for Optimal Power Flow Incorporating FACTS. *International Journal of Energy Optimization and Engineering*, *5*(2), 64–84. doi:10.4018/IJEOE.2016040104

Eloy, S., Dias, M. S., Lopes, P. F., & Vilar, E. (2016). Digital Technologies in Architecture and Engineering: Exploring an Engaged Interaction within Curricula. In D. Fonseca & E. Redondo (Eds.), *Handbook of Research on Applied E-Learning in Engineering and Architecture Education* (pp. 368–402). Hershey, PA: IGI Global. doi:10.4018/978-1-4666-8803-2.ch017

Elsayed, A. M., Dakkama, H. J., Mahmoud, S., Al-Dadah, R., & Kaialy, W. (2017). Sustainable Cooling Research Using Activated Carbon Adsorbents and Their Environmental Impact. In T. Kobayashi (Ed.), *Applied Environmental Materials Science for Sustainability* (pp. 186–221). Hershey, PA: IGI Global. doi:10.4018/978-1-5225-1971-3.ch009

Ercanoglu, M., & Sonmez, H. (2018). General Trends and New Perspectives on Landslide Mapping and Assessment Methods. In N. Ceryan (Ed.), *Handbook of Research on Trends and Digital Advances in Engineering Geology* (pp. 350–379). Hershey, PA: IGI Global. doi:10.4018/978-1-5225-2709-1.ch010

Erinosho, M. F., Akinlabi, E. T., & Pityana, S. (2016). Enhancement of Surface Integrity of Titanium Alloy with Copper by Means of Laser Metal Deposition Process. In E. Akinlabi, R. Mahamood, & S. Akinlabi (Eds.), *Advanced Manufacturing Techniques Using Laser Material Processing* (pp. 60–91). Hershey, PA: IGI Global. doi:10.4018/978-1-5225-0329-3.ch004

Farag, H., & Kishida, M. (2016). Kinetic Models for Complex Parallel–Consecutive Reactions Assessment of Reaction Network and Product Selectivity. In H. Al-Megren & T. Xiao (Eds.), *Petrochemical Catalyst Materials, Processes, and Emerging Technologies* (pp. 330–351). Hershey, PA: IGI Global. doi:10.4018/978-1-4666-9975-5.ch012

Faroz, S. A., Pujari, N. N., Rastogi, R., & Ghosh, S. (2017). Risk Analysis of Structural Engineering Systems Using Bayesian Inference. In P. Samui, S. Chakraborty, & D. Kim (Eds.), *Modeling and Simulation Techniques in Structural Engineering* (pp. 390–424). Hershey, PA: IGI Global. doi:10.4018/978-1-5225-0588-4.ch014

Fernando, P. R., Hamigah, T., Disne, S., Wickramasingha, G. G., & Sutharshan, A. (2018). The Evaluation of Engineering Properties of Low Cost Concrete Blocks by Partial Doping of Sand with Sawdust: Low Cost Sawdust Concrete Block. *International Journal of Strategic Engineering, 1*(2), 26–42. doi:10.4018/IJoSE.2018070103

Fragiadakis, M., Stefanou, I., & Psycharis, I. N. (2016). Vulnerability Assessment of Damaged Classical Multidrum Columns. In V. Sarhosis, K. Bagi, J. Lemos, & G. Milani (Eds.), *Computational Modeling of Masonry Structures Using the Discrete Element Method* (pp. 235–253). Hershey, PA: IGI Global. doi:10.4018/978-1-5225-0231-9.ch010

Gaines, T. W., Williams, K. R., & Wagener, K. B. (2016). ADMET: Functionalized Polyolefins. In H. Al-Megren & T. Xiao (Eds.), *Petrochemical Catalyst Materials, Processes, and Emerging Technologies* (pp. 1–21). Hershey, PA: IGI Global. doi:10.4018/978-1-4666-9975-5.ch001

Garg, H. (2016). Bi-Criteria Optimization for Finding the Optimal Replacement Interval for Maintaining the Performance of the Process Industries. In P. Vasant, G. Weber, & V. Dieu (Eds.), *Handbook of Research on Modern Optimization Algorithms and Applications in Engineering and Economics* (pp. 643–675). Hershey, PA: IGI Global. doi:10.4018/978-1-4666-9644-0.ch025

Gaspar, P. D., Dinho da Silva, P., Gonçalves, J. P., & Carneiro, R. (2016). Computational Modelling and Simulation to Assist the Improvement of Thermal Performance and Energy Efficiency in Industrial Engineering Systems: Application to Cold Stores. In F. Miranda & C. Abreu (Eds.), *Handbook of Research on Computational Simulation and Modeling in Engineering* (pp. 1–68). Hershey, PA: IGI Global. doi:10.4018/978-1-4666-8823-0.ch001

Ge, H., Tang, M., & Wen, X. (2016). Ni/ZnO Nano Sorbent for Reactive Adsorption Desulfurization of Refinery Oil Streams. In T. Saleh (Ed.), *Applying Nanotechnology to the Desulfurization Process in Petroleum Engineering* (pp. 216–239). Hershey, PA: IGI Global. doi:10.4018/978-1-4666-9545-0.ch007

Ghosh, S., Mitra, S., Ghosh, S., & Chakraborty, S. (2017). Seismic Reliability Analysis in the Framework of Metamodelling Based Monte Carlo Simulation. In P. Samui, S. Chakraborty, & D. Kim (Eds.), *Modeling and Simulation Techniques in Structural Engineering* (pp. 192–208). Hershey, PA: IGI Global. doi:10.4018/978-1-5225-0588-4.ch006

Gil, M., & Otero, B. (2017). Learning Engineering Skills through Creativity and Collaboration: A Game-Based Proposal. In R. Alexandre Peixoto de Queirós & M. Pinto (Eds.), *Gamification-Based E-Learning Strategies for Computer Programming Education* (pp. 14–29). Hershey, PA: IGI Global. doi:10.4018/978-1-5225-1034-5.ch002

Gill, J., Ayre, M., & Mills, J. (2017). Revisioning the Engineering Profession: How to Make It Happen! In M. Gray & K. Thomas (Eds.), *Strategies for Increasing Diversity in Engineering Majors and Careers* (pp. 156–175). Hershey, PA: IGI Global. doi:10.4018/978-1-5225-2212-6.ch008

Gopal, S., & Al-Hazmi, M. H. (2016). Advances in Catalytic Technologies for Selective Oxidation of Lower Alkanes. In H. Al-Megren & T. Xiao (Eds.), *Petrochemical Catalyst Materials, Processes, and Emerging Technologies* (pp. 22–52). Hershey, PA: IGI Global. doi:10.4018/978-1-4666-9975-5.ch002

Goyal, N., Ram, M., Bhardwaj, A., & Kumar, A. (2016). Thermal Power Plant Modelling with Fault Coverage Stochastically. *International Journal of Manufacturing, Materials, and Mechanical Engineering*, *6*(3), 28–44. doi:10.4018/IJMMME.2016070103

Goyal, N., Ram, M., & Kumar, P. (2017). Welding Process under Fault Coverage Approach for Reliability and MTTF. In M. Ram & J. Davim (Eds.), *Mathematical Concepts and Applications in Mechanical Engineering and Mechatronics* (pp. 222–245). Hershey, PA: IGI Global. doi:10.4018/978-1-5225-1639-2.ch011

Gray, M., & Lundy, C. (2017). Engineering Study Abroad: High Impact Strategy for Increasing Access. In M. Gray & K. Thomas (Eds.), *Strategies for Increasing Diversity in Engineering Majors and Careers* (pp. 42–59). Hershey, PA: IGI Global. doi:10.4018/978-1-5225-2212-6.ch003

Guha, D., Roy, P. K., & Banerjee, S. (2016). Application of Modified Biogeography Based Optimization in AGC of an Interconnected Multi-Unit Multi-Source AC-DC Linked Power System. *International Journal of Energy Optimization and Engineering*, *5*(3), 1–18. doi:10.4018/IJEOE.2016070101

Guha, D., Roy, P. K., & Banerjee, S. (2016). Grey Wolf Optimization to Solve Load Frequency Control of an Interconnected Power System: GWO Used to Solve LFC Problem. *International Journal of Energy Optimization and Engineering*, *5*(4), 62–83. doi:10.4018/IJEOE.2016100104

Gupta, A. K., Dey, A., & Mukhopadhyay, A. K. (2016). Micromechanical and Finite Element Modeling for Composites. In S. Datta & J. Davim (Eds.), *Computational Approaches to Materials Design: Theoretical and Practical Aspects* (pp. 101–162). Hershey, PA: IGI Global. doi:10.4018/978-1-5225-0290-6.ch005

Guraksin, G. E. (2018). Internet of Things and Nature-Inspired Intelligent Techniques for the Future of Biomedical Engineering. In U. Kose, G. Guraksin, & O. Deperlioglu (Eds.), *Nature-Inspired Intelligent Techniques for Solving Biomedical Engineering Problems* (pp. 263–282). Hershey, PA: IGI Global. doi:10.4018/978-1-5225-4769-3.ch013

Hansman, C. A. (2016). Developing Mentoring Programs in Engineering and Technology Education. *International Journal of Quality Assurance in Engineering and Technology Education*, *5*(2), 1–15. doi:10.4018/IJQAETE.2016040101

Hasan, U., Chegenizadeh, A., & Nikraz, H. (2016). Nanotechnology Future and Present in Construction Industry: Applications in Geotechnical Engineering. In A. Khitab & W. Anwar (Eds.), *Advanced Research on Nanotechnology for Civil Engineering Applications* (pp. 141–179). Hershey, PA: IGI Global. doi:10.4018/978-1-5225-0344-6.ch005

Hejazi, T., & Akbari, L. (2017). A Multiresponse Optimization Model for Statistical Design of Processes with Discrete Variables. In M. Ram & J. Davim (Eds.), *Mathematical Concepts and Applications in Mechanical Engineering and Mechatronics* (pp. 17–37). Hershey, PA: IGI Global. doi:10.4018/978-1-5225-1639-2.ch002

Hejazi, T., & Hejazi, A. (2017). Monte Carlo Simulation for Reliability-Based Design of Automotive Complex Subsystems. In M. Ram & J. Davim (Eds.), *Mathematical Concepts and Applications in Mechanical Engineering and Mechatronics* (pp. 177–200). Hershey, PA: IGI Global. doi:10.4018/978-1-5225-1639-2.ch009

Hejazi, T., & Poursabbagh, H. (2017). Reliability Analysis of Engineering Systems: An Accelerated Life Testing for Boiler Tubes. In M. Ram & J. Davim (Eds.), *Mathematical Concepts and Applications in Mechanical Engineering and Mechatronics* (pp. 154–176). Hershey, PA: IGI Global. doi:10.4018/978-1-5225-1639-2.ch008

Henao, J., & Sotelo, O. (2018). Surface Engineering at High Temperature: Thermal Cycling and Corrosion Resistance. In A. Pakseresht (Ed.), *Production, Properties, and Applications of High Temperature Coatings* (pp. 131–159). Hershey, PA: IGI Global. doi:10.4018/978-1-5225-4194-3.ch006

Huirache-Acuña, R., Alonso-Nuñez, G., Rivera-Muñoz, E. M., Gutierrez, O., & Pawelec, B. (2016). Trimetallic Sulfide Catalysts for Hydrodesulfurization. In T. Saleh (Ed.), *Applying Nanotechnology to the Desulfurization Process in Petroleum Engineering* (pp. 240–262). Hershey, PA: IGI Global. doi:10.4018/978-1-4666-9545-0.ch008

Ilori, O. O., Adetan, D. A., & Umoru, L. E. (2017). Effect of Cutting Parameters on the Surface Residual Stress of Face-Milled Pearlitic Ductile Iron. *International Journal of Materials Forming and Machining Processes*, *4*(1), 38–52. doi:10.4018/IJMFMP.2017010103

Imam, M. H., Tasadduq, I. A., Ahmad, A., Aldosari, F., & Khan, H. (2017). Automated Generation of Course Improvement Plans Using Expert System. *International Journal of Quality Assurance in Engineering and Technology Education*, *6*(1), 1–12. doi:10.4018/IJQAETE.2017010101

Injeti, S. K., & Kumar, T. V. (2018). A WDO Framework for Optimal Deployment of DGs and DSCs in a Radial Distribution System Under Daily Load Pattern to Improve Techno-Economic Benefits. *International Journal of Energy Optimization and Engineering*, *7*(2), 1–38. doi:10.4018/IJEOE.2018040101

Ishii, N., Anami, K., & Knisely, C. W. (2018). *Dynamic Stability of Hydraulic Gates and Engineering for Flood Prevention*. Hershey, PA: IGI Global. doi:10.4018/978-1-5225-3079-4

J., J., Chowdhury, S., Goyal, P., Samui, P., & Dalkiliç, Y. (2016). Determination of Bearing Capacity of Shallow Foundation Using Soft Computing. In P. Saxena, D. Singh, & M. Pant (Eds.), *Problem Solving and Uncertainty Modeling through Optimization and Soft Computing Applications* (pp. 292-328). Hershey, PA: IGI Global. doi:10.4018/978-1-4666-9885-7.ch014

Jagan, J., Gundlapalli, P., & Samui, P. (2016). Utilization of Classification Techniques for the Determination of Liquefaction Susceptibility of Soils. In S. Bhattacharyya, P. Banerjee, D. Majumdar, & P. Dutta (Eds.), *Handbook of Research on Advanced Hybrid Intelligent Techniques and Applications* (pp. 124–160). Hershey, PA: IGI Global. doi:10.4018/978-1-4666-9474-3.ch005

Jayapalan, S. (2018). A Review of Chemical Treatments on Natural Fibers-Based Hybrid Composites for Engineering Applications. In K. Kumar & J. Davim (Eds.), *Composites and Advanced Materials for Industrial Applications* (pp. 16–37). Hershey, PA: IGI Global. doi:10.4018/978-1-5225-5216-1.ch002

Jeet, K., & Dhir, R. (2016). Software Module Clustering Using Bio-Inspired Algorithms. In P. Vasant, G. Weber, & V. Dieu (Eds.), *Handbook of Research on Modern Optimization Algorithms and Applications in Engineering and Economics* (pp. 445–470). Hershey, PA: IGI Global. doi:10.4018/978-1-4666-9644-0.ch017

Joshi, S. D., & Talange, D. B. (2016). Fault Tolerant Control for a Fractional Order AUV System. *International Journal of Energy Optimization and Engineering*, *5*(2), 1–24. doi:10.4018/IJEOE.2016040101

Julião, D., Ribeiro, S., de Castro, B., Cunha-Silva, L., & Balula, S. S. (2016). Polyoxometalates-Based Nanocatalysts for Production of Sulfur-Free Diesel. In T. Saleh (Ed.), *Applying Nanotechnology to the Desulfurization Process in Petroleum Engineering* (pp. 426–458). Hershey, PA: IGI Global. doi:10.4018/978-1-4666-9545-0.ch014

Kamthan, P. (2016). On the Nature of Collaborations in Agile Software Engineering Course Projects. *International Journal of Quality Assurance in Engineering and Technology Education*, *5*(2), 42–59. doi:10.4018/IJQAETE.2016040104

Karaman, O., Celik, C., & Urkmez, A. S. (2016). Self-Assembled Biomimetic Scaffolds for Bone Tissue Engineering. In M. Bououdina (Ed.), *Emerging Research on Bioinspired Materials Engineering* (pp. 104–132). Hershey, PA: IGI Global. doi:10.4018/978-1-4666-9811-6.ch004

Karkalos, N. E., Markopoulos, A. P., & Dossis, M. F. (2017). Optimal Model Parameters of Inverse Kinematics Solution of a 3R Robotic Manipulator Using ANN Models. *International Journal of Manufacturing, Materials, and Mechanical Engineering*, *7*(3), 20–40. doi:10.4018/IJMMME.2017070102

Kesimal, A., Karaman, K., Cihangir, F., & Ercikdi, B. (2018). Excavatability Assessment of Rock Masses for Geotechnical Studies. In N. Ceryan (Ed.), *Handbook of Research on Trends and Digital Advances in Engineering Geology* (pp. 231–256). Hershey, PA: IGI Global. doi:10.4018/978-1-5225-2709-1.ch006

Khanh, D. V., Vasant, P. M., Elamvazuthi, I., & Dieu, V. N. (2016). Multi-Objective Optimization of Two-Stage Thermo-Electric Cooler Using Differential Evolution: MO Optimization of TEC Using DE. In F. Miranda & C. Abreu (Eds.), *Handbook of Research on Computational Simulation and Modeling in Engineering* (pp. 139–170). Hershey, PA: IGI Global. doi:10.4018/978-1-4666-8823-0.ch004

Kim, D., Hassan, M. K., Chang, S., & Bigdeli, Y. (2016). Nonlinear Vibration Control of 3D Irregular Structures Subjected to Seismic Loads. In P. Samui (Ed.), *Handbook of Research on Advanced Computational Techniques for Simulation-Based Engineering* (pp. 103–119). Hershey, PA: IGI Global. doi:10.4018/978-1-4666-9479-8.ch003

Knoflacher, H. (2017). The Role of Engineers and Their Tools in the Transport Sector after Paradigm Change: From Assumptions and Extrapolations to Science. In H. Knoflacher & E. Ocalir-Akunal (Eds.), *Engineering Tools and Solutions for Sustainable Transportation Planning* (pp. 1–29). Hershey, PA: IGI Global. doi:10.4018/978-1-5225-2116-7.ch001

Kose, U. (2018). Towards an Intelligent Biomedical Engineering With Nature-Inspired Artificial Intelligence Techniques. In U. Kose, G. Guraksin, & O. Deperlioglu (Eds.), *Nature-Inspired Intelligent Techniques for Solving Biomedical Engineering Problems* (pp. 1–26). Hershey, PA: IGI Global. doi:10.4018/978-1-5225-4769-3.ch001

Kostić, S. (2018). A Review on Enhanced Stability Analyses of Soil Slopes Using Statistical Design. In N. Ceryan (Ed.), *Handbook of Research on Trends and Digital Advances in Engineering Geology* (pp. 446–481). Hershey, PA: IGI Global. doi:10.4018/978-1-5225-2709-1.ch013

Kumar, A., Patil, P. P., & Prajapati, Y. K. (2018). *Advanced Numerical Simulations in Mechanical Engineering*. Hershey, PA: IGI Global. doi:10.4018/978-1-5225-3722-9

Kumar, G. R., Rajyalakshmi, G., & Manupati, V. K. (2017). Surface Micro Patterning of Aluminium Reinforced Composite through Laser Peening. *International Journal of Manufacturing, Materials, and Mechanical Engineering*, *7*(4), 15–27. doi:10.4018/IJMMME.2017100102

Kumari, N., & Kumar, K. (2018). Fabrication of Orthotic Calipers With Epoxy-Based Green Composite. In K. Kumar & J. Davim (Eds.), *Composites and Advanced Materials for Industrial Applications* (pp. 157–176). Hershey, PA: IGI Global. doi:10.4018/978-1-5225-5216-1.ch008

Kuppusamy, R. R. (2018). Development of Aerospace Composite Structures Through Vacuum-Enhanced Resin Transfer Moulding Technology (VERTMTy): Vacuum-Enhanced Resin Transfer Moulding. In K. Kumar & J. Davim (Eds.), *Composites and Advanced Materials for Industrial Applications* (pp. 99–111). Hershey, PA: IGI Global. doi:10.4018/978-1-5225-5216-1.ch005

Lemos, J. V. (2016). The Basis for Masonry Analysis with UDEC and 3DEC. In V. Sarhosis, K. Bagi, J. Lemos, & G. Milani (Eds.), *Computational Modeling of Masonry Structures Using the Discrete Element Method* (pp. 61–89). Hershey, PA: IGI Global. doi:10.4018/978-1-5225-0231-9.ch003

Loy, J., Howell, S., & Cooper, R. (2017). Engineering Teams: Supporting Diversity in Engineering Education. In M. Gray & K. Thomas (Eds.), *Strategies for Increasing Diversity in Engineering Majors and Careers* (pp. 106–129). Hershey, PA: IGI Global. doi:10.4018/978-1-5225-2212-6.ch006

Macher, G., Armengaud, E., Kreiner, C., Brenner, E., Schmittner, C., Ma, Z., ... Krammer, M. (2018). Integration of Security in the Development Lifecycle of Dependable Automotive CPS. In N. Druml, A. Genser, A. Krieg, M. Menghin, & A. Hoeller (Eds.), *Solutions for Cyber-Physical Systems Ubiquity* (pp. 383–423). Hershey, PA: IGI Global. doi:10.4018/978-1-5225-2845-6.ch015

Maghsoodlou, S., & Poreskandar, S. (2016). Controlling Electrospinning Jet Using Microscopic Model for Ideal Tissue Engineering Scaffolds. *International Journal of Chemoinformatics and Chemical Engineering*, *5*(2), 1–16. doi:10.4018/IJCCE.2016070101

Mahendramani, G., & Lakshmana Swamy, N. (2018). Effect of Weld Groove Area on Distortion of Butt Welded Joints in Submerged Arc Welding. *International Journal of Manufacturing, Materials, and Mechanical Engineering*, *8*(2), 33–44. doi:10.4018/IJMMME.2018040103

Maiti, S. (2016). Engineered Gellan Polysaccharides in the Design of Controlled Drug Delivery Systems. In M. Bououdina (Ed.), *Emerging Research on Bioinspired Materials Engineering* (pp. 266–293). Hershey, PA: IGI Global. doi:10.4018/978-1-4666-9811-6.ch010

Majumdar, J. D., Weisheit, A., & Manna, I. (2016). Laser Surface Processing for Tailoring of Properties by Optimization of Microstructure. In E. Akinlabi, R. Mahamood, & S. Akinlabi (Eds.), *Advanced Manufacturing Techniques Using Laser Material Processing* (pp. 121–171). Hershey, PA: IGI Global. doi:10.4018/978-1-5225-0329-3.ch006

Maldonado-Macías, A. A., García-Alcaraz, J. L., Hernández-Arellano, J. L., & Cortes-Robles, G. (2016). An Ergonomic Compatibility Perspective on the Selection of Advanced Manufacturing Technology: A Case Study for CNC Vertical Machining Centers. In G. Alor-Hernández, C. Sánchez-Ramírez, & J. García-Alcaraz (Eds.), *Handbook of Research on Managerial Strategies for Achieving Optimal Performance in Industrial Processes* (pp. 137–165). Hershey, PA: IGI Global. doi:10.4018/978-1-5225-0130-5.ch008

Mamaghani, I. H. (2016). Application of Discrete Finite Element Method for Analysis of Unreinforced Masonry Structures. In V. Sarhosis, K. Bagi, J. Lemos, & G. Milani (Eds.), *Computational Modeling of Masonry Structures Using the Discrete Element Method* (pp. 440–458). Hershey, PA: IGI Global. doi:10.4018/978-1-5225-0231-9.ch017

Mansor, M. R., Sapuan, S. M., Salim, M. A., Akop, M. Z., Musthafah, M. T., & Shaharuzaman, M. A. (2016). Concurrent Design of Green Composites. In D. Verma, S. Jain, X. Zhang, & P. Gope (Eds.), *Green Approaches to Biocomposite Materials Science and Engineering* (pp. 48–75). Hershey, PA: IGI Global. doi:10.4018/978-1-5225-0424-5.ch003

Mansouri, I., & Esmaeili, E. (2016). Nanotechnology Applications in the Construction Industry. In A. Khitab & W. Anwar (Eds.), *Advanced Research on Nanotechnology for Civil Engineering Applications* (pp. 111–140). Hershey, PA: IGI Global. doi:10.4018/978-1-5225-0344-6.ch004

Manzoor, A. (2016). MOOCs for Enhancing Engineering Education. In D. Fonseca & E. Redondo (Eds.), *Handbook of Research on Applied E-Learning in Engineering and Architecture Education* (pp. 204–223). Hershey, PA: IGI Global. doi:10.4018/978-1-4666-8803-2.ch010

Martin, A., Kalevaru, V. N., & Radnik, J. (2016). Palladium in Heterogeneous Oxidation Catalysis. In H. Al-Megren & T. Xiao (Eds.), *Petrochemical Catalyst Materials, Processes, and Emerging Technologies* (pp. 53–81). Hershey, PA: IGI Global. doi:10.4018/978-1-4666-9975-5.ch003

Melnyczuk, J. M., & Palchoudhury, S. (2016). Introduction to Bio-Inspired Hydrogel and Their Application: Hydrogels. In M. Bououdina (Ed.), *Emerging Research on Bioinspired Materials Engineering* (pp. 133–159). Hershey, PA: IGI Global. doi:10.4018/978-1-4666-9811-6.ch005

Mitra-Kirtley, S., Mullins, O. C., & Pomerantz, A. E. (2016). Sulfur and Nitrogen Chemical Speciation in Crude Oils and Related Carbonaceous Materials. In T. Saleh (Ed.), *Applying Nanotechnology to the Desulfurization Process in Petroleum Engineering* (pp. 53–83). Hershey, PA: IGI Global. doi:10.4018/978-1-4666-9545-0.ch002

Moalosi, R., Uziak, J., & Oladiran, M. T. (2016). Using Blended Learning Approach to Deliver Courses in An Engineering Programme. *International Journal of Quality Assurance in Engineering and Technology Education*, *5*(1), 23–39. doi:10.4018/IJQAETE.2016010103

Mohammadzadeh, S., & Kim, Y. (2017). Nonlinear System Identification of Smart Buildings. In P. Samui, S. Chakraborty, & D. Kim (Eds.), *Modeling and Simulation Techniques in Structural Engineering* (pp. 328–347). Hershey, PA: IGI Global. doi:10.4018/978-1-5225-0588-4.ch011

Mohanty, I., & Bhattacherjee, D. (2016). Artificial Neural Network and Its Application in Steel Industry. In S. Datta & J. Davim (Eds.), *Computational Approaches to Materials Design: Theoretical and Practical Aspects* (pp. 267–300). Hershey, PA: IGI Global. doi:10.4018/978-1-5225-0290-6.ch010

Mohebkhah, A., & Sarhosis, V. (2016). Discrete Element Modeling of Masonry-Infilled Frames. In V. Sarhosis, K. Bagi, J. Lemos, & G. Milani (Eds.), *Computational Modeling of Masonry Structures Using the Discrete Element Method* (pp. 200–234). Hershey, PA: IGI Global. doi:10.4018/978-1-5225-0231-9.ch009

Molina, G. J., Aktaruzzaman, F., Soloiu, V., & Rahman, M. (2017). Design and Testing of a Jet-Impingement Instrument to Study Surface-Modification Effects by Nanofluids. *International Journal of Surface Engineering and Interdisciplinary Materials Science*, *5*(2), 43–61. doi:10.4018/IJSEIMS.2017070104

Montalvan-Sorrosa, D., de los Cobos-Vasconcelos, D., & Gonzalez-Sanchez, A. (2016). Nanotechnology Applied to the Biodesulfurization of Fossil Fuels and Spent Caustic Streams. In T. Saleh (Ed.), *Applying Nanotechnology to the Desulfurization Process in Petroleum Engineering* (pp. 378–389). Hershey, PA: IGI Global. doi:10.4018/978-1-4666-9545-0.ch012

Montillet, J., Yu, K., Bonenberg, L. K., & Roberts, G. W. (2016). Optimization Algorithms in Local and Global Positioning. In P. Vasant, G. Weber, & V. Dieu (Eds.), *Handbook of Research on Modern Optimization Algorithms and Applications in Engineering and Economics* (pp. 1–53). Hershey, PA: IGI Global. doi:10.4018/978-1-4666-9644-0.ch001

Moreira, F., & Ferreira, M. J. (2016). Teaching and Learning Requirements Engineering Based on Mobile Devices and Cloud: A Case Study. In D. Fonseca & E. Redondo (Eds.), *Handbook of Research on Applied E-Learning in Engineering and Architecture Education* (pp. 237–262). Hershey, PA: IGI Global. doi:10.4018/978-1-4666-8803-2.ch012

Mukherjee, A., Saeed, R. A., Dutta, S., & Naskar, M. K. (2017). Fault Tracking Framework for Software-Defined Networking (SDN). In C. Singhal & S. De (Eds.), *Resource Allocation in Next-Generation Broadband Wireless Access Networks* (pp. 247–272). Hershey, PA: IGI Global. doi:10.4018/978-1-5225-2023-8.ch011

Mukhopadhyay, A., Barman, T. K., & Sahoo, P. (2018). Electroless Nickel Coatings for High Temperature Applications. In K. Kumar & J. Davim (Eds.), *Composites and Advanced Materials for Industrial Applications* (pp. 297–331). Hershey, PA: IGI Global. doi:10.4018/978-1-5225-5216-1.ch013

Náprstek, J., & Fischer, C. (2017). Dynamic Stability and Post-Critical Processes of Slender Auto-Parametric Systems. In V. Plevris, G. Kremmyda, & Y. Fahjan (Eds.), *Performance-Based Seismic Design of Concrete Structures and Infrastructures* (pp. 128–171). Hershey, PA: IGI Global. doi:10.4018/978-1-5225-2089-4.ch006

Nautiyal, L., Shivach, P., & Ram, M. (2018). Optimal Designs by Means of Genetic Algorithms. In M. Ram & J. Davim (Eds.), *Soft Computing Techniques and Applications in Mechanical Engineering* (pp. 151–161). Hershey, PA: IGI Global. doi:10.4018/978-1-5225-3035-0.ch007

Nazir, R. (2017). Advanced Nanomaterials for Water Engineering and Treatment: Nano-Metal Oxides and Their Nanocomposites. In T. Saleh (Ed.), *Advanced Nanomaterials for Water Engineering, Treatment, and Hydraulics* (pp. 84–126). Hershey, PA: IGI Global. doi:10.4018/978-1-5225-2136-5.ch005

Nogueira, A. F., Ribeiro, J. C., Fernández de Vega, F., & Zenha-Rela, M. A. (2018). Evolutionary Approaches to Test Data Generation for Object-Oriented Software: Overview of Techniques and Tools. In M. Khosrow-Pour, D.B.A. (Ed.), Incorporating Nature-Inspired Paradigms in Computational Applications (pp. 162-194). Hershey, PA: IGI Global. doi:10.4018/978-1-5225-5020-4.ch006

Nunes, J. F., Moreira, P. M., & Tavares, J. M. (2016). Human Motion Analysis and Simulation Tools: A Survey. In F. Miranda & C. Abreu (Eds.), *Handbook of Research on Computational Simulation and Modeling in Engineering* (pp. 359–388). Hershey, PA: IGI Global. doi:10.4018/978-1-4666-8823-0.ch012

Ogunlaja, A. S., & Tshentu, Z. R. (2016). Molecularly Imprinted Polymer Nanofibers for Adsorptive Desulfurization. In T. Saleh (Ed.), *Applying Nanotechnology to the Desulfurization Process in Petroleum Engineering* (pp. 281–336). Hershey, PA: IGI Global. doi:10.4018/978-1-4666-9545-0.ch010

Ong, P., & Kohshelan, S. (2016). Performances of Adaptive Cuckoo Search Algorithm in Engineering Optimization. In P. Vasant, G. Weber, & V. Dieu (Eds.), *Handbook of Research on Modern Optimization Algorithms and Applications in Engineering and Economics* (pp. 676–699). Hershey, PA: IGI Global. doi:10.4018/978-1-4666-9644-0.ch026

Osho, M. B. (2018). Industrial Enzyme Technology: Potential Applications. In S. Bharati & P. Chaurasia (Eds.), *Research Advancements in Pharmaceutical, Nutritional, and Industrial Enzymology* (pp. 375–394). Hershey, PA: IGI Global. doi:10.4018/978-1-5225-5237-6.ch017

Padmaja, P., & Marutheswar, G. (2017). Certain Investigation on Secured Data Transmission in Wireless Sensor Networks. *International Journal of Mobile Computing and Multimedia Communications, 8*(1), 48–61. doi:10.4018/IJMCMC.2017010104

Paixão, S. M., Silva, T. P., Arez, B. F., & Alves, L. (2016). Advances in the Reduction of the Costs Inherent to Fossil Fuels' Biodesulfurization towards Its Potential Industrial Application. In T. Saleh (Ed.), *Applying Nanotechnology to the Desulfurization Process in Petroleum Engineering* (pp. 390–425). Hershey, PA: IGI Global. doi:10.4018/978-1-4666-9545-0.ch013

Palmer, S., & Hall, W. (2017). An Evaluation of Group Work in First-Year Engineering Design Education. In R. Tucker (Ed.), *Collaboration and Student Engagement in Design Education* (pp. 145–168). Hershey, PA: IGI Global. doi:10.4018/978-1-5225-0726-0.ch007

Panneer, R. (2017). Effect of Composition of Fibers on Properties of Hybrid Composites. *International Journal of Manufacturing, Materials, and Mechanical Engineering*, *7*(4), 28–43. doi:10.4018/IJMMME.2017100103

Parker, J. (2016). Hubble's Expanding Universe: A Model for Quality in Technology Infused engineering and Technology Education. *International Journal of Quality Assurance in Engineering and Technology Education*, *5*(2), 16–29. doi:10.4018/IJQAETE.2016040102

Paul, S., & Roy, P. (2018). Optimal Design of Power System Stabilizer Using a Novel Evolutionary Algorithm. *International Journal of Energy Optimization and Engineering*, *7*(3), 24–46. doi:10.4018/IJEOE.2018070102

Pavaloiu, A. (2018). Artificial Intelligence Ethics in Biomedical-Engineering-Oriented Problems. In U. Kose, G. Guraksin, & O. Deperlioglu (Eds.), *Nature-Inspired Intelligent Techniques for Solving Biomedical Engineering Problems* (pp. 219–231). Hershey, PA: IGI Global. doi:10.4018/978-1-5225-4769-3.ch010

Peña, F. (2016). A Semi-Discrete Approach for the Numerical Simulation of Freestanding Blocks. In V. Sarhosis, K. Bagi, J. Lemos, & G. Milani (Eds.), *Computational Modeling of Masonry Structures Using the Discrete Element Method* (pp. 416–439). Hershey, PA: IGI Global. doi:10.4018/978-1-5225-0231-9.ch016

Penchovsky, R., & Traykovska, M. (2016). Synthetic Approaches to Biology: Engineering Gene Control Circuits, Synthesizing, and Editing Genomes. In M. Bououdina (Ed.), *Emerging Research on Bioinspired Materials Engineering* (pp. 323–351). Hershey, PA: IGI Global. doi:10.4018/978-1-4666-9811-6.ch012

Pieroni, A., & Iazeolla, G. (2016). Engineering QoS and Energy Saving in the Delivery of ICT Services. In P. Vasant & N. Voropai (Eds.), *Sustaining Power Resources through Energy Optimization and Engineering* (pp. 208–226). Hershey, PA: IGI Global. doi:10.4018/978-1-4666-9755-3.ch009

Pioro, I., Mahdi, M., & Popov, R. (2017). Application of Supercritical Pressures in Power Engineering. In L. Chen & Y. Iwamoto (Eds.), *Advanced Applications of Supercritical Fluids in Energy Systems* (pp. 404–457). Hershey, PA: IGI Global. doi:10.4018/978-1-5225-2047-4.ch013

Plaksina, T., & Gildin, E. (2017). Rigorous Integrated Evolutionary Workflow for Optimal Exploitation of Unconventional Gas Assets. *International Journal of Energy Optimization and Engineering*, *6*(1), 101–122. doi:10.4018/IJEOE.2017010106

Puppala, A. J., Bheemasetti, T. V., Zou, H., Yu, X., Pedarla, A., & Cai, G. (2016). Spatial Variability Analysis of Soil Properties using Geostatistics. In P. Samui (Ed.), *Handbook of Research on Advanced Computational Techniques for Simulation-Based Engineering* (pp. 195–226). Hershey, PA: IGI Global. doi:10.4018/978-1-4666-9479-8.ch008

Ramdani, N., & Azibi, M. (2018). Polymer Composite Materials for Microelectronics Packaging Applications: Composites for Microelectronics Packaging. In K. Kumar & J. Davim (Eds.), *Composites and Advanced Materials for Industrial Applications* (pp. 177–211). Hershey, PA: IGI Global. doi:10.4018/978-1-5225-5216-1.ch009

Ramesh, M., Garg, R., & Subrahmanyam, G. V. (2017). Investigation of Influence of Quenching and Annealing on the Plane Fracture Toughness and Brittle to Ductile Transition Temperature of the Zinc Coated Structural Steel Materials. *International Journal of Surface Engineering and Interdisciplinary Materials Science*, *5*(2), 33–42. doi:10.4018/IJSEIMS.2017070103

Razavi, A. M., & Ahmad, R. (2016). Agile Software Development Challenges in Implementation and Adoption: Focusing on Large and Distributed Settings – Past Experiences, Emergent Topics. In I. Ghani, D. Jawawi, S. Dorairaj, & A. Sidky (Eds.), *Emerging Innovations in Agile Software Development* (pp. 175–207). Hershey, PA: IGI Global. doi:10.4018/978-1-4666-9858-1.ch010

Reccia, E., Cecchi, A., & Milani, G. (2016). FEM/DEM Approach for the Analysis of Masonry Arch Bridges. In V. Sarhosis, K. Bagi, J. Lemos, & G. Milani (Eds.), *Computational Modeling of Masonry Structures Using the Discrete Element Method* (pp. 367–392). Hershey, PA: IGI Global. doi:10.4018/978-1-5225-0231-9.ch014

Ro, H. K., & McIntosh, K. (2016). Constructing Conducive Environment for Women of Color in Engineering Undergraduate Education. In U. Thomas & J. Drake (Eds.), *Critical Research on Sexism and Racism in STEM Fields* (pp. 23–48). Hershey, PA: IGI Global. doi:10.4018/978-1-5225-0174-9.ch002

Rodulfo-Baechler, S. M. (2016). Dual Role of Perovskite Hollow Fiber Membrane in the Methane Oxidation Reactions. In H. Al-Megren & T. Xiao (Eds.), *Petrochemical Catalyst Materials, Processes, and Emerging Technologies* (pp. 385–430). Hershey, PA: IGI Global. doi:10.4018/978-1-4666-9975-5.ch014

Rudolf, S., Biryuk, V. V., & Volov, V. (2018). Vortex Effect, Vortex Power: Technology of Vortex Power Engineering. In V. Kharchenko & P. Vasant (Eds.), *Handbook of Research on Renewable Energy and Electric Resources for Sustainable Rural Development* (pp. 500–533). Hershey, PA: IGI Global. doi:10.4018/978-1-5225-3867-7.ch021

Sah, A., Bhadula, S. J., Dumka, A., & Rawat, S. (2018). A Software Engineering Perspective for Development of Enterprise Applications. In A. Elçi (Ed.), *Handbook of Research on Contemporary Perspectives on Web-Based Systems* (pp. 1–23). Hershey, PA: IGI Global. doi:10.4018/978-1-5225-5384-7.ch001

Sahoo, P., & Roy, S. (2017). Tribological Behavior of Electroless Ni-P, Ni-P-W and Ni-P-Cu Coatings: A Comparison. *International Journal of Surface Engineering and Interdisciplinary Materials Science*, *5*(1), 1–15. doi:10.4018/IJSEIMS.2017010101

Sahoo, S. (2018). Laminated Composite Hypar Shells as Roofing Units: Static and Dynamic Behavior. In K. Kumar & J. Davim (Eds.), *Composites and Advanced Materials for Industrial Applications* (pp. 249–269). Hershey, PA: IGI Global. doi:10.4018/978-1-5225-5216-1.ch011

Sahu, H., & Hungyo, M. (2018). Introduction to SDN and NFV. In A. Dumka (Ed.), *Innovations in Software-Defined Networking and Network Functions Virtualization* (pp. 1–25). Hershey, PA: IGI Global. doi:10.4018/978-1-5225-3640-6.ch001

Saikia, P., Bharadwaj, S. K., & Miah, A. T. (2016). Peroxovanadates and Its Bio-Mimicking Relation with Vanadium Haloperoxidases. In M. Bououdina (Ed.), *Emerging Research on Bioinspired Materials Engineering* (pp. 197–219). Hershey, PA: IGI Global. doi:10.4018/978-1-4666-9811-6.ch007

Saladino, R., Botta, G., & Crucianelli, M. (2016). Advances in Nanotechnology Transition Metal Catalysts in Oxidative Desulfurization (ODS) Processes: Nanotechnology Applied to ODS Processing. In T. Saleh (Ed.), *Applying Nanotechnology to the Desulfurization Process in Petroleum Engineering* (pp. 180–215). Hershey, PA: IGI Global. doi:10.4018/978-1-4666-9545-0.ch006

Saleh, T. A., Danmaliki, G. I., & Shuaib, T. D. (2016). Nanocomposites and Hybrid Materials for Adsorptive Desulfurization. In T. Saleh (Ed.), *Applying Nanotechnology to the Desulfurization Process in Petroleum Engineering* (pp. 129–153). Hershey, PA: IGI Global. doi:10.4018/978-1-4666-9545-0.ch004

Saleh, T. A., Shuaib, T. D., Danmaliki, G. I., & Al-Daous, M. A. (2016). Carbon-Based Nanomaterials for Desulfurization: Classification, Preparation, and Evaluation. In T. Saleh (Ed.), *Applying Nanotechnology to the Desulfurization Process in Petroleum Engineering* (pp. 154–179). Hershey, PA: IGI Global. doi:10.4018/978-1-4666-9545-0.ch005

Salem, A. M., & Shmelova, T. (2018). Intelligent Expert Decision Support Systems: Methodologies, Applications, and Challenges. In T. Shmelova, Y. Sikirda, N. Rizun, A. Salem, & Y. Kovalyov (Eds.), *Socio-Technical Decision Support in Air Navigation Systems: Emerging Research and Opportunities* (pp. 215–242). Hershey, PA: IGI Global. doi:10.4018/978-1-5225-3108-1.ch007

Samal, M. (2017). FE Analysis and Experimental Investigation of Cracked and Un-Cracked Thin-Walled Tubular Components to Evaluate Mechanical and Fracture Properties. In P. Samui, S. Chakraborty, & D. Kim (Eds.), *Modeling and Simulation Techniques in Structural Engineering* (pp. 266–293). Hershey, PA: IGI Global. doi:10.4018/978-1-5225-0588-4.ch009

Samal, M., & Balakrishnan, K. (2017). Experiments on a Ring Tension Setup and FE Analysis to Evaluate Transverse Mechanical Properties of Tubular Components. In P. Samui, S. Chakraborty, & D. Kim (Eds.), *Modeling and Simulation Techniques in Structural Engineering* (pp. 91–115). Hershey, PA: IGI Global. doi:10.4018/978-1-5225-0588-4.ch004

Santhanakumar, M., Adalarasan, R., & Rajmohan, M. (2016). An Investigation in Abrasive Waterjet Cutting of Al6061/SiC/Al2O3 Composite Using Principal Component Based Response Surface Methodology. *International Journal of Manufacturing, Materials, and Mechanical Engineering, 6*(4), 30–47. doi:10.4018/IJMMME.2016100103

Sareen, N., & Bhattacharya, S. (2016). Cleaner Energy Fuels: Hydrodesulfurization and Beyond. In T. Saleh (Ed.), *Applying Nanotechnology to the Desulfurization Process in Petroleum Engineering* (pp. 84–128). Hershey, PA: IGI Global. doi:10.4018/978-1-4666-9545-0.ch003

Sarhosis, V. (2016). Micro-Modeling Options for Masonry. In V. Sarhosis, K. Bagi, J. Lemos, & G. Milani (Eds.), *Computational Modeling of Masonry Structures Using the Discrete Element Method* (pp. 28–60). Hershey, PA: IGI Global. doi:10.4018/978-1-5225-0231-9.ch002

Sarhosis, V., Oliveira, D. V., & Lourenco, P. B. (2016). On the Mechanical Behavior of Masonry. In V. Sarhosis, K. Bagi, J. Lemos, & G. Milani (Eds.), *Computational Modeling of Masonry Structures Using the Discrete Element Method* (pp. 1–27). Hershey, PA: IGI Global. doi:10.4018/978-1-5225-0231-9.ch001

Satyam, N. (2016). Liquefaction Modelling of Granular Soils using Discrete Element Method. In P. Samui (Ed.), *Handbook of Research on Advanced Computational Techniques for Simulation-Based Engineering* (pp. 381–441). Hershey, PA: IGI Global. doi:10.4018/978-1-4666-9479-8.ch015

Sawant, S. (2018). Deep Learning and Biomedical Engineering. In U. Kose, G. Guraksin, & O. Deperlioglu (Eds.), *Nature-Inspired Intelligent Techniques for Solving Biomedical Engineering Problems* (pp. 283–296). Hershey, PA: IGI Global. doi:10.4018/978-1-5225-4769-3.ch014

Sezgin, H., & Berkalp, O. B. (2018). Textile-Reinforced Composites for the Automotive Industry. In K. Kumar & J. Davim (Eds.), *Composites and Advanced Materials for Industrial Applications* (pp. 129–156). Hershey, PA: IGI Global. doi:10.4018/978-1-5225-5216-1.ch007

Shah, M. Z., Gazder, U., Bhatti, M. S., & Hussain, M. (2018). Comparative Performance Evaluation of Effects of Modifier in Asphaltic Concrete Mix. *International Journal of Strategic Engineering*, *1*(2), 13–25. doi:10.4018/IJoSE.2018070102

Shah, V. S., Shah, H. R., & Samui, P. (2016). Application of Meta-Models (MPMR and ELM) for Determining OMC, MDD and Soaked CBR Value of Soil. In S. Bhattacharyya, P. Banerjee, D. Majumdar, & P. Dutta (Eds.), *Handbook of Research on Advanced Hybrid Intelligent Techniques and Applications* (pp. 454–482). Hershey, PA: IGI Global. doi:10.4018/978-1-4666-9474-3.ch015

Sharma, N., & Kumar, K. (2018). Fabrication of Porous NiTi Alloy Using Organic Binders. In K. Kumar & J. Davim (Eds.), *Composites and Advanced Materials for Industrial Applications* (pp. 38–62). Hershey, PA: IGI Global. doi:10.4018/978-1-5225-5216-1.ch003

Sharma, T. K. (2016). Application of Shuffled Frog Leaping Algorithm in Software Project Scheduling. In P. Saxena, D. Singh, & M. Pant (Eds.), *Problem Solving and Uncertainty Modeling through Optimization and Soft Computing Applications* (pp. 225–238). Hershey, PA: IGI Global. doi:10.4018/978-1-4666-9885-7.ch011

Shivach, P., Nautiyal, L., & Ram, M. (2018). Applying Multi-Objective Optimization Algorithms to Mechanical Engineering. In M. Ram & J. Davim (Eds.), *Soft Computing Techniques and Applications in Mechanical Engineering* (pp. 287–301). Hershey, PA: IGI Global. doi:10.4018/978-1-5225-3035-0.ch014

Shmelova, T. (2018). Stochastic Methods for Estimation and Problem Solving in Engineering: Stochastic Methods of Decision Making in Aviation. In S. Kadry (Ed.), *Stochastic Methods for Estimation and Problem Solving in Engineering* (pp. 139–160). Hershey, PA: IGI Global. doi:10.4018/978-1-5225-5045-7.ch006

Shukla, R., Anapagaddi, R., Singh, A. K., Allen, J. K., Panchal, J. H., & Mistree, F. (2016). Integrated Computational Materials Engineering for Determining the Set Points of Unit Operations for Production of a Steel Product Mix. In S. Datta & J. Davim (Eds.), *Computational Approaches to Materials Design: Theoretical and Practical Aspects* (pp. 163–191). Hershey, PA: IGI Global. doi:10.4018/978-1-5225-0290-6.ch006

Siero González, L. R., & Romo Vázquez, A. (2017). Didactic Sequences Teaching Mathematics for Engineers With Focus on Differential Equations. In M. Ramírez-Montoya (Ed.), *Handbook of Research on Driving STEM Learning With Educational Technologies* (pp. 129–151). Hershey, PA: IGI Global. doi:10.4018/978-1-5225-2026-9.ch007

Singh, R., & Dutta, S. (2018). Visible Light Active Nanocomposites for Photocatalytic Applications. In K. Kumar & J. Davim (Eds.), *Composites and Advanced Materials for Industrial Applications* (pp. 270–296). Hershey, PA: IGI Global. doi:10.4018/978-1-5225-5216-1.ch012

Singh, R., & Lou, H. H. (2016). Safety and Efficiency Enhancement in LNG Terminals. In H. Al-Megren & T. Xiao (Eds.), *Petrochemical Catalyst Materials, Processes, and Emerging Technologies* (pp. 164–176). Hershey, PA: IGI Global. doi:10.4018/978-1-4666-9975-5.ch007

Sözbilir, H., Özkaymak, Ç., Uzel, B., & Sümer, Ö. (2018). Criteria for Surface Rupture Microzonation of Active Faults for Earthquake Hazards in Urban Areas. In N. Ceryan (Ed.), *Handbook of Research on Trends and Digital Advances in Engineering Geology* (pp. 187–230). Hershey, PA: IGI Global. doi:10.4018/978-1-5225-2709-1.ch005

Stanciu, I. (2018). Stochastic Methods in Microsystems Engineering. In S. Kadry (Ed.), *Stochastic Methods for Estimation and Problem Solving in Engineering* (pp. 161–176). Hershey, PA: IGI Global. doi:10.4018/978-1-5225-5045-7.ch007

Strebkov, D., Nekrasov, A., Trubnikov, V., & Nekrasov, A. (2018). Single-Wire Resonant Electric Power Systems for Renewable-Based Electric Grid. In V. Kharchenko & P. Vasant (Eds.), *Handbook of Research on Renewable Energy and Electric Resources for Sustainable Rural Development* (pp. 449–474). Hershey, PA: IGI Global. doi:10.4018/978-1-5225-3867-7.ch019

Subburaman, D., Jagan, J., Dalkiliç, Y., & Samui, P. (2016). Reliability Analysis of Slope Using MPMR, GRNN and GPR. In F. Miranda & C. Abreu (Eds.), *Handbook of Research on Computational Simulation and Modeling in Engineering* (pp. 208–224). Hershey, PA: IGI Global. doi:10.4018/978-1-4666-8823-0.ch007

Sun, J., Wan, S., Lin, J., & Wang, Y. (2016). Advances in Catalytic Conversion of Syngas to Ethanol and Higher Alcohols. In H. Al-Megren & T. Xiao (Eds.), *Petrochemical Catalyst Materials, Processes, and Emerging Technologies* (pp. 177–215). Hershey, PA: IGI Global. doi:10.4018/978-1-4666-9975-5.ch008

Tüdeş, Ş., Kumlu, K. B., & Ceryan, S. (2018). Integration Between Urban Planning and Natural Hazards For Resilient City. In N. Ceryan (Ed.), *Handbook of Research on Trends and Digital Advances in Engineering Geology* (pp. 591–630). Hershey, PA: IGI Global. doi:10.4018/978-1-5225-2709-1.ch017

Tyukhov, I., Rezk, H., & Vasant, P. (2016). Modern Optimization Algorithms and Applications in Solar Photovoltaic Engineering. In P. Vasant & N. Voropai (Eds.), *Sustaining Power Resources through Energy Optimization and Engineering* (pp. 390–445). Hershey, PA: IGI Global. doi:10.4018/978-1-4666-9755-3.ch016

Ulamis, K. (2018). Soil Liquefaction Assessment by Anisotropic Cyclic Triaxial Test. In N. Ceryan (Ed.), *Handbook of Research on Trends and Digital Advances in Engineering Geology* (pp. 631–664). Hershey, PA: IGI Global. doi:10.4018/978-1-5225-2709-1.ch018

Umar, M. A., Tenuche, S. S., Yusuf, S. A., Abdulsalami, A. O., & Kufena, A. M. (2016). Usability Engineering in Agile Software Development Processes. In I. Ghani, D. Jawawi, S. Dorairaj, & A. Sidky (Eds.), *Emerging Innovations in Agile Software Development* (pp. 208–221). Hershey, PA: IGI Global. doi:10.4018/978-1-4666-9858-1.ch011

Üzüm, O., & Çakır, Ö. A. (2016). A Bio-Inspired Phenomena in Cementitious Materials: Self-Healing. In M. Bououdina (Ed.), *Emerging Research on Bioinspired Materials Engineering* (pp. 220–246). Hershey, PA: IGI Global. doi:10.4018/978-1-4666-9811-6.ch008

Valente, M., & Milani, G. (2017). Seismic Assessment and Retrofitting of an Under-Designed RC Frame Through a Displacement-Based Approach. In V. Plevris, G. Kremmyda, & Y. Fahjan (Eds.), *Performance-Based Seismic Design of Concrete Structures and Infrastructures* (pp. 36–58). Hershey, PA: IGI Global. doi:10.4018/978-1-5225-2089-4.ch002

Vasant, P. (2018). A General Medical Diagnosis System Formed by Artificial Neural Networks and Swarm Intelligence Techniques. In U. Kose, G. Guraksin, & O. Deperlioglu (Eds.), *Nature-Inspired Intelligent Techniques for Solving Biomedical Engineering Problems* (pp. 130–145). Hershey, PA: IGI Global. doi:10.4018/978-1-5225-4769-3.ch006

Vergara, D., Lorenzo, M., & Rubio, M. (2016). On the Use of Virtual Environments in Engineering Education. *International Journal of Quality Assurance in Engineering and Technology Education*, *5*(2), 30–41. doi:10.4018/IJQAETE.2016040103

Verrollot, J., Tolonen, A., Harkonen, J., & Haapasalo, H. J. (2018). Challenges and Enablers for Rapid Product Development. *International Journal of Applied Industrial Engineering*, *5*(1), 25–49. doi:10.4018/IJAIE.2018010102

Wagner, C., & Ryan, C. (2016). Physical and Digital Integration Strategies of Electronic Device Supply Chains and Their Applicability to ETO Supply Chains. In R. Addo-Tenkorang, J. Kantola, P. Helo, & A. Shamsuzzoha (Eds.), *Supply Chain Strategies and the Engineer-to-Order Approach* (pp. 224–245). Hershey, PA: IGI Global. doi:10.4018/978-1-5225-0021-6.ch011

Wang, Z., Wu, P., Lan, L., & Ji, S. (2016). Preparation, Characterization and Desulfurization of the Supported Nickel Phosphide Catalysts. In H. Al-Megren & T. Xiao (Eds.), *Petrochemical Catalyst Materials, Processes, and Emerging Technologies* (pp. 431–458). Hershey, PA: IGI Global. doi:10.4018/978-1-4666-9975-5.ch015

Yardimci, A. G., & Karpuz, C. (2018). Fuzzy Rock Mass Rating: Soft-Computing-Aided Preliminary Stability Analysis of Weak Rock Slopes. In N. Ceryan (Ed.), *Handbook of Research on Trends and Digital Advances in Engineering Geology* (pp. 97–131). Hershey, PA: IGI Global. doi:10.4018/978-1-5225-2709-1.ch003

Zhang, L., Ding, S., Sun, S., Han, B., Yu, X., & Ou, J. (2016). Nano-Scale Behavior and Nano-Modification of Cement and Concrete Materials. In A. Khitab & W. Anwar (Eds.), *Advanced Research on Nanotechnology for Civil Engineering Applications* (pp. 28–79). Hershey, PA: IGI Global. doi:10.4018/978-1-5225-0344-6.ch002

Zindani, D., & Kumar, K. (2018). Industrial Applications of Polymer Composite Materials. In K. Kumar & J. Davim (Eds.), *Composites and Advanced Materials for Industrial Applications* (pp. 1–15). Hershey, PA: IGI Global. doi:10.4018/978-1-5225-5216-1.ch001

Zindani, D., Maity, S. R., & Bhowmik, S. (2018). A Decision-Making Approach for Material Selection of Polymeric Composite Bumper Beam. In K. Kumar & J. Davim (Eds.), *Composites and Advanced Materials for Industrial Applications* (pp. 112–128). Hershey, PA: IGI Global. doi:10.4018/978-1-5225-5216-1.ch006

About the Contributors

Kaushik Kumar, B.Tech (Mechanical Engineering, REC (Now NIT), Warangal), MBA (Marketing, IGNOU) and Ph.D (Engineering, Jadavpur University), is presently an Associate Professor in the Department of Mechanical Engineering, Birla Institute of Technology, Mesra, Ranchi, India. He has 17 years of Teaching & Research and over 11 years of industrial experience in a manufacturing unit of Global repute. His areas of teaching and research interest are Conventional and Non-conventional Quality Management Systems, Optimization, Non-conventional machining, CAD / CAM, Rapid Prototyping and Composites. He has 9 Patents, 28 books, 15 Edited Book Volume, 43 Book Chapters, 136 international Journal, 21 International and 8 National Conference publications to his credit. He is on the editorial board and review panel of 7 International and 1 National Journals of repute. He has been felicitated with many awards and honours.

Divya Zindani, (BE, Mechanical Engineering, Rajasthan Technical University, Kota), M.E. (Design of Mechanical Equipment, BIT Mesra), is presently pursuing PhD (National Institute of Technology, Silchar). He has over 2 years of Industrial experience. His areas of interests are Optimization, Product and Process Design, CAD/CAM/CAE, Rapid prototyping and Material Selection. He has 1 Patent, 4 Books, 6 Edited Books, 18 Book Chapters, 2 SCI journal, 7 Scopus Indexed international journal and 4 International Conference publications to his credit.

J. Paulo Davim received his Ph.D. degree in Mechanical Engineering in 1997, M.Sc. degree in Mechanical Engineering (materials and manufacturing processes) in 1991, Mechanical Engineering degree (5 years) in 1986, from the University of Porto (FEUP), the Aggregate title (Full Habilitation) from the University of Coimbra in 2005 and the D.Sc. from London Metropolitan University in 2013. He is Eur Ing by FEANI-Brussels and Senior Chartered

Engineer by the Portuguese Institution of Engineers with an MBA and Specialist title in Engineering and Industrial Management. Currently, he is Professor at the Department of Mechanical Engineering of the University of Aveiro, Portugal. He has more than 30 years of teaching and research experience in Manufacturing, Materials, Mechanical and Industrial Engineering, with special emphasis in Machining & Tribology. He has also interest in Management, Engineering Education and Higher Education for Sustainability. He has guided large numbers of postdoc, Ph.D. and master's students as well as coordinated & participated in several financed research projects. He has received several scientific awards. He has worked as evaluator of projects for international research agencies as well as examiner of Ph.D. thesis for many universities. He is the Editor in Chief of several international journals, Guest Editor of journals, books Editor, book Series Editor and Scientific Advisory for many international journals and conferences. Presently, he is an Editorial Board member of 25 international journals and acts as reviewer for more than 80 prestigious Web of Science journals. In addition, he has also published as editor (and co-editor) more than 100 books and as author (and co-author) more than 10 books, 80 book chapters and 400 articles in journals and conferences (more than 200 articles in journals indexed in Web of Science core collection/h-index 45+/6000+ citations and SCOPUS/h-index 53+/8500+ citations).

* * *

Esther Akinlabi is a full Professor in the Faculty of Engineering and the Built Environment, University of Johannesburg, with a demonstrated history of working in the higher education industry. Strong education professional skilled in AutoCAD, Analytical Skills, Research Design, Friction Stir Welding, and laser based additive manufacturing.

Hacene Ameddah, received his Doctorate in Mechanical Engineering from the University of BATNA 2, Algeria. Is a team leader of the laboratory of Innovation in Construction, Eco-design, and Seismic Engineering (LICEGS), member of Research Laboratory in Production (LRP). His research interests include NC tool path generation, curve and surface design, freeform fabrication process planning, the application of programing languages to problems in CAD/ CAM, medical imaging and functional design, computer-aided surgery, bio-manufacturing and rapid prototyping and optimizations techniques.

Krishna Mohan B., Assistant Professor of Department of Mechanical Engineering, completed his M. Tech in CAD/CAM from Godavari Institute of Engineering and Technology, Rajahmundry affiliated to JNTUK, is an astute professional with more than 14 years of experience in Teaching, Research and Industry. His M. Tech work was on Determination of Stress Intensity Factors in Welded Joints using FEM. He is associated with the department for the past 4 years. He possesses excellent modeling skills in CAD software like Solid Works and PTC Creo. He is a certified Solidworks Associate. His research interests include Additive Manufacturing, Composite Materials, and Simulation techniques using FEM. He is currently AICTE-CII Department Coordinator and CAD Lab in charge.

Sumit Bhowmik is an Assistant Professor in Department of Mechanical Engineering, National Institute of Technology Silchar, Assam. He received his B. E. in Mechanical Engineering from Tripura Engineering College (Tripura University), India, Master in Mechanical Engineering and PhD from Jadavpur University, Kolkata. He has over 15 years of teaching and research experience. He has more than 20 international journals and international Conference publications to his credit. His areas of interests are Fracture Mechanics, Materials properties, Composites, and Optimization, etc.

Dheeman Bhuyan is a graduate in Mechanical Engineering from CSVTU Bhilai and has a post graduate degree in mechanical design from Birla Institute of Technology, Mesra. He has over 4 years of experience in various fields of industry, academia and research along with multiple publications. He currently is serving as Assistant Professor in Girijananda Chowdhury Institute of Management and Technology, Guwahati while pursuing his doctoral research at the National Institute of Technology Meghalaya.

Azhar Equbal is an Assistant Professor in the Department of Mechanical Engineering at RTC Institute of Technology, Ormanjhi, Ranchi. He has Completed Ph.D in Manufacturing Engineering from National Institute of Foundry and Forge Technology, Ranchi, under the supervision of Dr. A.K. Sood .His area of specialization was Application of Fused deposition modelling process in Electrical Discharge machining. His current areas of interest include Process Parameters Optimization, Soft computing Approach and Artificial Intelligence. He has got more than 25 papers published in different reputed international journals.

Md. Asif Equbal is an Assistant Professor in the Department of Mechanical Engineering at Cambridge Institute of Technology, Tatisilwai Ranchi. He has Completed Ph.D in Forge Technology from National Institute of Foundry and Forge Technology, Ranchi, under the supervision of Dr. R. K. Ohdar. His area of specialization was Supply chain Management in Indian Context. His current areas of interest include Supply chain management, Process Optimization, and Artificial Intelligence. He has got more than 15 papers published in different reputed international journals.

Md Israr Equbal is an Associate Professor in the Department of Mechanical Engineering at JB Institute of Engineering and Technology, Moinabad, Hyderabad, India. He has completed his Ph.D in Forge Technology at National Institute of Foundry and Forge Technology, Ranchi, India. He holds an MTech in Manufacturing Engineering from National Institute of Foundry and Forge Technology, Ranchi, India. He has more than nine years of experience in teaching and research. His current areas of research include modelling and simulation, analysis and optimisation of parameters involving manufacturing processes and statistical process control and artificial intelligence. He has published more than 50 research articles in refereed journals and conferences.

Jagadish is an Assistant Professor at the Department of Mechanical Engineering, National Institute of Technology Raipur Chhattisgarh, India. He obtained his Ph.D. in Mechanical Engineering with specialization in Manufacturing Engineering from National Institute of Technology Silchar and Post graduation in Product Design and Development from National Institute of Technology Warangal, India. His areas of interests are advanced manufacturing process, polymer machining, green manufacturing, computer aided design, and applied soft computing techniques. He has over 3 years of industrial experience in the field of design and analysis and 5 years of teaching and research experience. He received an Institutional Award (Gold Medal) by institution of engineers India and best innovative award by springer for his outstanding research contribution and published more than 14 research papers and 6 book chapters. Currently, he is doing research in advanced machining process, green manufacturing, polymer machining, and decision making tools.

Hridayjit Kalita, (BE, Mechanical Engineering, SRM University, Kattankulathur), is presently pursuing M.E. (Design of Mechanical Equipment, BIT Mesra). His areas of interests are Conventional and advanced machining and manufacturing, Optimization, 3D printing. He has 2 Books, 4 Book Chapters, 2 Scopus Indexed international journal and 4 International Conference publications to his credit.

Emin Kececi is a Professor of Mechatronics at the Department of Mechanical Engineering at Abdullah Gul University, received his BS and MS in Mechanical Engineering from ITU and Duke University in 1996 and 1999 respectively. He obtained his PhD in Electrical Engineering from the University of Virginia in 2003. His current research focuses on mobile robots and medical devices as well as directing the AGU-MAKE, Maker Space. Dr. Kececi also served as a member of the Editorial Board of Robotica.

Satyanarayana Kosaraju, Associate Professor of Mechanical Engineering, completed his Ph.D from National Institute of Technology Warangal and has over Five years of academic and research experience in Indian universities. His Ph.D work was on Machinability characteristics: Modeling, Simulation, analysis and optimization of Ti-6Al-4V. Prior to PhD, he had earned Bachelors of Engineering in Mechanical Engineering from JayaPrakash Narayan college of Engineering, and Masters of Engineering with Product Design Development specialization in Mechanical Engineering from National Institute of Technology Warangal. Dr. Satyanarayana research interests include Material Testing & Characterization, single and multi-objective optimization techniques, Experimental, Analytical and Finite Element Investigations of various machining and forming processes in which he has more than 30 publications, in various journals and conferences. He is currently working on the uniaxial and biaxial deformation behavior of Titanium 21S alloy up to 4000C, anisotropic yield criteria for titanium 21S alloy, Forming Limit Diagrams (FLDs) at various temperatures and optimization. Dr Satyanarayana is a recipient of Bharath Vidhya Rathan Award. He is a member of ASME, IEANG and life member of ISTE.

Mark Liu is a fashion and textile designer advancing the application of scientific principles to traditional techniques used in the fashion industry. Noted for pioneering Zero-Waste Fashion design when exhibiting his fashion label in Estethica at London Fashion Week. His research develops the new field of Non-Euclidean Fashion Patternmaking. His current research explores: fabrics made from algae biotechnology, 3D printed garments, 3D scan to garment techniques and using fashion to teach STEM skills to school students.

Jennifer Loy is Professor of Additive Manufacturing Engineering at Deakin University in Victoria, Australia. Jennifer has a background in Industrial Design, with a specialisation in digital technologies. Her research interests are in design for additive manufacturing, with a particular focus on medical and health products. Jennifer is an adjunct professor with the Menzies Health Institute in Queensland. Her work includes product design and development, workforce evolution and training and supply chain management.

Daniel M. Madyira holds a BSc (Hons) degree in Mechanical Engineering from the University of Zimbabwe, an MSc. in the Design of Turbomachinery from Cranfield University and a PhD in Mechanical Engineering from the University of Johannesburg. He is currently a senior lecturer in the Department of Mechanical Engineering Science at the University of Johannesburg. His key research interests included high speed machining of titanium and its impact on fatigue performance. Some of his applied research work includes rotary kilns, structural integrity of buried pipelines and machining of metal powder green compacts. The fracture behavior and crack repairs is special importance of most of this work. He has supervised more that 15 MSc. and 6 PhD work and published more than 100 refereed publications.

Tawanda Marazani is a PhD Student at the University of Johannesburg in the Department of Mechanical Engineering Science under the supervision of Prof. Esther T Akinlabi and Dr. Daniel M Madyira. His areas of research interests include Additive Manufacturing, Repair of Cracks in Metals, Friction Stir Processing, Friction Stir Welding and Impingement Freezing Technology. He has so far published 2 conference papers, currently has 2 accepted journal papers and 1 accepted conference paper. He has 15 years combined academic and industrial experience.

James Novak is a Research Fellow in Additive Manufacturing at Deakin University, Victoria, Australia. He has previously been a postdoctoral researcher in product design and design for additive manufacturing at the University of Technology Sydney, and a lecturer and course convener at Griffith University. Novak is most well known for his full-size 3D printed bicycle frame which has been exhibited around the world at venues including the Red Dot Design Museum in Germany, and he has won international awards for his design and research contributions including the prestigious Dick Aubin Distinguished Paper Award in 2015.

Swadesh Kumar Singh, Professor & Dean R&D has done his M.Tech and Doctorate from IIT Delhi and specialized in Metal forming. He has worked in Indian Engineering Services(IES) as Assistant Executive Engineer. He has published 70 research papers in reputed international and national journals. He is awarded with "Career Award for young teachers by AICTE" and "Young Scientist award" by DST for his research projects. He authored the books on Production Engineering, Industrial Engineering, Reasoning and Aptitude for standard publications. He is the driving force of the latest technologies in material science which steer the students to adapt to the changes.

Anoop Sood is an Associate Professor in Manufacturing Engineering Department at NIFFT Ranchi. He has completed his PhD from NIT, Rourkela. He is author of many papers published in high-quality journals of repute His research areas are manufacturing processes, Additive manufacturing, scheduling, etc.

Index

S

T

U

V

W

www.ingramcontent.com/pod-product-compliance
Lightning Source LLC
LaVergne TN
LVHW081346300926
849161LV00008B/2086